William James at the Boundaries

# WILLIAM JAMES at the **Boundaries**

Philosophy, Science, and the Geography of Knowledge

FRANCESCA BORDOGNA

The University of Chicago Press
Chicago and London

Francesca Bordogna is assistant professor in the Department of History and the Science in Human Culture Program at Northwestern University.

The University of Chicago Press, Chicago 60637
The University of Chicago Press, Ltd., London
© 2008 by The University of Chicago
All rights reserved. Published 2008
Printed in the United States of America

17 16 15 14 13 12 11 10 09 08  1 2 3 4 5

ISBN-13: 978-0-226-06652-3    (cloth)
ISBN-10: 0-226-06652-5        (cloth)

Library of Congress Cataloging-in-Publication Data

Bordogna, Francesca.
    William James at the boundaries : philosophy, science, and the geography
of knowledge / Francesca Bordogna.
        p.   cm.
    Includes bibliographical references and index.
    ISBN-13: 978-0-226-06652-3 (cloth : alk. paper)
    ISBN-10: 0-226-06652-5 (cloth : alk. paper)    1. James, William, 1842–1910.
2. Philosophy and science.    I. Title.
    B945.J24B47    2008
    191—dc22

                                                    2008014908

∞ The paper used in this publication meets the minimum requirements of the American National Standard for Information Sciences—Permanence of Paper for Printed Library Materials, ANSI Z39.48–1992.

# CONTENTS

Acknowledgments      *vii*

Introduction: Mental Energy, Boundary Work, and the Geography
of Knowledge      *1*

1    PHILOSOPHY AND SCIENCE      *21*

2    PHILOSOPHY VERSUS THE NATURALISTIC SCIENCE OF MAN
James's Early Negotiations of Disciplinary and Pedagogical Boundaries      *59*

3    JAMES AND THE (IM)MORAL ECONOMY OF SCIENCE      *91*

4    MENTAL BOUNDARIES AND PRAGMATIC TRUTH      *137*

5    PRAGMATISM, PSYCHOLOGISM, AND A "SCIENCE OF MAN"      *155*

6    ECSTASY AND COMMUNITY
James and the Politics of the Self      *189*

7    THE PHILOSOPHER'S PLACE
James, Münsterberg, and Philosophical Trees      *219*

8    THE PHILOSOPHER'S MIND
     Routinists, Undisciplinables, and "The Energies of Men'"                    259

     Conclusions                                                                 269
     Notes                                                                       275
     Bibliography                                                                337
     Index                                                                       373

# ACKNOWLEDGMENTS

In the long years it took me to write this book I have accumulated debts to many individuals and institutions, and it is a pleasure to acknowledge them here.

This project began as a Ph.D. dissertation when I was a graduate student in the Committee on the Conceptual Foundations of Science (now the Committee on Conceptual and Historical Studies of Science) at the University of Chicago. There I had the good fortune to work with three exceptional advisers: Arnold I. Davidson, Lorraine J. Daston, and Robert J. Richards. All provided rigorous guidance and much food for thought with their pathbreaking work. Robert Richards and Lorraine Daston gracefully resigned themselves to the fact that occasionally they would have to act as personal coaches. I am deeply grateful to all of them.

I also acknowledge the researchers at the Max Planck Institute for the History of Science, where I carried on research as a graduate student first and, later, as a postdoctoral fellow. The intellectual community that Daston led made the field of the history of science unfold as a magic landscape. I remember with great pleasure the intense conversations the other fellows and I carried on in improbable spots dispersed through Berlin and in front of the tiny kitchen on the fifth floor—the true think tank of our Abteilung. My thanks also to the staff of the Max Planck Institute, especially Urs Schoepflin and Frau Kuntze, and to the other MPI fellows, especially Skuli Sigurdsson.

*William James at the Boundaries* grew into something very different from my Ph.D. dissertation. The metamorphosis was facilitated by a number of people and places. The Department of History and the Science in Human Culture program at Northwestern University provided an exciting intellectual environment. I would like to thank all of my colleagues, and especially Jock McLane for his institutional support. I am immensely grateful to Ken Alder, Thomas William Heyck, and Alex Owen, who read drafts of the entire manuscript and offered thoughtful comments at different stages of the project. Their own scholarly work enabled me to rethink the assumptions guiding mine and opened up new directions of research. My book would not have been quite the same without their good influence. I am also grateful to my former colleagues in the Program of Liberal Studies and in the History and Philosophy of Science program at the University of Notre Dame, where I was fortunate enough to teach before I moved to Northwestern, and especially to Don Howard.

I am most grateful to George Cotkin, David Leary, and Owen Flanagan, who read the whole manuscript and offered detailed and thought-provoking suggestions for improvement. I would also like to thank Charlene Haddock Seigfried, whose insights on William James have been important in shaping my own understanding of James and whose encouragement I deeply value.

I have benefited from conversations with, and criticism from, Daniela Barberis, Berna Eden, Sergio Franzese, Cheryce Kramer, Thomas Gieryn, David Joravsky, Andy Mendelsohn, Tara Nummedal, Alessandro Pagnini, Ignas Skrupskelis, and John Tresch, and I thank all of them for their insights. Parts of this work have been presented at many departments, conferences, and institutions, including the University of Chicago, Northwestern University, the Max Planck Institute for the History of Science, the University of Notre Dame, the University of South Carolina, Cambridge University, the University of Indiana at Bloomington, the University of California at Berkeley, and the National Humanities Center. Although I am unable here to acknowledge adequately the contributions of all the historians and sociologists of science, intellectual historians, and philosophers whom I met at these and many other institutions, I am grateful to all of them for their probing questions and valuable thoughts.

The book would not have been possible without financial support from several institutions. Grants from the Franke Institute for the Humanities at the University of Chicago, in 2000–2001, and the Alice Berline Kaplan Center for the Humanities (now Alice Kaplan Institute for the Humanities) at Northwestern University, in 2003–2004, gave me time off from teaching and allowed me to draft the book. The National Humanities Center in North Carolina, where I was a fellow in 2006–2007, offered a magnificent venue for the last revisions. The airy building, all transparent windows and doors facing onto the woods, made me think of William James's description of his summer place in New Hampshire—"the most delightful house you ever saw, . . . with fourteen doors, all opening outwards." The open lounge, the sunny niches with cozy chairs, and the paths in the woods created countless possibilities for exchange and discussion. My thanks to the Rockefeller Foundation, for sponsoring my stay at the center, and to the director of the center, Geoffrey Harpham, for creating such an intellectually vibrant environment. I am also grateful to Kent Mullikin, Marie Brubaker, Joel Elliott, James Getkin, the unforgettable Corbett Capps, and all the staff for making my stay so pleasant and profitable. The 2006–2007 fellows, especially Mimi Kim and Joe Viscomi, were always available for energizing intellectual exchange, and I recall with delight my conversations with them. Special thanks go to Karen Carroll, who did the first systematic copyediting of the entire manuscript. Without her help and implacable editorial drive, this book might have never been sent to press.

I am also grateful to the National Humanities Center's librarians (Eliza Robertson, Betsy Dain, and Jean Houston), and to the staff at other institutions where I did my research—the Houghton Library and the Pusey Library at Harvard University, the Massachusetts Public Library, the archive of the Society for Psychical Research at Cambridge University Library, the Wren Library at Trinity College, Duke University Library, and the University of Chicago Library.

A slightly different version of chapter 6 has been published in the *British Journal for the History of Science* ("Inner Division and Uncertain Contours: William James and the Politics of the Modern Self," 40, no. 147 [2007]: 505–536), while earlier drafts of sections of chapters 3 and 5 appeared in *History and Philosophy of the Biological and Biomedical Sciences* and the *Journal of the History of the Behavioral Sciences*, I am grateful to Simon Schaffer, Nick Jardine, and John C. Burnham and the anonymous reviewers for their advice and encouragement.

At the University of Chicago Press my thanks go to T. David Brent, executive editor, and Alan Thomas, editorial director of the humanities and science, for their help, advice, and support, as well as to Elizabeth Branch Dyson, Laura Avey, and especially to my copyeditor, Richard Audet, whose efforts have done much to improve the book. I would also like to thank former editor Catherine Rice, while I recall with gratitude the memory of Susan Abrams, who encouraged my work from the start.

My research assistants, especially Dana Rovang, Rachel Ponce, and Andrew Nelson, supplied invaluable research and editorial help.

My mother-in-law, Marianne Jurkowitz helped much, taking care of my child at critical junctures, while to her husband, Ned Alexander, I am indebted for technological help with the digital images reproduced in the book. My greatest debt, however, is to my husband Edward Jurkowitz, who learned more than he ever wanted to about William James. Taking time off from his own work as a scholar, he edited multiple drafts of all chapters, acted as an amazing intellectual sounding board, and provided intellectual stimulus and insights with his own work on the history of physics. To him goes my deepest gratitude and love—and a promise to pay off my debt. Our child Arianna has brought us much-needed mental perspective and balance. At six, she has already authored several "books" but is determined not to attend college, much less to become an academic.

My sisters Isabella and Giovanna gave me continued love and support. They are the best sisters in the world. To my father Giuseppe, who was able to give meaning to his life and ours as he emerged from deep coma, brain injury, and loss of memory, goes great admiration and immense love. This book is dedicated to him and to my mother Anna, who helped me in more ways than I can remember.

# Mental Energy, Boundary Work, and the Geography of Knowledge

On December 28, 1906, William James delivered his presidential speech before the American Philosophical Association (APA) assembled at Columbia University. A couple of years away from the peak of his international fame, James was a renowned, if no longer cutting-edge, psychologist and a prominent philosopher. His recent articles on radical empiricism and new work on pragmatism had created a stir in philosophical circles in America and abroad. The philosophers assembled at the meeting of the APA no doubt would have welcomed a discussion of some of the most contentious aspects of these provocative philosophical theories. They were likely disappointed—in his address James skirted all technical philosophical issues and concentrated instead on a practical problem of medical and social import: the question of how people could maximize their physical, mental, and moral energy. After discussing the merits and defects of the prevailing type of "scientific psychology" (structuralism) and its rival, functionalism, James described yoga breathing exercises, mental suggestion, consumption of brandy, and other techniques through which ordinary people struggling in extreme situations had been able to increase their levels of energy. He read aloud long testimonials from people who were at best marginal to

1

the interests of the American philosophical community, including one by a friend who suffered from what we would call bipolar disorder, and who had found peace in yoga. James commended the conception of philosophy articulated by the twenty-five-year-old Italian writer and journalist Giovanni Papini, an outcast in philosophical and academic circles, and concluded by inviting his listeners to take up the psychological and physiological problems of charting the "powers of men" and discovering new pathways to access human energy.[1]

As George Cotkin has shown in his important study *William James, Public Philosopher*, James's APA presidential address, titled "The Energies of Men," embodied the therapeutic of strenuosity and heroic life that a wide range of his contemporaries, including President Theodore Roosevelt, were urging on fellow citizens. The physical and spiritual exercises that James endorsed were a means both of reinvigorating the individual and of revitalizing America. Energy and mental energy were topics of great importance not only among progressives, but also to the many people who, across social classes, suffered from ennui, debilitation, and neurasthenia, and lamented the loss of self-directedness in the new social and economic order.[2] Yet why did James use his presidential address to the philosophers' association to promulgate his energizing message?

The question is worth pursuing, especially since many philosophers, from James's time up to today, have wondered what on earth James's address had to do with philosophy.[3] The immediate fortunes of "The Energies of Men" are revealing. Based on a talk that James had presented earlier that year at the Harvard Psychological Club, James's APA presidential address was received "with special interest" by the members of the American Psychological Association, who, that year, as in previous years, held their annual meeting in conjunction with the philosophers' meeting.[4] It was reprinted in magazines that reached a broad readership, and became hugely popular among mind curers, mental hygienists, and followers of the Emmanuel Church, which rushed to publish, in various formats, a modified version.[5] Occultists found that the paper sanctioned some of their goals (and means), and the Italian occult-modernist avant-garde, led for a short while by Papini and his pragmatist friends, enthusiastically presented the paper as an expression of a popular American spiritual and hygienic movement: the "New Thought." Yet, despite the waves of interest "The Energies of Men" produced in nonphilosophical circles, to all appearances among American philosophers it caused hardly a ripple. In fact, at a time when every philosophical paper that James published was endlessly reviewed or discussed in the leading philosophy journals, "The Energies of Men" seems to have gone nearly unnoticed by philosophers.

One does not have to look far to find the reasons behind this cool response. The APA had been founded only five years earlier, at the turn of the century, with the goal of transforming philosophy from a vague field without public recognition, and open to incursions from all sorts of amateurs, into a technical discipline, a specialized field, or, as many hoped, even a "special science," with a technically defined set of methods and problems. The founders of the philosophical association resented their apparent lack of academic and social prestige, and were especially annoyed that people with no philosophical training felt qualified to pronounce with authority on philosophical topics. In response to this situation, they mobilized the vocabularies of "specialization" and "professionalization," terms to which they attached a variety of meanings, and continually engaged in exercises of disciplinary self-definition.[6] American philosophers did not agree on the subject matter and methods of philosophy, or on the means by which their goals could be achieved. Yet many believed that they could successfully define their discipline and restore the intellectual and social authority it had enjoyed earlier in the nineteenth century only by enforcing a clear-cut distinction between professional philosophy and amateur dabbling, as well as between philosophy and other fields of inquiry, in particular the sciences.

"The Energies of Men" hardly contributed to those projects. It started with a discussion of the advantages and disadvantages of two dominant types of psychology, structural psychology and functional psychology. The first approach, James told his audience, relied on laboratory machinery, the customary "electric keys" and "revolving drums" of self-recording instruments. The second was instead premised on a much vaguer, hardly quantifiable, but, to James, overall more adequate type of observation: clinical observation. The central concept discussed in the paper, that of the "amount of energy available for running one's mental and moral operations," came from clinically oriented functional psychology. The current understanding of such energy was so vague that, James observed, "scientific psychologists" felt entitled to "ignore" it "altogether," and to leave it "to be treated by the moralists and mind-curers and doctors exclusively." Not so for James. He concluded his talk by inviting his audience to refine its conceptions of mental energy, and to tackle two problems: first, to produce a "topographic survey . . . of the limits of human power . . . , something like an ophthalmologist's chart of the limits of the human field of vision," and, second, to produce a "methodical inventory of the paths of access, or keys, differing with the diverse types of individual, to the different kinds of power." James envisioned this project as an "absolutely concrete" study, to be based mainly on "historical" and "biographical material," and rich with practical implications both for individual conduct and for society.[7]

James's APA address was "out of place" in several ways. It trespassed the emerging divides separating the discipline of philosophy from the so-called "special sciences," especially psychology and physiology, and from practical endeavors, such as medicine and psychiatry. It took a stab at "scientific psychology," singing the praises of a conception, that of mental energy, that was widely used by "common, practical men," but "never once mentioned or heard of in laboratory circles."[8] It violated established rules of decorum by broaching subjects that, James emphasized, were unmentionable in educated and polite circles. And it flirted with mental hygienists, mind curers, and popular spiritual philosophers, trespassing the boundaries that separated professional philosophy not only from popular philosophy, but also from pseudoscience and what many would have dubbed "quackery." Not surprisingly, the paper did not find favor with an audience of professionalizing and specializing philosophers. It probably confirmed for many of them what they already thought: James was a "psychologist trying to do philosophy and . . . failing."[9] Did James simply misjudge the interests of his audience? What was his aim?

*William James at the Boundaries* offers an image of William James and of late nineteenth-century philosophy and science that recasts James's APA performance. Rather than looking at James's presidential address as an oddity, just another of those bizarre things that James occasionally did, this book suggests that "The Energies of Men" was characteristic, even emblematic, of James's approach to philosophy and science. Indeed, it was one of many gestures and techniques through which James struggled to reconfigure the relationships between philosophy and the sciences, as well as professional and amateur discourses. Through these efforts, I contend, James reinterpreted the nature of philosophy and science and, by doing so, proposed a new vision for the intellectual and social order of knowledge.

This book engages variegated aspects of James's wide-ranging scientific and philosophical activities by examining what I argue is a central thrust in his work: an insistence on transgressing boundaries separating fields of knowledge, types of discourse, and groups of inquirers. James, for example, knocked through newly erected fences separating philosophy from psychology; he attempted to dismember the "Ph.D. octopus," which, he believed, would strangle American higher education, squeezing it into specialized studies; and he dared to bring psychology and metaphysics to the study of religion and mysticism. Throughout his life, as Cotkin noted, he blurred the boundaries separating professional philosophy from public and popular philosophy.[10] He mixed rhetorical registers and genres, and was notorious for engaging fields, discourses, and practices that newly established scientific and philosophical orthodoxies pushed outside the bounds of proper academic inquiry. His sustained interest in "supernormal" phenomena (such as telepathy and the trance of a medium) and his close study of the spectacular healings occasionally obtained by mind curers represent examples of repeated attempts to open the doors of science to phenomena ostracized by self-described "scientists." He questioned the social and epistemic barriers separating professionals from amateurs, and maintained ostentatious friendships with marginal and controversial figures, never missing a chance to advertise their feats in formal academic circles. James also made spectacular gestures meant to batter established codes of academic propriety, as he did when he invited American philosophers to explore how yoga, alcohol consumption, and other unorthodox techniques could help people attain higher levels of mental and physical energy. James, in short, was a "serial" transgressor of boundaries.[11]

This fact, clear to James's contemporaries, and irritating to many of them, has not escaped the attention of James scholars. James, a man who came from a wealthy and distinguished family, was long praised as a "friend of the underdogs," especially political minorities. Although plenty of scholars have questioned the exact nature of James's actual social sympathies, emphasizing, for example, the contradictions in his attitude toward workers, his "puerile ideas about the differences between men and women," his "patrician elitism," and the inadequacy of his social analyses, James continues to be seen as a figure who established connections across social barriers, and cultivated personal contacts with people with whom his social or professional peers would not have mingled.[12]

Scholars have also long noted that James did not stick to emerging disciplinary divides. Bruce Kuklick, in particular, has argued that within his pragmatist philosophy James transformed his psychology of cognition into a philosophical theory of truth. Gerald E. Myers, Graham Bird, Timothy Sprigge, Eugene Taylor, David C. Lamberth, and David Leary see profound continuities between James's epistemology (and metaphysics) and his psychology, specifically the philosophical psychology that he began articulating in the mid-1890s.[13] For Daniel J. Wilson, James was among the first in America to bring scientific training and methods into philosophy.[14] Charlene Seigfried similarly sees a close "linkage" between James's approach to questions of knowledge and truth and evolutionary biological theory. She maintains that James endeavored to "rejuvenate philosophy by bringing to it both the results and the methods of science," including especially a "natural history methodology," which, she shows, James applied to the study of human nature and the structures of human experience. In that way, Seigfried continues, James reframed philosophy as an instrument that people could use in order to understand and transform "the human condition."[15]

Other scholars have shown that James brought philosophy and science together with other modes of culture or realms of experience. David Leary's recent research, for example, unveils the ways in which literature, especially drama and poetry, combined with philosophy and psychology in James's work, while Jacques Barzun, Howard Feinstein, and others have examined James's psychological and philosophical work from the angle of James's artistic training and ambitions.[16] For David Hollinger, James's whole philosophical career was "largely controlled" by his attempt to reframe the notion of science so as to challenge the absolute authority that natural science's spokesmen advocated for it, and help find a via media between science and religion.[17] Paul J. Croce, studying debates surrounding doctrines of determinism and chance, submits that James was instrumental in bringing about a culture of uncertainty that eroded belief in science and religion, as well as in crafting intellectual tools for dealing with that culture.[18] Further exploring that nexus of science and religion, David Lamberth has shown that James's "late view of philosophy" was informed by James's work on religious experiences and psychic phenomena, a point that Richard Gale also sets forth in his novel interpretation of James's pluralistic metaphysics.[19] Lamberth further remarks that James operated without any clear distinction between "philosophy, the study of religion (philosophical or otherwise), and theological reflection," and observes that "the disciplinary distinctions that animate many twentieth-century discussions [of James's philosophy] prove foreign in many respects to an inherently interdisciplinary Jamesian perspective."[20]

Similarly, for Louis Menand James, properly speaking, was "not even a practitioner of a particular scholarly discipline,"[21] while Daniel Bjork contends that "vocational labels" do not lead to an adequate understanding of James's work, suggesting that the "center of James's vision" consisted in "a creative continuum that focused on art, natural science, medicine, aesthetics, psychology, and philosophy." That "relational space," Bjork continues, flowed directly from James's experiential metaphysics, a type of metaphysics that embedded all relations (including both conjunctive and disjunctive relations such as and-ness, with-ness, together-ness, and but-ness) in experience.[22]

Few scholars, in short, think it profitable to pigeonhole James's various activities into neatly bounded academic disciplines or fields of inquiry such as philosophy, psychology, clinical psychology, psychical research, and the psychology of religion. Yet what prompted James in the first place to transgress the divides increasingly separating disciplines, types of discourse, and groups of investigators?

This book seeks to answer this question. Following James as he plowed through those barriers, the book contends that James's trespassing of boundaries—far from being a peripheral sidelight as often portrayed, a consequence of his metaphysics, or a product of his legendary vocational uncertainties and his piecemeal, unsystematic education—represented an essential strategy within a broader intellectual and social project. I suggest that, by engaging in a studied form of transgressive "boundary work," James imagined both a new configuration of knowledge and a new vision for American society. Moreover, I argue that he redefined the boundaries separating individual citizens from one another and from the wider society, and offered matching portraits of the human self and of ideal scientific and philosophical inquirers. I contend that he mobilized his portraits of the self and of ideal knowledge producers so as to redefine, at once, the existing order of knowledge and the order of society.

I borrow the concept of "boundary work" from sociologists of science and historians of disciplinarity such as Thomas Gieryn, Steve Fuller, Sheila Jasanoff, Julie Klein, and James Good.[23] This type of activity includes creating, maintaining, and protecting, but also debunking, blurring, cracking, and crossing the boundaries that separate disciplines, fields of knowledge, kinds of discourse, and social groups. As Gieryn and Fuller have shown, the boundaries that demarcate elite science from pseudoscience, popular science, and nonscience, like the divisions separating disciplines, are never written in stone, but are constantly reconstituted and altered in local, culturally embedded negotiations. Thus boundary work, these and other scholars remind us, occurs both among and within disciplines. It is especially prevalent in the spaces at the "edges of science," in contested territories, and in the sites where "epistemic communities, institutions, methodologies, and object domains intersect."[24] Gieryn suggests that scientists typically engage in boundary work mainly for three purposes: "expelling" others from their discipline in order to secure professional authority over a particular domain; "expanding" their field and gaining control of territory previously controlled by a different group; or protecting a group's autonomy from external pressures (churches, the state, academic authorities, users of science).[25] However, boundary work is not always aimed at protecting the narrow professional or institutional interests of a particular group of people. As Good proposes, scientists may cross or blur boundaries to ensure circulation of statements and techniques. They may do that in order to "share common interests with members of other disciplines," to dispel "in-group partisanship" within disciplines or university departments, to enroll others by engaging them in "collaborative problem solving," or to develop transdisciplinary theories.[26]

Boundary work often functions as a means that scientists use to redirect specific disciplines. However, in some cases, knowledge producers can also use it to change an entire configuration of disciplines, and even rethink the meaning of disciplinarity. As Pierre Bourdieu stated, disciplines are "political structures" and "institutionalized formations for organizing schemes of perception, appreciation, and action, and for inculcating them as tools of cognition and communication."[27] From this perspective, boundary work ultimately is an attempt to refashion the persona of the scholar, the habits, virtues, and dispositions considered necessary to the production of knowledge in a particular discipline, as well as the "interpersonal styles" that should regulate social interactions between investigators working in the same discipline or within different disciplines.[28] I suggest that, by trespassing boundaries, James pursued all of these goals.

At the turn of the twentieth century, boundary work often involved the creation of spaces of knowledge. These included both physical spaces, such as the layout of a university campus or the construction of a building devoted to a specific discipline, and metaphorical spaces, such as those visualized by the many charts of the disciplines and trees of knowledge produced in that era. Historians and sociologists of science have shown that such spaces—especially physical spaces, such as laboratories, cabinets of curiosities, or botanical gardens—can perform a variety of functions. The layout of a natural history museum, for example, may visually display a scientific theory, and offer a persuasive argument for its acceptance. As Gieryn and others have illustrated, by virtue of their position, access, and arrangement, physical spaces of knowledge can also perform epistemic functions, including conferring epistemic authority upon a certain group of scientists vis-à-vis others.[29] Like other spaces, knowledge spaces often dictate social rules and regulate human interactions both within a scientific community and among scientists and their publics.[30] These epistemic and social functions are, in some cases, tightly interlinked, so that a space of knowledge may enforce epistemic norms precisely by regulating the social interactions that take place within that space.

I suggest that, like many of his contemporaries, James worked to set up spaces of knowledge, and argue that he mobilized those spaces not only to rearrange relationships among fields of knowledge, thereby reconfiguring an intellectual "geography of knowledge," but also to promote what may be called a "social geography of knowledge." By that phrase I refer not only to James's views of the proper distribution of intellectual/epistemic competences among different social groups, but also to the ways in which James mobilized different spatial arrangements of knowledge in order to promote different forms of sociability. In other words, I contend that the spaces that James assigned to investigators, like other knowledge spaces, served both intellectual and social functions.

In North America and in many European countries, natural scientists, social and human scientists, and other humanists abundantly resorted to boundary work to produce or reconfigure spaces of knowledge. Philosophers vigorously participated in such efforts. At the turn of the twentieth century, philosophers perceived that their discipline was going through a period of crisis, and located the causes of the crisis in the new prestige that the natural sciences and rapidly institutionalizing human and social sciences enjoyed both in the academy and in the broader society. In post–Civil War America natural scientists threatened philosophers' established roles as public moralists and, in some areas of the country, as educators of the new social and financial elites. In this new, complex situation many philosophers found that it was imperative to redefine the meaning of philosophy and its position vis-à-vis the sciences. Boundary work and spaces of knowledge, especially charts and trees of the sciences that offered synoptical arrangements of the disciplines, promised to be excellent tools in that respect.

Many among the philosophers and social scientists who addressed such questions found the emerging disciplinary divides to be profoundly empowering. They created barriers that would exclude amateurs and nonspecialists from their fields and prevent, or at least limit, exchanges among investigators working in different areas. Others, in contrast, worried about the fragmentation of knowledge brought about by professionalization and the institutionalization of inquiry, and strove to bring about new forms of unity. These unifiers, for the most part philosophers, believed that "modern" tendencies toward specialization had negative impact not only on knowledge, but also on the minds of citizens, their intellectual and moral faculties, and their ability to contribute to a well-ordered society.[31] These philosophers often reconceived philosophy as a *scientia scientiarum*, an "architectonic science" that alone was capable of properly training students' minds, forming citizens, providing a conceptual foundation for the sciences, ordering the disciplines, and bringing forth a unification of knowledge. In that way, they found a new mission for philosophy that made it once again a prestigious field of knowledge.

I paint James as a unifier, and argue that, like many of his colleagues, he promoted plans for a philosophical unification of knowledge. James shared with Herbert Spencer, a philosopher and polymath whom in other ways he rather disliked, the hope that some day it would be possible to create a "completely unified system of knowledge." This may sound surprising, since James had little sympathy for unitary schemes, and suspected that unitary narratives (of the cosmos, of the goal and origin of things, of moral purposes) would ultimately always become instruments for the suppression of differences. In what sense, then, can he be depicted as a "unifier"? I propose that, in contrast to many of his contemporaries, James pursued not an intellectual or conceptual or methodological or architectonic unity of knowledge, but a "social" unity of knowledge. By trespassing divides, he challenged the social effects of professionalization and the disciplinary compartmentalization of knowledge in order to promote the formation of pluralistic, yet cohesive, communities of inquirers. These communities would be open not only to professionals, but also to amateurs, consumers of knowledge, even the occasional "crank"; they would allow for a sense of common cause and for exchange between people who inhabited different intellectual and social worlds, and, in this way, they would bring forth a social unity of knowledge.

My answer, then, to the question of why James carried on transgressive boundary work weaves together James's views on knowledge production with the forms of social engagement that he considered ideal. Building on the work of other scholars, especially David H. Hollinger's and Lorraine Daston's work on the ethos of science, and Steven Shapin's and Edward P. Jurkowitz's analyses of the ways in which science practitioners combine social-political values, epistemic assumptions, and methodological preferences in their visions of proper scientific communities, I also suggest that much of James's boundary work aimed to foster social, moral, and epistemological attitudes—that is, a moral and social economy of knowledge—that would enable knowledge producers to create and maintain open, pluralistic, yet intimate communities of inquirers.[32]

The goal of enabling a social unification of knowledge and promoting proper communities of inquirers shaped in important ways James's scientific and philosophical work. It also shaped James's understanding of the practice of science and the practice of philosophy, and informed the new ideals of the scientist and the philosopher that he envisioned. As others have amply noted, James praised the virtues of tolerance and open-mindedness, while he condemned authoritarianism, dogmatism, and intolerance as "sins" against science.[33] I suggest that by painting the genuine scientific investigator as the embodiment of these social and epistemic virtues, he made the ability to transgress cognitive and social divides central to the practice of good science.

During the last decade of his life he also proposed a new understanding of philosophy, one that so far has received little, if any, attention. Introducing the notion of "general philosophy," a kind of endeavor that was distinct from technical philosophy, or "metaphysics," James expanded the goals and scope of philosophy. General philosophy, he explained, "include[d] all the sciences—logic, mathematics, physics, psychology, ethics, politics and metaphysics," and it aimed "at making of science what Herbert Spencer call[ed] a 'system of completely unified knowledge.'" It was philosophy in the ancient, and to James's mind, "more worthy sense" of the word.[34] By examining this conception of philosophy, I propose that James used it in order to allow for the formation of groups of inquirers in which people coming from a multiplicity of academic fields, professions, and other callings could participate in "cross-disciplinary" conversations and work together to pursue common goals.[35] Moreover, I argue, James refashioned the image of the ideal philosopher, making him into a paragon of just the mental and social sensibilities necessary to maintain that kind of community. "Genuine" philosophers (as distinguished from the "professional" philosophers) were, for James, those individuals who, because of their mental breadth and truly "encyclopedic" attitude—especially their ability to imagine other people's mental states and look at things from different perspectives—could function as "cross-roads" of truth, that is, as sites of intersection between different disciplinary and intellectual paths. He identified the philosophers "in the 'great' sense of the word" both by means of their location—boundary spaces, interstitial sites, and sites of intersection—and by means of special mental qualities that would enable them to act as "go-betweens," and made philosophers' main task the facilitating of exchanges and encounters among people who traveled along different disciplinary, professional, and social roads.[36] The philosopher, then, became an enabler of encounters within and across different academic and social domains, by means of which all the participants would work to unify knowledge. That, I argue, was James's plan for the "social" unification of knowledge.

One project that James pursued was the creation of a new "science of man." As Charlene Seigfried, Sergio Franzese, and other scholars have noted, like other philosophers before him James perceived that the human subject was uniquely located in the domains of nature and culture, and animated by desires that went beyond the given order of nature. Throughout his life, these scholars have argued, James worked at a "pragmatic anthropology," a philosophical inquiry into human nature that would not only explore human nature and "man's relationships to his environments," but also help people transform and improve their lives.[37] For James, a proper pragmatic inquiry into human nature would have to embrace a range of approaches, including philosophical, psychological, naturalistic, religious, and medical ones, and draw from a variety of practices. Not only did James challenge the supposed deterministic implications of the natural sciences, arguing that they need not bring "messages of death" to all human desires and aspirations,[38] but he also questioned the basic modes of organization of knowledge that made the articulation of an integrative study of human nature difficult to pursue. James also feared that the increasing institutional divergence between disparate approaches to the study of man would prevent scholars and others from appreciating and *cultivating* the wholeness of the human being. Thus, I suggest, James conceived the science of human nature as a project that would require the cooperation of a wide range of people, expanding beyond philosophers, physiologists, and psychologists to encompass physicians, psychiatrists, and educators, and even the proponents and followers of regimes of popular science and popular philosophy. The science of human nature that James envisioned would have to be a cross-disciplinary endeavor, and it required the systematic trespassing of divides.

James's account of the human self was central to this project. Itself a product of a cross-divisional and multivocal inquiry that straddled the divides separating metaphysics, normal adult psychology, psychical research, clinical psychology, the psychology of religion, and discussion of mystical and hygienic practices, the Jamesian self, I suggest, supported the formation of heterogeneous, open, cross-divisional epistemic-social communities. *William James at the Boundaries* explores the links between self and community both by probing the social valences of James's account of the self and by examining James's approach to a metaphysical problem that he took to be of paramount importance, namely the problem of the "Many and the One." Was the universe "one" or "many"? What kind of relationships existed between the one and the many? "If things be individual (as common sense affirms)," James in addition asked, "how can they interact at all?—for how can what is separate communicate? And what is meant at any rate by interaction?" [39] By exploring James's approach to this profound and widely debated metaphysical problem, I suggest that he simultaneously framed a practical conception of the human self and a social vision that would make possible, and underscore, the cross-divisional modes of social relationships that were so important to him.

James attacked the problem of the one and the many from a variety of angles. One central way of framing it asked how different "consciousnesses," different "pulses" or "bits" of experience, as well as different "selves," could "freely" combine into larger wholes, without losing their individualities and defining traits. James labeled this question the "problem of the self-compounding of consciousnesses" or also "the compounding of selves." [40] Building on the work of scholars like Charlene Seigfried, James Kloppenberg, Richard Gale, Gerald Myers, and Timothy Sprigge, I argue that to James this problem was not only a technical metaphysical and psychological problem but also a social problem. For James (as for other philosophers and psychologists of the time) it translated into the question of whether it was possible to rethink both society and the nature of the individual, and whether it was possible to imagine a type of society in which individuals could fully participate in a profoundly communal life, and nevertheless retain their individual autonomy and agency.

Inflected in psychological, metaphysical, epistemological, cosmological, psychiatric, biological, and zoological terms, the question of how the individual related to the wider society had great momentum in the late nineteenth century. In the United States it found great resonance especially among those who perceived that the country was headed toward a kind of economy dominated by giant corporations over which individuals, including owners, had little or no control. As a rich historiography has shown, intellectuals, artisans, workers, and middle-class men and women found that the new economic and social circumstances endangered the sovereignty and self-directedness of the individual citizen. Many wondered to what extent the autonomy and agency of the individual could be compatible with the powerful economic, political, and social institutions that imposed an ever-growing socialization on life. James, like many others, worried about the challenges that the new economic order posed to the autonomous self; yet, like other pragmatists, as James Livingston proposes, he also perceived that the new social order disclosed new possibilities, and explored new forms of subjectivity.[41] I suggest that, by doing so, James offered a conception of the self that enabled him to reframe the nature of the relationships of the individual with society, as well as the relationships among individuals in society.

James experienced the modern self as pluralistic and weak. Like many other psychologists, and also novelists and social commentators, he located inner division at the core of the self and opened up its boundaries. Carrying on yet another form of "boundary work"—this time of a psychological and metaphysical nature—James negotiated the borderlines that separated different pulses of consciousness that follow one another within the mind of an individual, and redefined the nature of the borders that divide different selves, or people, both from one another and from larger social and mental wholes. In particular, by allowing for "telescoping" and "interpenetration" along the margins of the individual self, he rooted the individual in community. At the same time, by prompting a range of techniques of self-cultivation and unification of the self, he worked to endow the multistranded, porous, and weak individual he drew with strength and agency. In short, even while avoiding all a priori metaphysical and ontological foundations for a unitary, closed, sovereign self, he bolstered the autonomy of the leaky self and the individual against the overwhelming homogenizing forces of capitalism. Thus, although the Jamesian individual might not be able to—or even ultimately be designed to—defeat big political and economic institutions, he/she could mount resistance against them and, through example, persuade others to follow. Above all, the form of selfhood that James construed would challenge the ways in which people interacted with each other in a capitalistic society—an absolute social and political priority for James. In lieu of the distant, anonymous, and purely "legal" business relationships that, according to James, characterized corporations and other bureaucratic organizations, he envisioned a completely different mode of social engagement, a deeply "sympathetic understanding of the other," even an intimate, almost mystical, interpenetration with others, one overrunning the boundaries of the open self.[42] James believed that such deep personal relationships would foster the formation of "concatenations" of individuals, and pluralistic, diverse, "intimate" communities within which individuals could retain their particularities, autonomy, and agency, yet cooperate for common goals. This, I submit, was James's "social" solution to the problem of the one and the many, and one of the chief goals of James's cross-divisional science of human nature.

To return to James's 1906 address before the American Philosophical Association, "The Energies of Men," I propose that in presenting it James aimed to instantiate the kind of inquiry into human nature that he hoped would result from cooperative, cross-divisional communities. With "The Energies of Men" James also aimed to alter the way that many American philosophers understood their discipline and challenge the professional self that many of them championed. The notion of energy, and especially the idea that by breaking through one's rigid routines and ingrained habits a person could tap unused reserves of energy, was central to James's program for the promotion of diverse, pluralistic communities of inquirers, as well as to his plans for the revitalization of philosophy and for the transformation of increasingly inflexible philosophers. Many of his friends, James told his audience, were so tied up in their disciplinary routines and so concerned about their scientific respectability that they were unable to engage unorthodox topics of conversation—including the bizarre subjects that James pointedly broached in his lecture. Their paralyzing intellectual inhibitions, James subtly implied, were not dissimilar from pathological cases of "habit-neuroses," functional diseases of the nervous system that resulted from the habitual repetition of an action or from the exclusive use of one part of the body, and made it impossible for patients to carry on other activities. With his unorthodox paper James challenged American philosophers to break through their disciplinary routines and professional restrictions, "unlock unused reservoirs of investigating power," and engage exciting topics outside the bounds of professional propriety.[43]

He also invited them to interact with the many heterogeneous groups of people who were greatly interested in the problem of increasing human energy and improving productivity and life. Indeed, with his address James enacted before a philosophical audience his novel understanding of the nature of philosophy and the function of the philosopher. In the address James read long excerpts of other people's work, making room for voices that belonged not to recognized professional philosophers, but to marginal characters whose concerns would otherwise never have been heard at a philosophical meeting. By transforming his "solo" presidential lecture into a multivocal performance, James illustrated by example the kinds of cross-divisional exchanges and conversations that he believed philosophers ought to keep alive, and that he regarded as necessary to the articulation of a properly integrative, pragmatic science of man. For an hour, he rendered the podium of the American Philosophical Association the kind of social-intellectual crossroads of knowledge he considered essential for the vitality of the academy and society.

To conclude, this book argues that many of James's scientific and philosophical interventions were part and parcel of a broader project that aimed to redefine American academic culture and society by making possible new modes of social relationships and intellectual exchange both in the academy and in society. James negotiated the divides separating academic disciplines and reconfigured the boundaries of the human self. In doing so, he offered new images of the human self as well as of ideal scientific and philosophical inquirers. These individuals would trespass intellectual and social boundaries, act as sites of intersection between different roads of inquiry, and engage in intimate exchanges across the divides of the self. They would keep alive cross-disciplinary cooperation and promote the formation of intimate yet heterogeneous communities that James deemed essential both to the academy and to social life. In the end, by redefining the boundaries of the self and those of knowledge James promoted a new intellectual-social configuration of knowledge and a new vision for American society.

The book is organized around topics rather than in chronological order. Chapter 1, "Philosophy and Science," sets the background for the rest of the book and introduces many of the book's secondary characters. It traces the fluid relationships between philosophy and science at the turn of the twentieth century. Focusing chiefly on episodes that James witnessed and people whom he encountered in the United States and during his many European travels, it examines some of the negotiations through which philosophers and scientists redefined their fields of inquiry and their public images. One such episode was the inauguration of Emerson Hall, a philosophy building at Harvard University. My examination of an early proposal for the architecture and location of the building suggests that the initiator of the project, the German-born psychologist and philosopher Hugo Münsterberg, understood Emerson Hall to provide an emblem depicting the subordination of all the scientific disciplines to philosophy and expressing the philosopher's leading role within the academy and in society.

Chapter 2, "Philosophy versus the Naturalistic Science of Man," examines James's early efforts to mediate the growing tensions between a more traditional philosophical inquiry into human nature as articulated in the context of "moral philosophy" and a newer "naturalistic," empirical "science of man" premised on brain physiology, evolutionary theory, and other natural sciences. After briefly examining the British and American debates that pitted moral philosophers against men of science in a fight for jurisdiction over the study of human nature and the mind, and also over higher education and public morality, I show how James negotiated the intellectual and pedagogical divides that separated the philosophical and the naturalistic sciences of man. Building on Ignas Skrupskelis's work, the chapter details how James endeavored to project a new image of his own academic self as "a man of the two disciplines," namely introspective philosophy and physiology.[44] It also argues that it was through his encounters with the "disputed boundary" increasingly dividing those two "lots" in the "field" of knowledge that James framed an early synthetic understanding of philosophy. Philosophy, which in the 1870s James defined as the "reflection of man on his relations with the universe," was to be an inductive, "architectonic," enterprise wherein, based on the facts accumulated by the "special sciences," philosophers (or philosopher-scientists) could infer general principles.[45]

The last section of the chapter briefly discusses James's famous attempt in The Principles of Psychology (1890) to recast psychology as a natural science, separated from metaphysics. As Skrupskelis has shown, that separation was meant to be only provisional, and James revisited it later in that decade.[46] This section concludes that during this period James began drawing a distinction between two ways of thinking about philosophy: first, as metaphysics—the discipline in charge of scrutinizing the assumptions made by the special sciences and solving the metaphysical puzzles they raised—and, second, as the "total body of truth," an overarching framework within which the disparate sciences could find unity.

Chapter 3, "James and the (Im-)moral Economy of Science," argues that James offered an alternative social and moral economy of science, as well as a new image of the genuine scientific investigator. From early on, James resisted both the prevailing scientific ethos of emotional detachment, self-restraint, and self-effacement, and the instrumentation of scientific objectivity—including, for example, self-recording or graphing machines, which represented to him and many others the material embodiment of that ethos. He rejected the idea that scientists should act as passive self-recording instruments, on which reality would inscribe its own definition, and emphasized instead the importance played by personal factors in the production of knowledge, including scientific knowledge. Meanwhile, as David Hollinger, Deborah J. Coon, and others have noted, he made the virtues of openness, anti-absolutism, and tolerance essential to scientific and intellectual life, offering a distinctive kind of scientific and intellectual ethos. Those moral, social, and epistemic virtues also ensured that the act of breaching boundaries would be central to the practice of good science.

This chapter studies the links between James's epistemology and his social and moral economy of science, focusing on James's activities as a psychical researcher, especially his approach to questions regarding the validity of evidence and testimony for putative supernormal phenomena (such as telepathy, or the "return" of spirits of the dead through the trance of a medium). It shows that James's epistemology of evidence was closely intertwined with the modes of social relationship that he promoted among researchers and with amateur practitioners during his investigations of psychic phenomena.

Chapters 4 and 5 deal with James's pragmatist account of truth and the controversy that surrounded it. In these chapters I contend that James's most influential philosophical theory represented not only an intervention within technical philosophical debates, but also an attempt to break the emerging barriers that separated philosophy from psychology and other human sciences. Chapter 4 ("Mental Boundaries and Pragmatic Truth") briefly shows how James transported a range of psychological and even physiological notions (such as the theory of ideo-motor action) into the philosophical concept of truth. Chapter 5 ("Pragmatism, Psychologism, and a 'Science of Man'") locates the early twentieth-century controversy over pragmatist theories of truth within the context of broader debates over psychologism and within the tense, indeed adversarial, relationships between philosophers and investigators who perceived themselves as psychologists. The chapter shows that James's account of truth deliberately violated the newly promulgated requirements around which the American Philosophical Association was striving to build a disciplinary identity. It reinterprets the controversy over pragmatism as a fundamental clash over the scope and character of philosophy and its future as a discipline. The concluding section of this chapter recasts James's pragmatist account of truth, and other central aspects of his pragmatism, as part and parcel of James's proposed science of man. There I suggest that James's attempts to create a properly inclusive practice of philosophy went hand in glove with his efforts to create an integrated image of the human subject and to frame a cross-disciplinary inquiry into human nature.

The account of the human self that James articulated, especially in the last decade or so of his life, was part of that project. I examine it in Chapter 6,

"Ecstasy and Community: James and the Politics of the Self." James's conception of the self drew heavily from psychology, psychopathology, psychical research, mysticism, and metaphysics, as well as from popular philosophy, mental hygiene, and a range of popular projects for the cultivation of the self. Revisiting the question of the social valences of James's account of the self, this chapter argues that James's conception of the self and also the techniques of self-cultivation that he advocated were elements of a broader rethinking of the relationship between individual and community. By rethinking self and society, James aimed to promote new modes of social interaction that would allow for community and solidarity, but also for individual agency, spontaneity, and autonomy.

Chapter 7, "The Philosopher's Place: James, Münsterberg, and Philosophical Trees," maintains that the question of the proper place for the philosopher and for philosophy was central to James's conception of "general philosophy" and to the ways in which, in the last decade of his life, he reframed the relationships between philosophy and the sciences. James depicted "general" philosophy as the "trunk" of the "tree" of knowledge, from which the sciences had branched out. The chapter locates this conception of philosophy amid the widespread period debates over the issue of the unity of knowledge and the question of the proper ordering—especially the right "spatial" ordering—of the sciences. It begins by discussing the ambitious plans of James's colleague Hugo Münsterberg for the classification of the disciplines and the unification of knowledge, as they were evidenced in one of his pet projects, the organization of the International Congress of Arts and Sciences that accompanied the St. Louis exhibition of 1904. I show that Münsterberg's conference program embodied a meticulous hierarchy of knowledge, one that located philosophy atop the academy and elite philosophers near the pinnacle of a highly differentiated social order. The chapter then examines other charts and trees of the disciplines intended to represent the true order of knowledge, and traces related proposals for redefining philosophy as a *scientia scientiarum*. I suggest that these taxonomical and unifying plans, including Münsterberg's, were often aimed at accomplishing two tasks: they served as a means for repositioning philosophy in an advantageous position vis-à-vis the special sciences, and they conveyed their creators' visions for the organization of the academy and the broader society.

Next I turn to William James, approaching him from the angle of his disagreements with Hugo Münsterberg on the question of the proper place of philosophy and the philosopher. While Münsterberg located the philosophers in metaphorical, exclusive "temples of philosophy," which found a material embodiment in "his" Emerson Hall at Harvard, James articulated a quite different philosophical space. He wanted philosophers to conduct their trade in liminal places, which would function as sites of cross-disciplinary and cross-divisional encounters and discussions. Here I argue that James and Münsterberg's divergences on the issue of philosophical spaces were central to their profound disagreements about the intellectual and social geography of knowledge, and over the ways in which knowledge should be unified. In this way, I suggest, through their conflicts they evinced and promoted two incompatible conceptions of the nature of philosophy and its function.

James not only assigned philosophers certain places, but also identified the mental sensibilities that characterize the true philosopher. Chapter 8, "The Philosopher's Mind: Routinists, Undisciplinables, and 'The Energies of Men,'" unveils those qualities by tracing James's attacks on disciplinary "routines" and professional "habits." Like other contemporaries, James perceived disciplines and professions as shaping the minds and bodies of those who followed their routines by engendering habits. All habits, however, were not necessarily good, and James believed that disciplinary habits could produce certain undesirable cognitive and social effects: namely, rutted modes of inquiry, tunnel vision, inability to engage with topics that fall beyond the purview of a discipline, and strict obeisance to norms regulating style and conduct that James summed up with the terms "academic respectability" and "decorum." The chapter underlines that James consciously challenged these norms by employing untechnical writing and rhetorical styles, and by praising, indeed even setting forth as ideals, "undisciplinable" intellectuals, such as his friend Thomas Davidson, a man who offered stern resistance against every possible habit. The chapter concludes by reconsidering James's APA presidential paper, "The Energies of Men," and reading it as an enactment of the traits that James assigned to the proper philosophical mind.

# 1

# Philosophy and Science

Philosophy is a queer pursuit, reckoned, as it sometimes is, to be the most sublime, and sometimes to be the most trivial of human occupations.
WILLIAM JAMES

## A New Home for Harvard Philosophers and Psychologists

On the December 27, 1905, Hugo Münsterberg, chairman of the Division of Philosophy at Harvard University, proudly began the dedication ceremonies for a new philosophy building, Emerson Hall. The three-story construction would house the entire Division of Philosophy, with the exception of the Education Department.[1] The first floor held a large lecture hall and several small seminar rooms. A "monumental staircase" led to the second floor, which housed a "large philosophical library" (with a special psychological wing). The third floor was entirely dedicated to the psychological laboratory. This consisted of twenty-four rooms, including a lecture room for the courses in comparative psychology, dark rooms, a battery room, a sound-proof room, several light-proof rooms, a photography room, and "a large, well-lighted and well-ventilated" room for animals, containing aquaria and cages for birds, rodents, guinea pigs, monkeys, and other medium-sized mammals. (See fig. 1.1.) A further unheated room in the attic above the laboratory would house amphibians, reptiles, and in the winter, hibernating animals.[2]

Born in Germany to a converted Jewish family, Münsterberg had moved to Harvard University in 1892 from Freiburg, initially with the task of directing the psychological laboratory. After some vacillation, he decided to stay at Harvard, and had quickly climbed the university hierarchy.[3] The idea of creating a new building entirely dedicated to philosophy had been brought up earlier, but Münsterberg made it his own and brought it to fruition. Münsterberg's long-coveted project started to take shape in 1901, when a Visiting Committee supported his demand for a "worthy monumental building" exclusively devoted to philosophy, to be erected in a "quiet central spot of the Harvard yard."[4] Four years later, after an intense fund-raising campaign conducted personally by members of the department, the building was ready for use.

The inauguration of Emerson Hall signaled the opening of the annual meetings of the American Philosophical Association and the American Psychological Association, both of which took place in the new facility. The event was immediately followed by a joint session of the two associations, on the theme "The Affiliation of Psychology: With Philosophy or with the Natural Sciences?"—a theme specifically chosen by Münsterberg. The German philosopher-psychologist opened the joint session with a paper on "The Place of Experimental Psychology," which hammered home several points. First, the new building provided the long-desired solution to a problem that had plagued the Harvard psychology laboratory ever since William James had founded it back in the mid-1870s: lack of space.[5] At the same time, Emerson Hall offered a palpable answer to the question of the relationship between psychology and philosophy. By lodging the psychological laboratory right inside the philosophy building, Münsterberg argued, Harvard made the clear statement that experimental psychology belonged neither to the biological nor to the physical sciences; psychology was part and parcel of philosophy. Münsterberg rhetorically asked his audience:

> Have I been right in housing psychology under this roof? I might have gone to the avenue yonder and might have begged for a psychological laboratory in the spacious quarters of the Agassiz Museum, to live there in peaceful company with the biologists; or I might have persuaded our benefactors to build for me a new wing of the physical laboratory. But I insisted that the experimental psychologist feels at home only where logic and ethics, metaphysics and epistemology keep house on the next floor.[6]

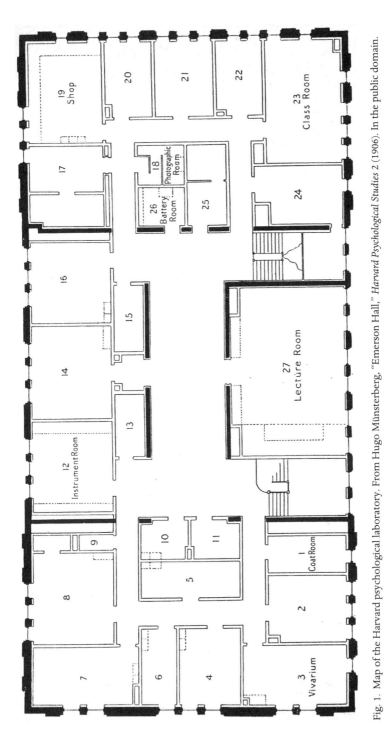

Fig. 1. Map of the Harvard psychological laboratory. From Hugo Münsterberg, "Emerson Hall," *Harvard Psychological Studies 2* (1906). In the public domain.

For Münsterberg, the similarity between a psychological and a physical laboratory was solely a matter of "externalities." While recognizing that experimental psychology was "an independent exact science," he insisted that philosophy was the "true basis" for psychology, and that only the psychologist who possessed a "philosophical background" could extricate the significance of a psychological experiment.[7] Had the psychological laboratory moved to the "naturalistic headquarters," Münsterberg continued, the consequences for psychology would have been quite negative. Inevitably, the laboratory would have narrowed its range of experiments to a few topics—sensation, perception, and reaction—ones already covered by physiology laboratories, thereby losing its very right to exist. Without training in philosophy and working contacts with philosophers, moreover, psychologists risked becoming mere "dilettantes," "somnambulists" walking on the edge of a roof, and failing to appreciate the height of it.[8]

Yet, to Münsterberg, Emerson Hall fulfilled even more important functions. The building put an end to a situation that had been inconvenient for philosophers, providing them now with a single location in which they could meet with colleagues, store their books and papers, and teach philosophy courses—all activities that were formerly "scattered under many roofs" throughout the campus.[9] Indeed, as Münsterberg remarked, Emerson Hall symbolized the unity of philosophy, collecting in one building all its various "branches"—ethics, logic, metaphysics, aesthetics, epistemology, psychology, and social ethics (the building hosted on the second floor a "museum of social ethics"). In addition, the building would encourage daily interactions and "mutual assistance" among philosophers working in the area, helping create a sense of community.[10]

Finally, and most important, Emerson Hall was to embody Münsterberg's answer to a question that was to him of the greatest importance: the question of philosophy's role and position with respect to the sciences. In 1901, in his letter to the Visiting Committee, Münsterberg had expressed his hope that the building would occupy a "central" position on the Harvard campus. That strategic position, he had argued, would symbolize both an ancient and new function for philosophy: giving unity to the university and restoring the unity of the sciences. At a time of growing specialization and division of labor, keeping alive the ideal of the unity of knowledge was an essential goal. The university had somewhat lost sight of the fact that philosophy was "more than one science among other sciences." Philosophy was indeed "the central science which alone [had] the power to give inner unity to the whole university work." The old title of "Ph.D." awarded every year to students in the sciences expressed symbolically that "all the special sciences"—that is, in the terminology of the time, all the scientific disciplines that, in contrast with philosophy's proverbial interest in supreme generalities, concentrated on a particular area of knowledge—"[were] ultimately only branches of philosophy." Unfortunately, the "truth of that symbol [had] faded away . . . in the academic community."[11] Nevertheless, with the end of the days of "antiphilosophical naturalism" and "unphilosophical positivism," days dominated by "realistic energies" and "the triumph of natural science and technique," a new philosophical age was imminent. Philosophy was now ready again to fulfill "its old historical mission." Today, Münsterberg stated, "the mathematicians and physicists, the chemists and biologists, the historians and economists eagerly turn again and again to philosophy, and on the borderland between philosophy and the empirical sciences they seek their most important problems and discussions."[12] He went on:

> The world begins to feel once more that all knowledge is empty if it has no inner unity, and that the inner unity can be brought about only by that science which enquires into the fundamental conceptions and methods . . . [of] the special sciences, . . . into [their] presuppositions and ultimate axioms, into the laws of mental life which lie at the basis of every experience, into the ways of teaching the truth, and above all into the value of human knowledge. . . . To foster this spirit of the twentieth century in the life of our University there is no more direct way possible than to give a dignified home to the philosophical work.[13]

Emerson Hall would effectively symbolize the coming of the new age, and would display "in stone" the new order of knowledge.

Münsterberg's efforts to create Emerson Hall and the events he organized for the inauguration of the building in December 1905 represented a visible intervention in a widespread debate over the value of philosophy, its nature, and its position in regards to the natural and human sciences in the modern academic landscape. Reconfiguring the spatial distribution of knowledge and positioning philosophy at its center (or, in other spatial arrangements, at its top, as we will see in chapter 7) was one effective way of addressing the issue and presenting a powerful answer to it. Indeed, as Münsterberg had clearly seen, the creation of a space where philosophers could conduct their work was essential to determining philosophers' self-understanding and public image, as well as to constructing a community of philosophers. For him, informal and private spaces, such as individual offices and libraries in one's home, would no longer suffice. Philosophers needed official spaces, and perhaps because of that Münsterberg made his own home (a "house that lent itself picturesquely to entertaining") into such a space, by collecting "rare antiques" and beautiful artcrafts from Japan, India, Mexico, Russia, and other countries.[14] Emerson Hall, a "dignified," "noble," "imposing" building, with its "beautifully crafted wood furniture and library," was to be an emblem of the dignity, poise, and decorum proper for the philosopher. The classical style of the construction made it look like a "Greek building," and pointed to the ancient genealogy of philosophy, signifying that Harvard philosophers belonged to a noble community going all the way back to the Greek philosophers.[15] The life-sized bronze statue of Ralph Waldo Emerson, unveiled by Emerson's son Edward during the dedication ceremony, provided an authoritative exemplar: it would remind visitors and the broader public, including the members of the American Philosophical Association and the American Psychological Association, that Harvard philosophers did not think of themselves as narrow specialists or secluded investigators.[16] Like Emerson, they intended to take on a role of social and moral leadership.

### Philosophers and Scientists

In a classical sociological study, Andrew Abbot argued that the professions, together with the disciplines, constitute a "system." The emergence or the institutionalization of a new profession or a new discipline is never an isolated event; it always involves messy negotiations, confrontations, and struggles over turf and jurisdiction with practitioners working in other professions or disciplines. These processes often result in shifts or alterations of the entire system of disciplines.[17] More recently, discussing the current state of anthropology and its "disciplinary borderlands," James Clifford put the point nicely: "There are no natural or intrinsic disciplines. . . . Disciplines define and redefine themselves interactively and competitively."[18] Philosophy at the turn of the twentieth century provides a perfect illustration of such claims. Beginning in the latter half of the nineteenth century, philosophers in many Western countries perceived that philosophy was going through a deep crisis, and found it imperative to reconfigure their field and redefine their public roles. They realized that to achieve those goals, they could not limit themselves to offering claims regarding the methods, topics, and epistemological rules proper to philosophy. Instead, like Münsterberg, they soon found that they would have to confront the claims advanced by other knowledge producers and social actors, define themselves interactively with rival groups, and clarify the position that philosophy ought to occupy in a rapidly changing disciplinary system. In other words, philosophers would have to change the entire geography of knowledge, that is, in the context of this book, the entire arrangement of the disciplines, and the configuration of their conceptual and methodological relationships as well as the "social" relationships among practitioners of different disciplines. Philosophers might labor to extricate philosophy from literature, art, and theology, or to clarify the relationships between those various endeavors; for the most part, however, they found that they must concentrate their efforts on the sciences. These efforts are the focus of this chapter and this book.

Most frequently philosophers' attempts to clearly redefine their discipline and public personae intersected and reacted against—in some cases bumped straight into—similar efforts made by practitioners of the natural and human sciences. As many recent studies in the history of science have demonstrated, in the mid- and late nineteenth century, scientists (or "natural philosophers" or "men of science," as investigators of nature often preferred to label themselves) were busy clarifying the scope, position, and authority of science, and creating new identities. By the turn of the twentieth century, recognized communities of "orthodox" scientists had emerged in many places, but throughout the preceding century scientists still comprised a variety of groups that competed over legitimacy and public recognition. For example, as Alison Winter has shown, in early Victorian England scientific practitioners were divided along lines of political orientation and class, but also along other lines that separated figures within the same social class. These various groups engaged in rival projects, and offered different understandings of nature and the rules governing scientific work. They portrayed their activities as legitimate science and those of their antagonists as "heterodox," and they often endeavored to edge out members of other groups. (In Britain scientific investigators such as Francis Galton worked to exclude clergymen scientists, on the grounds that they were incapable of behaving like men of science.)[19]

In the wider public sphere, men of science moving in a variety of circles sought to develop the public's "appreciation" for science, increase their "occupational opportunities," and position scientific studies as foundational in curricula of higher education.[20] They also endeavored to expand the boundaries of their emerging disciplines, and to clearly demarcate "science" from other activities, especially those that, by the turn of the twentieth century, established science practitioners construed as "popular science," "pseudo-science," and nonscience. Scientists' boundary work often involved the creation of material and social spaces of knowledge. These included, for example, geological and zoological museums, laboratories, and scientific institutes arranged in ways that would define scientific work, regulate access to science, "consolidate" divisions of labor and identity, promote forms of interaction, and, more generally, "guide the behavior of inhabitants."[21] Not only could these spaces make visible specific scientific theories,[22] but by virtue of their location, arrangement, and architectural style, also often enabled scientists to stake their claims to authority with respect to competitors and audiences, thus functioning, to borrow a phrase coined by Thomas Gieryn, as "truth spots."[23]

Meanwhile, scientific practitioners of various sorts struggled to redefine their public and academic roles. They often did so by differentiating themselves from scores of other people who engaged some of the same questions, including, depending on the circumstances, gentlemanly dilettantes, low-class amateurs, mechanics, artisans, and engineers, as well as men of letters, theologians, public moralists, and philosophers. At times, however, science practitioners strategically borrowed characteristics from other groups' self-images and identities (that of the man of letters, for example) in order to attain their goals.[24] They also pursued multiple models of the scientific investigator, ranging from the figure of the secluded specialist, insulated from social influences and free to determine the course of his inquiries on purely intellectual grounds, to the figure of the broadly educated man who, by dint of his universal capacities and representative character, believed himself to be qualified to engage in public affairs and broader social questions.[25]

Even though they have not received the same degree of attention, in the late nineteenth and early twentieth century self-titled "philosophers" similarly deployed a range of social, material, and rhetorical devices to achieve autonomy, attain public visibility, and redefine their intellectual pursuits and public personae. Centuries of conscious reinforcement of the ancient image of the philosopher as a pure, disincarnate, rational mind, single-mindedly occupied in the pursuit of Truth, have often translated into enduring historiographical practices that narrate the history of philosophy as a timeless dialogue among "resurrected philosophers," taking place in that most intangible of places, the philosophers' paradise.[26] Yet, as other historians remind us, philosophers in every era not only trade in ideas but also carry on all sorts of practical, even bodily, activities, ones that require much more than a pure mind, a pencil, and perhaps a library.[27] Philosophers at work in Western countries at the turn of the twentieth century provide a case in point. Like practitioners busy "producing" many other disciplines, philosophers created disciplinary genealogies, canons, and exemplary models in an attempt to specify cores of acceptable questions, topics, and methods.[28] They endeavored to build communities of inquirers and new audiences, often perceiving that in the changing configurations of the disciplines they needed to work together and create group alliances and identities. Like scientists, they also worked to shape material or metaphorical spaces where knowledge could be produced, assembled, divided up, ordered, showcased, or classified. Philosophers delineated the boundaries of their "field," and simultaneously defined "centers" and "peripheries," "insides" and "outsides." They policed borderline areas (such as the contentious "territory" of the study of the human mind), and they worked to shift the position of philosophy in the geography of knowledge by abundantly producing, as examined in detail in chapter 7, new taxonomies of the disciplines and maps of knowledge in which philosophy occupied a carefully chosen location.

To these ends they deployed all the main tactics that Gieryn finds to be associated with boundary work: expulsion (or exclusion), defense of autonomy, and expansion. Thus many philosophers set up and put in motion the machineries of "professionalization" and "specialization," terms laden with a variety of different meanings, in order to prevent outsiders from stepping into the field, and to restrict some types of inquirers to the role of amateurs or silent audiences.[29] Some narrowly defined the field and gave up large territories, seeking that way to protect their autonomy. Others struggled instead to expand philosophy's jurisdiction and to achieve some form of control over the sciences. Still others resorted to strategies that creatively engaged scientists and other rival groups. Some of them, including Hugo Münsterberg, as we will see in chapter 7, even staged complex public performances that symbolically assigned to scientists, philosophers, and other "performers" carefully preestablished roles that typified the proper arrangement of the disciplines.

Regardless of the specific tactics that philosophers chose to deploy, their attempts to redefine their field of inquiry and their social and cultural identities almost always required the presence of the scientist as spectator, friend, exemplary model, foe, negative ideal, or coperformer in a complicated and carefully choreographed pas de deux. This chapter looks at philosophers' efforts of self-definition by examining the ways in which they sought to construe the relationships, broadly understood, between philosophy and the sciences.[30] Viewing such efforts as part of the complex set of negotiations through which scientists and philosophers simultaneously came to define themselves and their fields allows for a new perspective on William James and the turn-of-the-century context in which he worked.

### William James and His Zigzagging Career

During the inauguration of Emerson Hall, William James, then in his early sixties, was sitting silently in the audience—fuming. Münsterberg, whom James had attracted from Freiburg to Harvard back in 1892 in the hopes of being relieved of the duty of directing the growing Harvard psychology laboratory, had become one of his most vexing enemies at home. The two could hardly agree on anything.

James had long been aware of Münsterberg's plans for Emerson Hall. As a good citizen of the department, he had even helped to raise the money needed for it (occasionally lamenting that his "victims"—the wealthy potential donors—did not "bleed well"). James, however, was "not very sanguine about 1000's coming in."[31] Indeed, he had expressed reservations about the whole business from the beginning. The need for space for the laboratory was real, but was it really necessary to create a home for Harvard philosophers? James did not think so. As soon as he was informed about the plans for the new building, he confessed to Münsterberg that he was not "sure" that he "shouldn't be personally a little ashamed of a philosophy Hall," even though, of course, he had promised to "express no such sentiments in public."[32] In 1903, when it looked like the project might actually materialize thanks to a $50,000 donation, James sent a concerned letter to his colleague George Herbert Palmer questioning the initial choice of the architect and the site.[33] James urged the committee build the new philosophy hall in a different, preferably more peripheral, spot (James suggested the "Holmes' field region"), or to design Emerson Hall as an "almost identical mate" to a neighboring building.[34] The Harvard corporation ended up assigning the job to a different architect. Other than that, however, neither of James's recommendations was followed, for both struck at the heart of Münsterberg's vision for the centrality, uniqueness, and supremacy of philosophy.

During the inauguration of the building, given that the only other speakers had been President Eliot and Ralph Waldo Emerson's son, Münsterberg's ostentatious intervention went beyond the limits of what James could bear.[35] Münsterberg had delivered "five speeches in one hour"; he had "bossed the show";[36] he had obviously "committed a fault of taste." His decision to play "the chief part" in what, after all, was a "Yankee" affair was unacceptable.[37] James complained to Eliot, who then notified Münsterberg, with the result that Münsterberg threatened to quit his job, and James had to apologize.

This skirmish, only one of many between James and Münsterberg, was more than a conflict between two famous members of the department whose philosophical Weltanschauungen were polar opposites. It was indicative of a deep divergence concerning the nature of philosophy and its function in the academy and in society. As we will see in chapter 7, it was a disagreement about distribution of power and forms of sociability, as well as the moral and social sensitivities that should inform the activity of philosophers and scientists. In the end it was a deep conflict between two radically different social arrangements of knowledge, and two radically opposite visions of what an ideal society should look like. This conflict, as I will suggest, revolved crucially around the issue of the proper location of philosophy and of the philosopher in material and symbolic spaces of knowledge. Thus, it will be important for this analysis to pay attention to the knowledge-spaces to which James and Münsterberg gave shape with their activities and knowledge practices.

Since the beginning of his career James had found that the issue of philosophy's relationship vis-à-vis the sciences had a profound personal relevance. As a young man, Louis Menand tells us, William James was "vocationally impaired." He constantly vacillated over the direction that his career should take. In the late 1850s James was studying art in hopes of becoming a painter. As Linda Simon has reconstructed, James also loved to walk in the fields and woods, to collect "specimens and pond water" that he would "examine under his microscope."[38] At his parents' home in Newport, Rhode Island, where he was studying painting under the guidance of William Morris Hunt, James arranged a makeshift laboratory where he performed chemistry experiments and other tests that included "imbibing" chemicals and experimenting on himself with galvanic batteries.[39] A few years later, when the family moved to Cambridge, he put together another small laboratory.[40] In those years James was debating a serious question. Should he continue studying art, or should he turn to the natural sciences?[41] He ended up enrolling at the Lawrence Scientific School at Harvard, where he studied, with only moderate success, chemistry, anatomy, and physiology with Eliot, and with the naturalists Jeffries Wyman, Asa Gray, and Louis Agassiz. However, James never graduated from the Lawrence Scientific School. In January 1864 he was hesitating between four alternatives: "Natural History, Medicine, Printing, Beggary."[42] One month later he enrolled in Harvard Medical School, but shortly afterward he interrupted his medical studies, and embarked on a naturalistic adventure, joining Louis Agassiz in an expedition to South America.[43] Plagued by mosquitoes, fleas, and even a mild form of smallpox, after a few weeks James "knew" that he did not want to become a naturalist, and wrote home that he "was going to study philosophy all [his] days."[44] He managed to receive his MD only in 1868, after many second thoughts and interruptions, caused by depression, back pain, and other physical ailments. That was the only nonhonorary degree he ever earned. As a medical student James worked for a short time at Massachusetts General Hospital, but quickly determined that medicine would also not be his profession.[45]

His continuing depression, insomnia, and time devoted to unsuccessful therapies, and a complicated relationship with his father, took a toll on his proposed courses of study, and caused (and perhaps also mirrored) much uncertainty over what he would like to do. James was still very attracted to natural history, biology, and physiology, and was quite pleased when in 1872 Eliot hired him to teach physiology at Harvard. At this time, however, he was also attracted to philosophy, even though he had reservations about the state of perennial doubt and "instability" in which philosophy seemed to keep its adepts. In 1873 he confided to his diary that "philosophical activity as a *business* is not normal for most men, and not for me." He feared that philosophy "breeds hypochondria." He craved "for some stable reality to lean upon," and thought he could find it in biology. Nevertheless, he would regard philosophy as his "vocation." His interest, he jotted down, would always be for "general problems," that is, in accord with contemporary usage of the term, "philosophical problems."[46] He suspected that such problems ought not be attacked "directly" and "in their abstract form," but instead should be addressed by solving "minor concrete questions" (possibly the questions raised by biology and physiology) as well as just "by living."[47]

From the late 1860s through the 1880s, James, who lacked formal training in philosophy (including the training in "moral" and "mental" philosophy obligatory at most liberal art colleges), gave himself a thorough philosophical education. He also read avidly, and critically, works in scientific philosophy, and devoured studies in anthropology, geography, physiology, clinical psychology, biology, and even botany.[48] At home his philosophical interests were spurred by informal discussions such as those he found at the "Club"—a "junior edition of the Saturday Club" (of which James's father was a member)—and the short-lived "Metaphysical Club," both examples of the "private philosophical and literary societies" where "intellectual work" was carried on in America prior to the emergence and establishment of the research university.[49] The Metaphysical Club famously included, among others, James, Charles S. Peirce, and Chauncey Wright (the group's "boxing master"), all of whom Peirce described as three young "men of science" with strong interests in philosophical and scientific questions.[50]

At Harvard, James initially offered courses in physiology, anatomy, and zoology.[51] In 1875 he added a graduate course in physiological psychology and, in 1876–77, an undergraduate course in psychology, which was listed within the Division of Natural Sciences.[52] In 1876 he was appointed assistant professor of physiology, and in 1880, after significant finagling, he managed to become assistant professor of *philosophy*.[53] As James confessed to his friend Ferdinand Canning Scott Schiller, the "first philosophical lecture that I ever listened to was when I began to lecture on philosophy myself!"[54] He began offering philosophical courses in addition to his courses in psychology, both groups of courses now listed in the philosophy department.[55] In 1889 James's title changed again, and he became professor of psychology, also in the Department of Philosophy.[56] Shortly after publishing his *Principles of Psychology* (1890)—the mammoth opus that both summarized the state of the field of psychology up to the time of its publication and offered many of James's original insights into a variety of areas—James started to complain about his teaching duties, and especially about his obligation to work in the psychological laboratory, for which, as he made amply clear, he had little inclination and insufficient skills.

In 1892, at the time when he sought to persuade Münsterberg to come to Harvard and "take charge of the Psychological Laboratory," James wrote to his friend Shadworth Hodgson, a British metaphysician, that, after devoting some ten years to psychology, he had "a longing for *erkenntnistheorie* [*sic*], and even cosmology."[57] He famously confided to the German psychologist and philosopher Carl Stumpf in 1895 that he "wish[ed] to get relieved of psychology as soon as possible," and the following year he asserted that he felt as if he "had bought the right to say good-bye to Psychology for the present, and turn to more speculative directions."[58] Although James stopped teaching his physiological psychology course in the mid-1890s, for a few years he continued teaching courses and offering popular lectures in the psychology of religion, the psychology of education, and abnormal psychology. In 1897 he reassumed the title of professor of philosophy, and concentrated on elaborating the philosophical programs for which he became most famous: pragmatism and radical empiricism. Despite the intentions he had expressed to his friends, James did not stop practicing science or cultivating his scientific interests. In particular, as Eugene Taylor and others have detailed, James continued carrying out studies in abnormal and supernormal psychology, or "psychical research" (the study of such phenomena as the trance of a medium, automatic writing, or the projection of the double), and kept abreast of literature on mental illness, child psychology, mental hygiene, and normal adult psychology.[59]

In a classic sociological study on the origins of scientific psychology, Joseph Ben-David and Randall Collins painted William James as an "idea-hybrid," a person who combined ideas taken from different fields into a new "intellectual synthesis."[60] According to their classification, James failed to be a "role-hybrid"—that is, a scholar who consciously mixed different "professional" or "academic" roles in an attempt to create a new "scientific role"—chiefly because, despite his work in psychology, he "finally decided on the traditional role of philosopher rather than the new role of scientific psychologist." This book posits instead that both the notion of "idea-hybrid" or that of "role-hybrid" do little to explain James's activities. James did much more than combine ideas and techniques from different emerging disciplines. And he did more than simply embrace an (unspecified) "traditional role of [the] philosopher" or the emerging identity of the "scientific psychologist" or combine those two identities (assuming, for a second, that such identities existed ready-made).

Many of James's endeavors involved a transgression of institutional, intellectual, and social divisions. He questioned both the boundaries that increasingly separated philosophy from the special sciences, and those that separated one special science from another. He also challenged the divides that separated orthodox science from what many of his colleagues construed as superstition, and crisscrossed the line that divided professional from amateur constituencies. This book argues that through such boundary work, James reinvented the role and the function of the philosopher at the same time as he reimagined the mental, moral, and social attitudes that would befit the genuine scientific investigator, thus redefining both the practice of philosophy and that of science. In doing so, he imagined a new social arrangement of knowledge, and linked it tightly to a new vision of society.

## James's British Circles: Philosophy Wars

James's early vocational and career uncertainties may have been the initial spur for his attentive consideration of the question of the relationships (most broadly understood) between the sciences and philosophy. However, during his frequent and protracted European travels and visits to foreign laboratories, learned societies, philosophical clubs, and scientific institutes, he quickly realized that that question preoccupied a great many men of science and self-described philosophers. Inflected in many different ways, questions concerning the mutual definition of philosophy and science and the extent to which scientists and philosophers should cooperate, kept popping up in the learned circles that James frequented in Britain, France, and the German-speaking lands. Through his exchanges with European and American colleagues, James became aware not only of the urgency that those questions carried for all those involved, but also of the existence of a wide range of possible answers and tactics to implement them. In an era of U.S. history that, as Daniel Rodgers has argued, was marked by an intense transatlantic flow of ideas and policies (as well as of commodities and people) and by an unusual American receptiveness to European models and "imported ideas," James—a compulsive traveler who crossed the Atlantic first when he was two years old and, the last time, just weeks before his death—looked abroad for new ways of framing a problem or new possible solutions.[61] Certainly James cannot be reduced to the figure of a "broker" of the transatlantic commerce of ideas and was never entirely comfortable with a cosmopolitan identity, insisting instead on the value of the "tender plant" of his "Americanism" and American culture.[62] Nevertheless, like many other travelers of the time (including educators, social scientists, social reformers, scientists, artists, and writers), he resorted to encounters abroad and tourism, both cultural and "natural," as a means to spin international scientific and philosophical networks, and to ensure the effective circulation of ideas, techniques, instruments, and practices.

This and the next two sections in this chapter trace, without any attempt at completeness, several of the debates regarding the question of philosophy's position vis-à-vis the sciences to which James was exposed during his sojourns abroad, following him through some of his travels to England, France, and the German-speaking lands. The last two sections return full circle to the United States and to Harvard's Emerson Hall.

In a letter of May 1878, the British philosopher-psychologist George Croom Robertson asked James, then in his mid-thirties, if he had any suggestions about how to put together an article on the state of philosophy in the United States.[63] The paper was to be part of a series of articles that Croom Robertson had commissioned to various scholars, some of which had already been published in the journal that he edited, Mind—"the first English journal devoted to Psychology and Philosophy."[64] James did not take up Croom Robertson's implicit invitation, and the report ended up being written by James's former student G. Stanley Hall (who apparently wrote it quickly, and just for the sake of making some additional money).[65] James read Hall's article and most likely some of the other articles in the series; after all, Mind was a journal that he regularly read, and where he hoped he would soon be published.

In the editorial that had opened the first issue of the journal in 1876, Croom Robertson had lamented the desultory state of philosophy in England. The country lacked an academic philosophical tradition, he wrote, and philosophy failed to enjoy any "scientific consideration." The authors of the articles on philosophy in London, Oxford, and Cambridge joined this jeremiad. Thus Henry Sidgwick (1838–1900), Cambridge's most renowned moral philosopher in the last quarter of the century, complained that at Cambridge the Moral Sciences Tripos (which included studies in "philosophical" fields such as psychology, ethics, metaphysics, political economy, and politics) was far less prestigious than the Mathematical Sciences Tripos and the Classics Tripos. Sidgwick was notorious for discouraging Cambridge students from sitting in the Moral Sciences Tripos, on the grounds that "the study of the moral sciences" did not prepare students for any profession, except for that of teaching them—hardly the type of attitude that could produce a local expert community of philosophers.[66]

Although in mid- and late Victorian Britain philosophy had an institutional basis, research in philosophy was pursued mostly outside of the universities by public moralists and "men of letters," a heterogeneous group mostly of

middle class people who, as T. W. Heyck has shown, included novelists, po-
ets, and journalists.[67] Essayists, jurists, politicians, magazine editors, scientific
investigators, science spokespersons, and scientific journalists, as well as people who were many of these things at once, could contribute to philosophy,
as illustrated by the cases of John Stuart Mill and of George Henry Lewes
(George Eliot's partner), a critic and literary essayist, editor, novelist, "expos-
itor of philosophy," and (apparently rather mediocre) dramatic actor.[68] Until
the 1860s there was no clear distinction between the man of letters and the
man of science.[69] Thus, for example, Herbert Spencer, who had never been
interested in philosophy or psychology until he was thirty, once the interest
arose (almost accidentally) could quickly train himself and eventually be
considered one of Britain's "most distinguished philosophers" by many of
his contemporaries, including philosophers with academic credentials.[70]

Until late in the century, philosophical writers did not address specialized
audiences. They often published in literary magazines or journals of opinion
for the broad readership that such magazines had helped to create.[71] Because
there was no clearly recognizable professional community of philosophers,
self-styled philosophers for the most part talked about philosophy at dinners
in restaurants or at their private residences, at meetings of informal philosophical clubs or small philosophical societies, during walks, and, occasionally, at larger dinner parties and soirées.[72]

James's British philosophical and scientific exchanges took place on such
occasions. In the summer of 1880, during a visit to London, he met Croom
Robertson and the metaphysician Shadworth Hodgson—a "bespectacled,"
"bashful & amiable philosopher . . . charming in the extreme," with whom
James had exchanged photographs over a year earlier, and whom James,
for a while, considered a sort of mentor.[73] James also met the philosopher-
psychologists James Ward and Alexander Bain, whose demeanor James
likened to "a snapping turtle," and who struck him as "utterly dogmatic
and charmless," and devoid of any "atmosphere" in "his mind."[74] He dined
once with Herbert Spencer (whose work James had already publicly criticized), but their conversation was "rambling and trivial," as James cautiously
avoided touching "the subject of philosophy," while Spencer "seemed quite
contented [sic] to gossip."[75] He met the neurologist Hughlings-Jackson, the
brain physiologist David Ferrier, and the Darwins.[76] A couple of years later
James returned to London to renew and extend his scholarly contacts. During his stay his social life "[spread] its wings amazingly," as he reported with
satisfaction to his wife Alice.[77] He became acquainted with Francis Galton,
who, "blooming & blushing & mincing-speeched," took James to the Royal
Society first, and then to the Athenaeum, for a dinner of the Royal Society's
"inexpressibly august Philosophical Club" (a scientific rather than a philo-

sophical club). James met other fellows of the Royal Society, including the physiological psychologist William Benjamin Carpenter and "Darwin's bulldog," Thomas H. Huxley, who, years later, James would present as the negative model of the man of science with metaphysical pretensions.[78] James was also introduced to two recently founded philosophical societies, groups "which existed in England at this time for the purpose of uniting good fellowship and philosophy": the Aristotelian Society and the "Scratch Eight."[79]

A small informal philosophical club, the Scratch Eight revolved around the literary essayist, critic, and editor Leslie Stephen, and included eight members who met over dinner (James was admitted as the "ninth").[80] Members included men of broad interests, such as the jurist Frederick Pollock, the philosophical and psychological writer James Sully, and Edmund Gurney, a "magnificent Adonis, 6 feet 4 in height with an extremely handsome face, voice & general air of distinction about him."[81] Gurney put James in touch with Henry Sidgwick and with the Society for Psychical Research, an institution established in 1882 by investigators interested in the "scientific" study of "phenomena designated by such terms as mesmeric, psychical, and Spiritualistic."[82] James joined the society, and became very close to both Sidgwick and Gurney. The author of a few metaphysical papers and a book on aesthetics and acoustics (which he wrote before he switched to psychical research full time), Gurney has no place in today's histories of philosophy. Yet James and other contemporaries had no qualms about his status as a philosopher, although Gurney's good looks struck James as "altogether the exact opposite of the classical idea of a philosopher."[83] When James left England, Croom Robertson kept him posted about the club and its vicissitudes. In the winter of 1884 he wrote that the Scratch Eight had temporarily stopped meeting, since people were pulled off in different directions by their various engagements—Stephen "teaching his Literature lectures at Cambridge" and Pollock "engulfed in Law-professoring at Oxford."[84] Shortly afterwards, the club dissolved for good.

The Aristotelian Society, founded in 1879–80 and still flourishing today, proved to be much more enduring. James was introduced to the society by Shadworth Hodgson, its founder and first president. James attended a few meetings and presented two papers, one in absentia. In February 1883, James was made a corresponding member of the society.[85] As such, he received a copy of most of Hodgson's papers and addresses. Reading those pamphlets James could not have been ignorant of an issue that, evidently, was a great source of concern to Hodgson and the Aristotelian Society: the issue of philosophy's status as a learned pursuit. The Aristotelian Society, Hodgson wrote, was "the only [British] Society . . . which has arisen spontaneously,

and unconnected with any College, University, or other public body, 'for
the systematic study of Philosophy.'" That kind of pursuit, however, in
England was clearly "not... recognized." "Let us not conceal it from ourselves," Hodgson stated, that "we are a nondescript tribe; a small tribe.... To
the ordinary Englishman of culture we appear as a rare and inexplicable
variety of the *dilettante* species."[86] That was not surprising, since at the time,
properly speaking, there was no "philosophy," only "a wrangling of sects"
and a multiplicity of competing philosophies. "The aim of this Society,"
Hodgson continued, was to alter that state of affairs and to make philosophy
into a recognizable, distinct "learned pursuit."[87]

Hodgson's concerns over the status of philosophy were amply shared.
Back in the 1850s, the Scottish metaphysician James Frederick Ferrier had
compared British philosophers to a loosely related group of people each
busy playing his own game: one played chess, whereas his adversary played
billiards, and so on. Since philosophers played different games, there was no
chance that they could really engage or understand one another; they could
not even meaningfully argue against each other, and a victory of one over
the other was simply impossible.[88] In 1879 Thomas Hill Green, professor of
moral philosophy at Oxford and perhaps the most influential philosopher in
late nineteenth-century Britain, complained about the absence of a type of
philosophy that could stand as autonomous both from theology and from
natural science.[89] Green lamented that, according to the British public, it
was "generally very doubtful, to say the least, whether Moral Philosophy, in
any distinctive sense of the term, represent[ed] any true object of human
inquiry."[90] Proving him correct, at his death in 1882, the Master of Balliol
College, who delivered the funeral sermon, acidly commiserated with Green
for having "chosen to devote himself to an important, if unfortunately
precarious, branch of knowledge."[91]

For many of the self-avowed philosophers who were troubled by philosophy's lack of status and even lack of a precise definition, the natural sciences
were a source of great concern. In the last decades of the nineteenth century,
British philosophers certainly worried a good deal about the materialistic
and deterministic worldview championed by men of science busy drawing
metaphysical and moral lessons from the theory of evolution, the generalized notion of the reflex arc, or the principle of conservation of energy.
However, in the last two decades of the century, philosophers were no less
concerned about the new images that men of science sought to project. Not
only did scientific inquirers feel entitled to write about metaphysics, but they
also publicly ridiculed philosophers and claimed for themselves the roles of
educators and public moralists, a role that had traditionally pertained to
philosophers and clergymen. The naturalist and science spokesman T. H.

Huxley was a case in point.[92] Sometimes mistakenly depicted as the prototype of the modern figure of the professional, specialized scientist, throughout his life Huxley carefully cultivated the persona of a man of liberal education and broad cultural interests. As Paul White has shown, Huxley sought to portray himself not as a narrow specialist (a person who would consequently only be able to enjoy limited social and cultural authority), but rather as a person of broad scientific and literary standing entitled to speak with authority on matters of cultural, social, and political import.[93] Huxley made it clear that scientific work was a prerequisite for philosophical inquiry: "the laboratory," he stated in 1895, "is the fore-court of the temple of philosophy; and whoso has not offered sacrifices and purification there, has little chance of admission into the sanctuary."[94] He challenged late nineteenth-century philosophical writers by offering a type of secular morality founded on the theory of evolution, which he advertised as more fit for modern times than metaphysical or religious ethics.

Lewes provides another example. He reported to the British public that philosophy, "once the pride and glory of the greatest intellects," had "everywhere in Europe fallen into discredit." Few continued to "believe in [philosophy's] large promises." Science was progressive, Lewes continued in a Comtian vein; it was "cosmopolitan" and it enjoyed consensus among its practitioners. Philosophy, instead, moved in circles and kept coming back to its errors.[95] It was national and fraught with dissent. In his two popular histories of philosophy Lewes attempted to demonstrate the "impossibility of Philosophy," the "incompetence" of the "philosophical method," and the inadequacy of the metaphysicians, whom, in his biographical sketches, he sometimes caricatured as pathetic embodiments of the traditional philosophical traits of absentmindedness, impracticality, and ineptitude.[96] Philosophy was to be practiced by a new kind of man: active and well trained in the sciences. Metaphysics was to be replaced, once and for all, by Positive Philosophy, a new type of philosophy that, as Comte had indicated, was to be based exclusively on the methods and results of the natural sciences. This new scientific philosophy alone, Lewes intoned, could unify the sciences and put a remedy to the "evil of specialty," an evil that "affect[ed] the very highest condition of Science, namely, its capability of instructing and directing society."[97] Any attempt to philosophize by other means was bound to end in failure.

British philosophers responded in a variety of ways to these threats. Among James's friends, James Ward counterattacked the "vilifiers" of metaphysics, stating that the progress of the special sciences was due precisely to metaphysics.[98] For Henry Sidgwick and Shadworth Hodgson, on the other hand, the only way of defining philosophy and showing its legitimacy as a distinct field of inquiry was by clearly demarcating the boundaries that separated it from the sciences.[99] Surveying the most widely held opinions about the nature of philosophy and of science, Sidgwick found that, in contrast with the sciences, which always dealt with "some special department of fact always marked off and separated from other departments," philosophy dealt with the "systematic knowledge of things in general."[100] Sidgwick shared with many others (including positivists like Comte, Lewes, and Spencer), an "architectonic" vision of philosophy, according to which philosophy was an attempt to combine the "special disciplines" and the various sciences into a systematic, rational system.[101] In its systematic efforts, Sidgwick wrote, philosophy was both the "crown" of the sciences and the "germ" from which they stemmed—acting as the propelling force behind the budding of new sciences from the tree of knowledge. (Sidgwick, however, had to concede that unfortunately no such systematic philosophy was yet in sight, and that science was "not yet crowned.")[102] For Hodgson, however, painting philosophy as an architectonic science and reducing it to a science of sciences, a *"scientia scientiarum,"* missed precisely the point. Hodgson attacked the "positivists" twice: once, for circulating a dark public image of the philosophers as "ignorant and obscurantist" conspirators busy spreading a false wisdom for hidden social purposes,[103] and, a second time, for conflating philosophy with science, thus losing the right to claim for philosophy any distinct *"locus standi* in the intellectual world."[104] In sum, for many British philosophers, including some of James's philosophical friends, the question of the relationships between philosophy and the sciences (as well as that of the interactions between cultivators of philosophical studies and men of science) was important and inescapable. Indeed, it rested very much at the center of their philosophical engagements.

British philosophers provided James not only with a set of answers to that question but also with different embodiments of the persona of the philosopher, including that of the "professional" philosopher. This was a term the meaning of which even those who saw themselves as "professional philosophers" in late Victorian Britain could barely agree on. Green, Sidgwick, and Hodgson are three cases in point. Green, in line with his attempts to reform Oxford University, worked hard to transform philosophy into a "specialized" field of inquiry—a vision that ran against Oxford's traditional understanding of philosophy as a technique for the mental and moral cultivation of its pupils.[105] Thus, for Green, who clearly viewed himself as a professional philosopher, professionalization went hand in hand with specialization and with the making of philosophy into an academic discipline. Sidgwick likewise presented himself as a "professional philosopher," but by that he meant something very different. For him philosophy, in contrast to the sciences, was not and could never become a specialized discipline. Philosophy did not investigate any well-demarcated area of knowledge. It was the "study of things in general."[106] Students learned from Sidgwick that, by definition, the philosopher could be no specialist, and that what characterized the philosopher instead was a frame of mind, an attitude of "candor, self-criticism, and regard for truth."[107] As for Hodgson, whom James once described as "a *gentleman* to his finger tips & a professional philosopher as well . . . a rare combination," he was concerned to make philosophy into a distinct "learned pursuit"—rather than an academic discipline—one recognizable as such by the cultivated Englishman.[108] Hodgson never held any academic position, and was proud that very few among the members of his Aristotelian Society came "from the ranks of those who are or have been professionally employed in the teaching [of philosophy]."[109] Apparently he took the study of philosophy to require seclusion, in glaring contrast with Green and Sidgwick, both of whom amply engaged in political life and in all sorts of public projects, obviously armed with the belief that the philosopher had an important role to play in shaping citizens.[110]

These three cases reveal a broader point: late nineteenth-century efforts to redefine philosophy and make it into an established field of inquiry were not necessarily linked with projects of specialization, and specialization or discipline formation was not seen as strictly tied to professionalization. These processes, which sometimes have been conflated, proceeded at different paces. Furthermore, participants showed no agreement on what kind of specialization and professionalization should take place. A perceptive observer, William James sensed something odd about the British that touched on this issue. Writing in 1883 from London to the British philosopher Thomas Davidson, James confessed quite frankly that there was something strange in "the english [*sic*] race." The British, James wrote, including presumably the philosophers and men of science with whom he had recently interacted,

> seem to be voluntarily clogging their lives with so many superfluities that one would suppose to stand in the way . . . of pure intellection. They drag down through the centuries, smeared with the fog & the smoke, their great muffler of respectabilities, insisting on keeping the amateurish attitude, & not talking after the manner of professionals about anything.[111]

Many of James's British friends, even those interested in establishing philosophy in the academy or in shaping it as a well-defined type of inquiry, often balked at professionalization and at the institutionalization of inquiry (Davidson himself being, as we will see, the most glaring example), and preferred to behave in ways that James, busily fashioning his own academic career, found "amateurish."[112] And yet these British acquaintances gave James food for thought and provided examples showing that the refashioning of philosophy did not need to proceed through the narrow path of "professionalization" that, at the turn of the century, many American philosophers would pursue.

## Philosophers of the Third Republic

The philosophical and scientific network that James began to weave in the early 1880s extended to the other side of the English Channel, where James met many philosophers, physiologists, neurologists, and psychologists. Among them was Charles Renouvier, the editor of the journal *La Critique Philosophique*. "The queerest looking old boy you ever saw," as James wrote to his wife, Renouvier was, according to James's first biographer, "the greatest single influence upon James's thought."[113]

In 1882 James met the self-described "experimental psychologist" Théodule Ribot, founding editor of the *Revue Philosophique*, "an insignificant little fellow, but thoroughly bon-enfant, & . . . probably as useful a man to philosophy as there is in France."[114] James presumably had read Ribot's article on the state of philosophy in France in the 1877 issue of *Mind*, part of the journal's efforts to chart the state of philosophy in the last quarter of the century. In that partisan essay Ribot identified four main philosophical positions. The first was a belated, "torpid" form of Cousinian eclecticism, which Ribot blamed for having "divorced [itself] . . . completely and shamefully from the sciences."[115] The second was an "obscure" form of "mystic spiritualism," which, Ribot charged, accepted the "data of science" but "interpreted [them] at fancy," subordinating facts to its moral and metaphysical goals.[116] The third was "positivism," both of the Comtian variety and an "experiential" variety—the latter one being a type of philosophy most fit to men "possessing scientific culture" and "distrusting metaphysics," and to investigators busy transforming former philosophical disciplines (such as psychology) into natural or biological sciences.[117] The fourth was Renouvier's unorthodox neo-Kantianism.[118]

Despite Ribot's dismissal of all types of philosophy, except for his own experiential positivism, as unscientific, philosophers working within each of the four approaches identified by Ribot engaged the question of philosophy's relationship to the sciences, and many argued for the necessity of some form of interaction.[119] Thus, as Daniela Barberis notes, the undisputed leader of Third Republic eclectics, Paul Janet (whose *Principes de métaphysique et de psychologie* James carefully read), was far less dismissive of science than

Ribot implied.[120] And third-generation spiritualists like Henri Bergson and Émile Boutroux thoroughly engaged the sciences in their work. "Beautiful Bergson," the philosopher whom in the last years of his life James famously credited with the central ideas of his own radical empiricism, made the question of science central to his attempt to define philosophy.[121] Although Bergson contrasted philosophy, especially his "nouvelle philosophie," with the quantitative and spatialized thinking of the sciences, he packed his philosophical work with references to mathematics, physics, biology, psychology, and psychiatry, as well as new technological domains, in particular vision and motion photography.[122] Like Bergson, Boutroux offered a detailed critique of science, specifically its mechanistic and deterministic assumptions. He was very concerned that, due to the mounting tendency to "specialism," philosophy—a "universal" type of knowledge essentially different from all the specialized disciplines—would fall apart and be replaced by a multiplicity of disconnected sciences.[123] Yet this "gentlest and most modest of philosophers," as James years later described him when the two became great friends, firmly believed in the need for interaction between philosophers and scientists, and worked to make that happen.[124] As for Renouvier, he incorporated a discussion of mathematical notions into the heart of his ethics, with the result that his philosophy, as the purist Shadworth Hodgson complained to James, was an unsavory "Mittelding, half science, half philosophy."[125]

As Jean-Jacques Fabiani has shown, French academic philosophers were greatly concerned with the challenges that scientists, especially double-identity practitioners such as the philosophy-trained experimental psychologist Ribot, posed to philosophy.[126] In their efforts to defend philosophy (and themselves) from such "insiders-outsiders," these philosophers resorted to a variety of maneuvers, ranging from furious attempts to place scientists outside of the confines of philosophy, to the creation of professional societies,[127] to armed battle against education modernizers. Modernizers challenged the privileged position of philosophy in the state system of higher education and campaigned to institute scientific and vocational curricula, with full access to the baccalaureate and, as a consequence, to governmental positions.[128] Against them, philosophers strove to reassert their traditional roles of educators and moral arbiters of society, hammering the point that the classical curriculum alone (especially the philosophy course) could help shape citizens, and even sustain France's "influence abroad."[129] Others depicted philosophy as the "source," "the goal," and the "conscience" of the sciences, arguing that philosophy alone could direct the sciences and give them unity. Others still worked to enforce epistemological and metaphysical classifications that placed philosophy at the top of the hierarchy of the disciplines.[130]

Demarcating philosophy from the natural and social sciences, and reenvisioning its position in a broader geography of knowledge, was central to the practice of philosophy in late nineteenth- and early twentieth-century France. French philosophers found in such endeavors an effective strategy for redefining their field and reasserting their public roles.

## "German philosophy is dead."

"German philosophy is dead," Charles Augustus Strong, a young American philosopher, wrote William James in January 1891.[131] Strong, a student of James's at Harvard, had moved to Germany to study philosophy. He quickly figured out that, rather than "stick[ing] to [his] old trade of abstract philosophy," it would be more profitable for him to prepare for philosophy by studying German psychology and the natural sciences. Few German philosophers, of course, would have agreed with Strong's blunt claim, but many had shared for years the feeling that academic philosophy was going through a deep crisis or, more optimistically, had just begun to recover from one.[132] Philosophy, however, seemed to prosper outside of the academy through the popular forms of scientific (often materialistic) metaphysics devised by biologists, physiologists, zoologists, and physicians such as Ernst Haeckel, Henrick Czolbe, Jacob Moleschott, Karl Vogt, and Ludwig Büchner.[133] These natural scientists felt entitled not only to take up philosophical questions, but also to weigh in on the nature of philosophy, and to speak with authority on the question of who ought to be considered a true philosopher. For example, Haeckel, the most visible proponent of Darwinism in the German-speaking lands, mobilized his biological investigations both to articulate a new, monistic type of metaphysics (a form of hylozoism) and to redefine the field of philosophy. Philosophy, he stated, was by no means a "separate science," nor could it exist independently of "the common empirical sciences." The task of the "Queen of the sciences" was that "of combining the general results of the other sciences, and of bringing their rays of light to a focus as in a concave mirror." Haeckel challenged academic philosophers' definition of the "philosopher." "It is a sure sign of a philosopher that he is not a professor of philosophy," he provocatively stated. "Every educated and thoughtful man who str[ove] to form a definite view of life" was entitled to be considered "a philosopher."[134] Haeckel took a stab at academic philosophers busily "watch[ing] with jealous eyes" every "door" that gave access to philosophy, entirely oblivious to the fact that the natural sciences, especially biology, offered a more than legitimate "ticket of admission" to philosophy.[135]

This was precisely the kind of response that Strong encountered in Germany when he consulted with the experimental psychologist Hermann Ebbinghaus and the philosopher and educational reformer Friedrich Paulsen about the best way to prepare for a career in philosophy.[136] Ebbinghaus never engaged traditional philosophical areas of inquiry, but he was a philosopher by training, and felt entitled to offer authoritative advice concerning the requirements necessary for the correct practice of philosophy. He lectured Strong that one must "prepare for philosophy by working at science." Paulsen agreed: nobody would be "fit" for the philosophical "business . . . who [had] not been drilled in the school of positive science and served his apprenticeship there."[137] Strong dutifully reported both conversations to James.

James found that Ebbinghaus's laboratory work on memory (for which the German psychologist was famous) was as "heroic" and equal (in its tediousness) to the meaningless and useless feats of saints.[138] He was much more attracted to Paulsen's views. Two years later he eagerly read Paulsen's *Einleitung in der Philosophie* (1893), in which Paulsen presented his position concerning the interaction between philosophers and scientists. German academic philosophers had dismissed the book as too "imprecise and dilettantistic."[139] James, however, enjoyed it thoroughly, and praised precisely the book's popular tone, its lack of technicality, its unpretentiousness, and even its inexactitude.[140] The two philosophers started a correspondence that continued until 1902, and James sponsored the English translation of Paulsen's *Einleitung*.[141]

Paulsen challenged all attempts to make philosophy a distinct discipline (for example, by assigning it a special method or a special subject matter). What Paulsen did not like about these attempts was that ultimately they aimed to draw "a line around philosophy" and to separate it from the sciences. To him that was wrong: no boundaries divided philosophy from the sciences.[142] Philosophy was "simply *the sum-total of all scientific knowledge*," the "science of the sciences." It was a "uniform"—if never-completed— "system," a *"universitas scientiarum."*[143] While each particular science investigated a "portion of reality," philosophy's subject matter was "the whole of reality."[144] Thus the sciences did not exist "outside of philosophy." Rather, they all "belonged to philosophy"; they were "parts" of it.[145]

Paulsen believed in keeping the field open. He refused to write for supposed professionals and always endeavored to reach a broad, unspecialized audience. Not surprisingly, for him philosophy could not be a "profession." Properly speaking, "professional philosophers"—people who, according to Paulsen's definition of the "philosopher," could grasp all of science—did not exist. Philosophy was more than anything else a mental disposition. Anyone who kept an eye on the goal of the unity of knowledge was a philosopher, regardless of his particular topic of investigation. Thus naturalists such as Charles Darwin and Alexander von Humboldt provided excellent examples of the "philosophical mind."[146] Paulsen stressed that, in contrast with the "modern *Philosophieprofessor*," the most famous philosophers of the past (from the Pre-Socratics to Aristotle, Thomas Aquinas, Descartes, Leibniz, and Wolff) were people endowed with "encyclopedic minds," general savants, scientists who combined the study of metaphysics and theology with that of mathematics, physics, optics, and psychology.[147] As we will see in chapter 7, later in his life James appropriated this viewpoint (and Paulsen's image of the philosopher) in his own attempts to refashion philosophy and reconfigure the geography of knowledge.

James was well aware of the great variety of opinions that late nineteenth-century German philosophers embraced concerning the proper relationships between philosophy and the sciences, and the identity of the philosopher or the psychologist. In 1882 James traveled to Prague, where he met the "good & sharp nosed" Carl Stumpf (1838–1916), a "pale and anxious" man, with whom he soon made friends.[148] Both a philosopher and a psychologist, Stumpf, as Martin Kusch writes, was "eager, or willing, to fuse the role of the philosopher with that of the natural scientist," and to blur the boundaries separating philosophy from the natural sciences.[149] Like James, Stumpf was personally rather ambivalent toward experimental work.[150] Nevertheless, he stressed the importance of experimentation and maintained that the recent "renaissance" ("Wiedergeburt") of philosophy was mainly due to "a psychology which [had] been carried out in the spirit of the natural sciences."[151] As a contemporary American observer commented, for Stumpf philosophy was the most general and highest of the sciences, but it was still a science: the philosophy that would "endure" was the philosophy that "[grew] out of the other sciences," used their "methods" and rules of evidence, and spoke "their dialect."[152] In Prague James also "walked and supped" with Ernst Mach, a professor of physics and a "genius of all trades." James admired Mach's broad interests: the man had "read everything & thought about everything," and in Mach's physiological works James detected the presence of a "masterful . . . philosopher."[153]

A few weeks later, James was in Leipzig, where he met Wilhelm Wundt, the founder of what many contemporaries considered the first psychology laboratory in the world. A psychologist and professor of philosophy, Wundt had received training in medicine and had years of experience in physiology laboratories. In the mid-1870s James had depicted Wundt as the embodiment of a new type of philosopher—a person who could command both science and philosophy—and credited him, and others like him, with a "serious revival of philosophical inquiry."[154] Thanks to the work of Mitchell Ash, Martin Kusch, and others, we now know that, in the mid-1870s and afterwards, Wundt was much less eager to combine the personae of the physiologist and the philosopher than James would have imagined. Wundt's reluctance depended on reasons that had much to do with academic politics and with his desire to be accepted among his philosophical colleagues.[155] Not surprisingly then, Wundt kept revising his opinion concerning the issue of philosophy's relationship to the sciences. At times he contended that philosophy rested on the empirical findings of the sciences.[156] At other times he prophesied that philosophy would regain its influence over the empirical sciences, or stressed that philosophy alone could indicate the conditions necessary for the production of scientific knowledge and unify the sciences into a system of knowledge.[157] Initially favorably impressed by Wundt's "refined elocution," "agreeable voice & ready tooth showing smile," James later revised his opinion and condemned Wundt's pretension to occupy "a sort of [infallible] pontifical position" as "shameful."[158]

A decade later, during one of his European vacations, James bumped into yet another German philosopher with strong opinions concerning philosophy's role with respect to the sciences: Richard Avenarius (1843–96). In 1877, in cooperation with Wundt and others, Avenarius had founded the *Vierteljahrsschrift für wissenschaftliche Philosophie* (Quarterly for Scientific Philosophy), a journal that was premised on the assumption that "the time [was] past when philosophy [could] hope to live apart from the other sciences."[159] James subscribed to the journal from 1877 to 1898. He met Avenarius by chance in the spring of 1893, when the two philosophers and their wives happened to be guests at the same pension in Luzern, on Lago Maggiore.[160] James found him to be "a very good natured creature,"[161] but he could not reconcile himself to Avenarius's plans for philosophy. Avenarius strove to transform philosophy into an "exact" or "strict" science ("strenge Wissenschaft"), James reported to Stumpf. James had serious misgivings about that: "I have an *apriori* distrust of all attempts at making philosophy systematically exact just now.... The frequency with which a man loves to use the words 'streng wissenschaftlich' [strictly scientific] is beginning to be for me a measure of the shallowness of his sense of the truth. Altogether, the less we have to say about 'strenge' the better, I think, in the present condition of speculations."[162]

Ebbinghaus, Paulsen, Stumpf, Mach, Wundt, and Avenarius all believed in the necessity of some form of close interaction between philosophy and the sciences. Many disagreed. Neo-Kantians and neo-Hegelians, in particular, were often quite hostile to all attempts to blur the role of the philosopher with that of the scientist.[163] At the turn of the twentieth century, these philosophers engaged in impassioned and ultimately successful battles that aimed to separate philosophy from the sciences, purify the persona of the philosopher, and make philosophy once again into a true foundation for all the sciences. James was well aware of their endeavors, especially since, as we will see in chapter 5, many arguments developed by these purists provided handy weapons that James's critics deployed against his philosophical work.

To sum up, issues about philosophy's position vis-à-vis the sciences were central to the work of many of James's British, French, and German-speaking philosophical and psychological acquaintances. Their views reflected the peculiarities of national contexts and were strictly linked to particularities of local institutions and systems of education. Yet participants in debates about those issues also communicated across national divides through international networks such as the ones that James endeavored to create through his travels. James posted in his photograph album the pictures of many of the philosophers and philosopher-scientists whom we have encountered in this chapter, as if to display the possibility for such transnational conversations.[164] These investigators provided James with a wide range of examples (positive and negative) of the ways in which the question of the proper relationships between philosophy and science (and philosophers and scientists) could be framed, and the tactics by means of which particular answers could be enforced. Most of them, as we saw, carried out some form of boundary work, which, indeed, was central to the practice of philosophy (and neighboring sciences) in all those countries. They all supplied a vocabulary and a stock of arguments to which James would resort, or which he would attack or modify, in his own attempts to redefine philosophy, the persona of the philosopher and that of the scientist, and the academy.

### America: From "Moral Philosophy" to the APA

The question of philosophy's position toward the sciences loomed large in the United States as well as in Europe. Historian of philosophy Daniel J. Wilson has argued that this specific issue propelled the creation in 1901 of the first national professional association, the American Philosophical Association, and more generally the professionalization of the field at the turn of the century. He suggests that those events resulted mainly from American philosophers' worries about the increasing professionalization, specialization, and institutionalization of the sciences.[165]

From the late eighteenth century until roughly the Civil War, philosophy in America meant "mental" and "moral philosophy." Mental philosophy included logic, metaphysics, epistemology, and the study of the human mind, whereas moral philosophy was concerned with ethics and normative questions.[166] Moral philosophy, in combination with mental philosophy, was sometimes construed as an all-embracing study of human nature, institutions, and society. This philosophical "science of man," as it was sometimes labeled, culminated in the formulation of rules that should regulate conduct in private life as in public affairs. Philosophy was very important in the colleges, where, as Bruce Kuklick wrote, moral philosophers "emerged as the guardians of the character of the youth," and "guarantor[s]"—especially in the Northeast—of middle- and upper-class values.[167] By the 1870s, however, moral philosophers found their claims to educational leadership challenged on many fronts. Naturalists, self-appointed science spokesmen, and other proponents of a "modern" education were ready to replace philosophers in their social and pedagogical capacities. These men promised to bring new, more effective tools to the task of "reconstruction."

By the late 1870s, the academic and professional rise of American science, together with the metamorphosis of the science amateur into a specialized researcher, had become visible in many fields. Graduate schools and research-based graduate training in various sciences, including mathematics, physics, chemistry, and psychology, along with specialized journals and nationally based scientific societies, had been established at several universities in a process that contributed to the overall transformation of the American university.[168] The institutional growth of the natural sciences and disciplines that had been formerly regarded as philosophical—whether it can be taken as a parameter of their "professionalization" in "our" sense of the word or not—went hand in hand with the increased prestige surrounding the special sciences. Scientific methods and the values associated with them were increasingly viewed as norms for research in the universities. Natural scientists could impugn philosophy as an unscientific pursuit, inferior to the sciences. By the mid-1890s and the beginning of the twentieth century, social scientists occasionally claimed for their own disciplines the overarching position of a *scientia scientiarum*, or *regina scientiarum*, that philosophers had traditionally ascribed to philosophy (and theologians to theology).[169] G. Stanley Hall, for example, proclaimed the beginning of an era that would "be known hereafter as the *psychological* era of scientific thought." Psychology, he declared, would become the "long hoped for . . . science of man," and would thus replace moral philosophy once and for all.[170] Chicago sociologist Albion Small painted sociology, rather than philosophy, as the architectonic science in charge of planning the building of knowledge and coordinating the efforts of political economists, historians, moralists, psychologists, anthropologists, and all the scholars interested in finding out "the meaning of human experience."[171]

Philosophy, according to many American thinkers, was going through a serious crisis. In 1892 Jacob Gould Schurman, the editor of the newly established *Philosophical Review*, observed that there was "scarcely a province of the entire realm of science and scholarship which is now without an official organ in America... except for philosophy.... There is neither an organ nor an organization of philosophical activity in America."[172] Almost ten years later, E. G. Creighton, an idealist philosopher at Cornell University, complained that "in many colleges and universities the place of philosophy" was "only grudgingly conceded," and that philosophy was still regarded as "a more or less useful handmaid to theology, or perhaps to education." Philosophy, Creighton lamented, lacked the "general recognition, even among the public, that is accorded to many of the other sciences," and the "philosopher" was not "universally conceded to be a specially trained scholar whose opinions in his own field [were] as much entitled to respect as those of the physicist or biologist in his special domain."[173] As a result "men wholly unschooled in the subject... frequently [felt] themselves competent... to write philosophical books and articles," and exhibited not infrequently "the greatest contempt for professional philosophers."[174]

It was in response to such attacks that Creighton and others established the American Philosophical Association in 1901.[175] In the early years of the century, many members of the Association realized that any attempt to improve philosophy's public standing and academic prestige would have to address the question of philosophy's relationship to science. Was philosophy a science? Ought it to become one? While a few rejected that idea, the majority of philosophers insisted that philosophy ought to transform itself into a science if it ever wanted to escape, as C. S. Peirce put it, from its "crude condition."[176] But how could philosophy become "scientific"? And what did that exactly mean? Here philosophers could not agree. That is not surprising since in the United States there was no consensus on what exactly constituted science. For some scientists and some philosophers, science was marked by a rejection of any kind of metaphysical claims, by a strictly empiricist epistemology, by a specific "standpoint," or by the use of the "hypothetical deductive method" or other quantitative, empirical, and experimental methods. However, they had sharply divergent understandings of how these epistemological or methodological requirements could be satisfied. For others, the defining feature of science consisted in specialization and professionalization, in sharp contrast with philosophy's proverbial generality and fuzzy professional requirements. For still others, science consisted in the first place in a "way of life" or in a certain "ethos"—that is, a group of epistemic, cognitive, moral, and social "attitudes" that informed scientific behavior and were taken to be necessary for the production of scientific knowledge.

But here again opinions multiplied both among scientists and philosophers, since different types of scientific ethos were available. According to many, the practice of science demanded first of all exactitude, accuracy, and precision. This Weberian "ideal type," as Kathryn M. Olesko has shown, "embodied a particular conception of scientific truth," creating high, almost mathematical, "standards for certainty."[177] In some sciences it translated into an understanding of the scientist as a piece of apparatus that should be carefully calibrated in order to eliminate personal error and observation discrepancies. A closely related, but different, cluster of "scientific" attitudes centered around emotional detachment, self-restraint, and disinterestedness.[178] Those who associated those virtues with science sometimes drew rigid boundaries between the "scientist" and the "technologist," and made the scientist into an isolated researcher who worked in the secluded, pure, uncontaminated space of the laboratory, free from outside influence and devoid of practical concerns.[179] In contrast to this ethos of disinterestedness and seclusion, a different moral economy of science valued the practical and public utility of science, and configured the scientist first of all as a "practical man," even as a social leader.[180] Still others depicted the scientist as a frontier man, a pioneer, an explorer, while a different variety demanded of the scientist, first of all, heroism, courage, fortitude, and the ability to accept without blinking what a mechanical and meaningless nature had in store for humans. The psychologist G. Stanley Hall, in 1920, combined all these traits in a composite portrait of the scientist that drew from several models, including the Emersonian "man of action," the Romantic hero, and the Nietzschean superman. All these attitudes could be variously combined in moral economies of science, within which the meaning of any single trait (for example, scientific purity or courage) could vary greatly.[181]

As Wilson notes, in the early years of the American Philosophical Association (APA), American philosophers were sometimes seduced by a scientific spirit of community, as opposed to the perceived "individualism" and individual idiosyncracies of the philosopher.[182] Thus, in his presidential address in 1902 before the APA, Creighton contended that philosophy could become a science only by adopting the "cooperative" attitude that distinguished modern science.[183] Science was chiefly "the result of conscious cooperation between a number of individuals," Creighton told his philosophical audience.[184] He had severe doubts that philosophers who wrote "in isolation from their contemporaries" could produce more than idiosyncratic results.[185] It was his hope that the newly founded Association would foster a genuine spirit of comradeship and communication among philosophers, and would thus contribute to making philosophy into a scientific enterprise. James's friend Charles Sanders Peirce would have agreed, even though by the first decade of the twentieth century his odd behavior had managed to completely isolate him from most American philosophers. To Peirce, the "most vital factors" of modern science were "the moral factors," and especially that the "method" of science had "been made social."[186] Back in the late 1860s and 1870s, Peirce had imagined a cooperative community of inquirers that, over the long run, through a social process of cancellation of errors, would attain the truth.[187] Peirce, who worked for thirty years at the U.S. Coast Survey and was an "all-around prodigy of science, mathematics, and philosophy,"[188] urged his "fellow-students of philosophy" to absorb the "spirit" of science, expressly its spirit of community.[189] Philosophy could not escape its "infantile condition," so long as philosophers remained unable to "come to agreement upon scarce a single principle."[190] It could best imitate science by becoming "architectonic," insofar as architecture—in contrast, for example, with sculpture and painting—was the product not of one "single artist" but of a whole "army" of people, and was to be enjoyed not "by a few" but by "the whole people."[191]

The ethos of community was not monolithic, however, since philosophers, like scientists and scores of social commentators, understood "community" in widely different ways. For Peirce, the proper scientific community was marked by "solidarity," even a kind of "scientific Eros." This scientific love, however, was deindividualizing. As a result the scientific community could best be compared to a "colony of insects," in which "the individual strives to produce that which he himself cannot hope to enjoy."[192] Some scientists and philosophers championed rigidly hierarchical modes of social relationships, such as those institutionalized in Wundt's Leipzig psychological laboratory.[193] Others still envisioned bureaucratic scientific (and philosophical) communities, characterized by impersonal, formal, distant relationships. A few, including, as we will see, William James and some of his pragmatist friends, instead worked to create scientific-philosophical communities premised on sympathetic, even "intimate," friendships among equal participants. These different ways of understanding the meaning of community were reflected in scientific, and philosophical, practice.

Most philosophers, like Peirce, felt that, regardless of whether philosophy should be configured as a community of "cooperating" inquirers, philosophy ought to "cooperate" with the sciences. For a few cooperation was (almost) a "fait accompli." Thus the philosopher and psychologist George Trumbull Ladd was relieved to note in 1895 that the time when "the young Hercules of science" was praised for "strangling the snakes of metaphysics" was over. Metaphysics was no longer a "taboo" subject among students of physics. Indeed, "there [was] no rigid demarcation between science and philosophy"; each could "flourish only in dependence on the other."[194] For other philosophers, however, cooperation required a good deal of effort.

Once again, philosophers understood "cooperation" in a variety of ways, which conveyed different assumptions about actual relationships and distribution of power. At Harvard, Josiah Royce, the philosopher of "loyalty," promoted interactions *inter pares* between philosophers and scientists, organizing a series of seminars in which philosophers as well as chemists, neuropathologists, and embryologists were invited to work together on topics of common interest.[195] For others, the scale of power leaned on the side of philosophy. This assumption was most frequently embodied in very widespread architectonic conceptions of philosophy, according to which philosophy alone could erect the edifice of knowledge and arrange the particular facts discovered by the special sciences into a coherent whole. In the 1880s, for example, the idealist (and future pragmatist) John Dewey depicted philosophy as "the highest of all sciences." Philosophy completed the sciences; it was also their "basis." While each special science investigated a "particular sphere," philosophy alone studied the relation of those "particulars" to the whole. Because of that, philosophy was not only "*a science*" but it was also "all Science taken in its organic systematic wholeness."[196] In the 1890s the Cornell philosopher Jacob Gould Schurman depicted philosophy as the "circle of knowledge," into which all the special sciences would one day converge and find unity.[197] His colleague Creighton and the young philosopher Morris R. Cohen (a former philosophy student at Harvard) deployed the vocabulary of synthesis, suggesting a priority of philosophy over the sciences (with Cohen stressing philosophy's function as a guide to social and moral reform).[198]

More aggressive philosophers announced the beginning of a new era in which the sciences would recognize their subordination to philosophy. Thus, for example, the Berkeley "pluralistic" idealist George Holmes Howison, one of James's philosophical friends, outlined a hierarchy of the sciences that placed philosophy at the "summit of the disciplines" in a "well-organized university."[199] In 1903 Howison (then chair of the Department of Philosophy at the University of California) was invited to give a lecture to the assembled body of Berkeley scientists. Pretending he was a "soon to be tortured prisoner" in the "enemy's camp" (probably one of the campus's scientific buildings), Howison invited scientists and philosophers to put aside any arrogance and enmity. At the same time he reiterated what he took to be the correct distribution of power. Scientists, Howison ventured, saw philosophy as "a vast interrogation point," a vain enterprise steeped in "conjectures and dreams,"[200] but philosophers could safely "smile" at scientists' "patronizing self-satisfaction." True, the science of nature was indispensable to human life; yet, as philosophers very well knew, science was neither sufficient nor ultimate: only philosophy could deal with the real aims of human life. The ultimate, truly supreme knowledge was philosophical, and science was at best secondary and ancillary to it.[201] Howison used his speech to the Berkeley scientists to emphasize another important point. Although philosophers, he continued, had nothing against the "scientific man" per se, they would not tolerate scientists who invaded the philosophers' turf and who claimed for their metaphysical insights "the peculiar prestige of science." "Maturer philosophers," Howison told his scientific audience, had every right to denounce such "trespasser[s]" and to "demand the credentials." They had every right to discredit not science, but those "votaries" of science who "entering into the foreign boundaries," mistook "the frontier outworks for the interior centre and citadel of the region."[202]

Howison's position was very similar to that, as we saw, Münsterberg trumpeted on the occasion of the inauguration of the new philosophy building at Harvard—and the two joined efforts in order to achieve their common goal. It also closely resembled the position that Münsterberg took in the first years of the century when he laid out the general program for the International Congress of Arts and Sciences, associated with the St. Louis World's Fair. This grandiose project, as we will see in chapter 7, arranged all the sciences into a system unified under the aegis of idealistic philosophy, and made philosophy into the supreme kind of knowledge and the philosopher into a social and political leader.

### Philosophy and Psychology: Back to Emerson Hall

The science that worried American philosophers the most was psychology. In transparently self-celebrating disciplinary histories written around the turn of the twentieth century, psychologists narrated the story of the emergence of a "new," "scientific psychology," independent of philosophy. These narratives often equated the origin of the new psychology with the establishment of laboratories—secluded spaces with controlled access that symbolized psychology's scientific status and signified its autonomy from social, religious, and metaphysical demands alike.[203] The authors of these early accounts identified suitable "founding fathers," usually W. Wundt, G. T. Ladd, G. Stanley Hall, and William James, and proudly quantified what they took to be the marks of professionalization: the growing number of psychological laboratories, the variety of specialized journals available in the field, and the increasing numbers of PhDs.[204] Above all, narratives featuring the success story of the new psychology sought to shift psychology's location in the geography of knowledge, and to position it in the region of the natural sciences at a safe distance from theology and "that ill-defined medley which has passed by the name of philosophy."[205] The authors of such disciplinary histories emphasized the enormous distance that separated the new psychology from the "careless observation" and "flimsy guess work" of the old, unscientific, philosophical psychologists (the proponents of "rational," "theoretical," or "speculative psychology") and boldly claimed that "philosophy, in modern times" had "contributed nothing but stumbling-blocks in aid of psychology."[206]

As Michael Sokal, John O'Donnell, Mitchell Ash, William Woodward, and others have shown, psychologists' quest for autonomy was far from a linear process.[207] It was much contended both by many of those who considered themselves philosophers in a new, more technical, sense, and by those who would have described themselves as primarily psychologists. In no place were those tensions more visible than in the psychologists' professional society, the American Psychological Association. Established in 1892 under the guidance of G. Stanley Hall by a small group of scholars, the society was modeled after the American Physiological Association, and it soon established links with other scientific organizations.[208] In the first several years, the society was open to self-described "philosophers," who at the time, as we saw, lacked a professional society. Philosophers, however, soon found that they were "ill-suffered," and that their work was regarded as old-fashioned, even "medieval."[209] Years later, an observer recalled that many psychologists obviously resented "the intrusion of the philosophical camel into the psychological tent."[210] These psychologists worked to reduce philosophers' presence in the American Psychological Association, and made plans to promote the formation of an "American Philosophical or Metaphysical Association," a suggestion that, as some commented, would have amounted to eliminating philosophers "entirely from the [Psychological] Association."[211] The foundation of the American Philosophical Association in 1901, however, only temporarily alleviated the tensions. While the two societies for several years held joint meetings, members of the American Psychological Association worked to make the society's membership policy more stringent, so as to exclude more philosophical investigators. Philosophers also sensed that psychology journals, which had formerly provided venues for publication, no longer welcomed their contributions.[212]

It comes as no surprise, therefore, that, in turn-of-the-twentieth-century America, the issue of psychology's relation to philosophy should periodically give rise to heated discussions. Many philosophers and psychologists

endeavored to make their fields autonomous from each other, and worked
to create rigid dividing "boundaries" (their term) that would preclude psychologists from stepping into the domain of philosophy, and vice versa.
However, many others welcomed one form or another of interaction. Ladd,
for example, in his presidential address before the American Psychological
Association, urged psychologists to retain intimate relationships with philosophy, and warned them that any attempt to split "sovereignty between the
two" was bound to be "unsatisfactory."[213] Similarly, not a few philosophers
worried over psychology's desertion from the philosophical camp. Other
participants framed the issue in terms of control and subordination. Among
philosophers, some attempted to reestablish the control of philosophy (that
"federal executive") over its "provincial center," psychology.[214] Symmet-
rically, among self-proclaimed "scientific" psychologists, a few contended
that philosophy was now subordinated to psychology. For example, in 1895
James McKeen Cattell, then president of the American Psychological Association, declared the "twilight of philosophy" could be changed into "dawn"
only by the "light" of psychology. Again, in a typical survey of the field published in 1899, Edward F. Buchner, comparing the traditional psychology of
old with the "new" psychology, stated: *"Then* one's psychology grew out of,
and was dependent on, one's philosophy. *Now,* one's philosophy depends
on one's psychology."[215]

All of these issues and positions were brought to the attention of philosophers and psychologists once again on December 27, 1905, during the joint
session of the American Philosophical Association and the American Psychological Association that followed the inauguration of Emerson Hall at
Harvard University. As we have seen, the session was organized around
the question of whether psychology's "affiliation" was with philosophy or
with the "natural sciences." Münsterberg opened the session with a paper
in which he defended Harvard's decision to house psychology in the philosophy building: the "place" of psychology was with philosophy.[216] The other
panelists included the psychologists G. Stanley Hall and J. R. Angell and the
philosophers A. E. Taylor and Frank Thilly. Thilly, who had been the founder
of a Western Association of Philosophy in 1900, months before the foundation of the American Philosophical Association, expressed his concerns
about psychology's separatist tendencies. He sadly noted that "philosophy,
the sometimes queen, has become a dowager; her children have deserted
her, all but a few barren daughters, we are often told, for whom nobody
cares. . . . And now the demand is frequently heard that psychology has cut
loose from her old-fashioned sisters, and set up an establishment of her
own or gone to live with the natural sciences."[217] He argued that psychology was a natural science neither in its subject matter nor in its method,

and tried to persuade his mixed audience that "affiliation with philosophy [was] in the interest of both fields." Since a separation would amount to a "mutilation," Harvard had done well to house philosophy and psychology (including the psychological laboratory) in one and the same building. The second philosopher participating in the panel, the Plato scholar A. E. Taylor of McGill University, could not disagree more strongly. Taylor demanded a sharp separation between philosophy and psychology. A sworn enemy of all forms of psychologisms—that is, as I use the term in this book, of all attempts to approach philosophical problems by means of psychological notions, methods, or techniques— Taylor was eager to locate psychology with the natural sciences ("the empirical sciences of physical nature") and to confine it to the "laboratory," so as to minimize psychological incursions into the domain of philosophy.[218]

On the more psychological front, Hall and Angell also took opposite positions. Hall argued that psychology was a branch of natural science and had nothing to do with metaphysics. The psychologist's "business" was "to examine the physical and physiological conditions of mental states, and this it [could] do only by employing the methods of the natural scientist." (This stance actually conflicted with Hall's own psychological practice, and he confessed that he had taken the position mostly because he had been asked to do so by the organizers of the session. He was embarrassed to present such ideas right inside Emerson Hall, the symbolic structure of the "marriage" of psychology and philosophy.) In contrast, the Chicago psychologist J. R. Angell was of the opinion that psychology was "in too delicate a position" to alienate either scientists or philosophers. Instead, it should continue to cultivate friendly relationships with both. Indeed, psychology had the important mission of "bringing together the interests of philosophers and natural scientists." Angell found that the "spiritual intimacy" between Harvard philosophers and psychologists would make the location of the new Harvard psychological laboratory within Emerson Hall profitable to all.[219]

William James, who had attended the inauguration of Emerson Hall, stayed for the joint session. He had long been familiar with all the opinions voiced during the discussion. He had known Hall for years and had grown accustomed to regarding him as a former friend, as well as a former student.[220] His relationship with Angell, a former Harvard student, was quite cordial: James admired Angell, who, in turn, regarded James as the founder of functionalism in psychology. James had also long been on friendly terms with Frank Thilly, but recently had learned that he needed to regard Taylor as a rabid enemy. Even though none of the panelists mentioned James (at least judging from the published version of their papers), James was most likely on their

minds, and no doubt on the minds of many among the audience. His fluctuating positions about the proper ways of arranging the relationships between
philosophy and psychology had given rise to much discussion. His attempt
in *Principles of Psychology* (1890) to construe psychology as a natural science
clearly distinct from philosophy, and averse to the discussion of metaphysical
issues, had framed much of the discussion in the early 1890s. By the mid-
1890s, however, James had complicated his position, and turned to a kind of
philosophical psychology that paved the way for radically empiricist metaphysics. Both radical empiricism and James's pragmatist account of truth
were taken by many to provide examples of philosophical doctrines that combined (or, as critics complained, conflated) philosophy with psychology, and
months before the inauguration of Emerson Hall, A. E. Taylor, one of the
panelists, had attacked the pragmatists exactly on that point.

If James had long thought about the question of the proper relationships between psychology and philosophy, he had also long been concerned
with the broader questions to which, as Münsterberg proudly emphasized,
Emerson Hall had provided a visible and to James, distasteful, answer: the
questions of the meaning of philosophy, its relationships to the sciences, its
position in the geography of knowledge, and the place of philosophers in the
academy and society. James, as we will see, kept returning to these questions
throughout his career, and his attempts to address these issues shaped both
his philosophical and his scientific work, at the same time as they reflected
James's concrete scientific practices and the distinctive features that marked
his metaphysical craft. Like other participants, in confronting these issues
James shifted from the epistemological and the methodological to the moral
and the social, and the other way around. Neither his solutions nor the tools
that he deployed—including his transgressive boundary work and the construction of spaces of knowledge—were unique. We will see that, as for other
contemporaries, those tools were for James a means of challenging the prevailing epistemological regime, reshaping existing disciplines, and changing the geography of knowledge. They were also instruments that, like others, James deployed in order to remake the philosopher and the scientist,
and promote a new ethos of science, new modes of social engagement, and
a new vision of society. What was different and unique was the specific
nexus between James's vision of knowledge production, the scientific and
philosophical personae that he promoted, the forms of interaction that he
sought to inculcate, and his vision of the human self and of society. It is this
nexus that this book explores.

# 2

# Philosophy versus the Naturalistic Science of Man

## James's Early Negotiations of Disciplinary and Pedagogical Boundaries

In December 1867 William James, who was then in Berlin, sent a letter to his father requesting money to cover his "trifling expenses." Perhaps anticipating objections, he also sent word of his "future plans." He intended to travel to Heidelberg and, health permitting, do some laboratory work with "two Professors . . . strong on the physiology of the senses": Hermann von Helmholtz, one of the most famous physiologists in Europe, and Wilhelm Wundt. James observed that "as a central point of study . . . the *border ground* of physiology and psychology, overlapping both, wd. be as fruitful as any," and announced that he was "now working on it."[1] Five years later, Wilhelm Wundt (whom James had not yet managed to meet) resorted to a similar metaphor, announcing to his fiancée that he would "direct his aspirations to 'a somewhat suspect borderland between physiology and philosophy.'"[2]

Both James and Wundt referred to the study of the mind, which they proposed to approach from a physiological perspective. In the 1870s, as later in the century, that was a highly contentious topic. Truly a "borderland," it lay at the intersection of different epistemic communities and diverse research practices, including, among many other approaches, classical psychologies based on "introspection" (a term to which a variety of meanings were attached), prescriptive

doctrines about the proper disciplining of the mind, and emerging experimental approaches derived from a range of different, even incompatible,
research and laboratory traditions. The study of the mind, according to many,
belonged to a broader study of human nature. It was perceived as crucial
to the maintenance, or reform, of the social order, and a variety of actors,
ranging from public moralists to educators, social reformers, and politicians, staked strong claims on it.[3] It represented a terrain on which not only
different conceptions of what it meant to be human but also conflicting
conceptions of the nature of science and of philosophy clashed.

   The first part of this chapter discusses James's preoccupations with that
"border ground" in the 1870s, tracing his interventions in two debates that
revolved around the science of human nature and the study of the mind:
intellectual debates about the direction in which the field should go, and pedagogical debates about higher education and the proper training of students'
minds. The second part of the chapter turns to the early 1890s, and briefly
discusses the approach to psychology that James defended in *Principles of
Psychology* (1890), while examining psychological debates in which James
participated in those years. During both periods James abundantly resorted
to boundary work, negotiating and renegotiating the divides that separated
naturalistic approaches to the study of human nature and the mind from
philosophical ones. However, as James engaged those boundaries, the order
among those various regimes of inquiry was changing, with different "fields"
of knowledge shifting their positions in relation to other enterprises or "territories" of research. James's position vis-à-vis these fields, and his aims, also
shifted. He adjusted his tactics skillfully (at times even opportunistically),
using boundary work to achieve his changing goals. The chapter traces the
conceptions of philosophy and psychology—and the corresponding, sometimes implicit, ideals of the philosopher and the scientist—that James forged
through his negotiations of disciplinary and pedagogical divides.


### Invading the Secluded Gardens of "Philosophy"

In 1856 Francis Bowen, the Alford Professor of Natural Religion, Moral Philosophy, and Civil Polity at Harvard College, stated that there is "a general
science of human Nature, of which the special sciences of Ethics, Psychology, Aesthetics, Politics, and Political Economy are so many departments, all
founded upon the essential unity of the human mind and character, and the
consequent similarity of its manifestations under similar circumstances."[4]
This declaration appeared in the introduction to Bowen's textbook, *The Pri-*

*nciples of Political Economy*, and it expressed what, as Julie Reuben suggests, Bowen took to be a matter of fact: that there was a general science of the human subject, that this science was philosophical, and that the investigation of the nature of man and social institutions pertained to moral philosophy. Bowen's claim reflected an established tradition. For example, in the late eighteenth century an American teacher had similarly stated that the philosophical science of man was "the science that teaches men their duty and the reasons of it," and that it did so on the basis of the study of "the physical and mental equipment of man," of economic, social, and political institutions, as well as of religion and the evidences of Christianity.[5]

To be sure, this conception of moral philosophy was not universal, but throughout the nineteenth century a few regarded it as the science that provided the framework for a general inquiry into human nature. As late as the 1890s, Henry Sidgwick, the Cambridge moral philosopher who appeared in chapter 1, observed that the prevailing opinion among British thinkers of the previous generation was that "the science of Mind" or "the science of Man" was the main task of philosophy.[6] Sidgwick reported that even Sir William Hamilton (the Scottish philosopher whose work was very influential in America in the first half of the century) and J. S. Mill, two philosophers who had disagreed on almost everything else, had agreed on that point. To Hamilton, the science of mind, "with its suite of dependent sciences," including logic, ethics, and politics, constituted Philosophy itself.[7] To Mill, "the proper meaning of Philosophy" was "the scientific knowledge of Man, as an intellectual, moral and social being."[8] "And this view—which seems to blend Philosophy indistinguishably with Psychology or Sociology or both," Sidgwick continued, "still survives among us" and is implied by the term "Mental Philosophy."

In the United States before the Civil War, "moral philosophy," understood as a broad philosophical science of human nature, and "mental philosophy" (the study of the human mind) represented the pinnacle of college education.[9] Taught by the president of the college (oftentimes a minister) or by religiously and morally minded instructors (such as the conservative Bowen at Harvard), the canonical senior-year course in mental and moral philosophy provided a framework that synthesized for the student all the previous college instruction, putting it in a safe Christian perspective, and supplying rules of conduct for private life and public affairs. Thus moral philosophers, who in communities like Boston and Cambridge enjoyed a close alliance with the economic and political elites, could represent themselves as moral leaders, guardians of the moral and social order.[10] They were in charge of educating the future generations and preparing them for statesmanship.

In the second half of the century, however, that privileged—real or imagined—position of philosophy was challenged on many counts, as the social and economic framework that moral philosophers sanctioned began to be questioned. Two closely interlinked sets of debates are relevant here: the debates revolving around the issue of who was entitled to articulate a science of man, and those over the reform of college education.

As we saw in chapter 1, in the late nineteenth century the growing success of the natural sciences, along with the attendant transformation of the amateur naturalist into a specialized, professional figure, threatened philosophers' academic prestige and power. Thus in 1879 the psychologist Granville Stanley Hall, fresh from his laboratory experiences in Berlin and Leipzig, described the philosophical instruction offered at some three hundred American colleges as "rudimentary," "medieval," doctrinaire, and highly dependent on the theological views of the founders, trustees, or presidents of the various institutions. In the "American college," Hall concluded, philosophy could not hold its own when compared with the sciences.[11]

In the last three decades of the century, naturalists, biologists, and science writers stepped right into the domain of moral philosophy, taking up the questions of the nature of "man" and of his relation to the universe—questions that, by widespread agreement, defined the domain of "moral and mental philosophy" as opposed to "natural philosophy."[12] As William James put it, a "deluge" of scientific facts—concerning such widely different topics as "the polyp's tentacles, the throat of the pitcher-plant, the nest of the bower-bird, the illuminated hind-quarters of the baboon, and the manners and customs of the Dyaks and Andamanese"—"have swept . . . into the decent gardens in which, with her disciples refined Philosophy was wont to pace," leaving "but little of their human and academic scenery erect."[13]

The investigation of human nature became a contested "territory," to use a metaphor that was often deployed by period practitioners. As James also wrote, "A real science of man is now being built up out of the theory of evolution and the facts of archeology, the nervous system and the senses."[14] That "real science of man," as James understood it, was premised on some of the hottest scientific advances: Hermann von Helmholtz's laboratory studies of the physiology of sensation; evolutionary approaches to psychology, particularly Herbert Spencer's; the flourishing research on "mental physiology," that is, the physiology of the brain and the nervous system (as articulated, for example, by William Benjamin Carpenter and by the physiologist Henry Maudsley); and the study of animal fossils and human fossils, or "anthropology."[15] In developing a naturalistic (physical, medical, physiological) discourse of the human being, men of science questioned philosophers' exclusive right to the study of man and challenged the subjective methods that philosophy traditionally brought to bear on that topic (especially intro-spection). In the words of Edward Youmans, the founder and editor of *Popular Science Monthly*, one of the organs for the diffusion of the new "scientific" approach to inquiry and education in the United States, the "scientific study of Human Nature" had successfully turned upside down the traditional philosophical study of man.[16] By firmly placing man within, rather than above, nature, natural scientists had shown once and for all that human nature should be studied with the methods of the natural sciences. In par-ticular, the study of the human mind fell into the domain of physiology. Youmans's views, of course, were not original. He simply expressed and made public the antagonism between the philosophical inquiry of human nature and the new naturalistic science of man.

But naturalists threatened to do more than step into philosophers' intel-lectual territory and question their methods; they challenged moral philoso-phers' claims to moral and social leadership and used their naturalistic science of man as a platform to advertise their visions about morality, religion, pol-itics, and society. Thus in Britain the evolutionist T. H. Huxley, a scientific investigator who felt entitled to lecture to broad audiences on such themes as education and biblical evidence, contended with men of letters for ed-ucational and civic leadership. Not only did "natural knowledge" provide "increased comfort" but the naturalistic study of man and of "man's place in nature" also laid the foundations for "the laws of conduct" and a "new morality."[17] Similarly, Herbert Spencer, taking up the philosophical topic of ethics, had clearly indicated that his main goal was to find a "scientific basis" for the "principles of right and wrong in conduct at large."[18] In Germany the physiologist Ludwig Büchner made the same claim: the physiological and anthropological study of human nature would provide a platform for a new public morality, one associated with the liberal, even radical, social project that he supported.[19]

Moral philosophers' opposition to the naturalistic inquiry into human nature focused on just that point. Thus, for example, the Oxford idealist and public moralist T. H. Green, in discussing whether it was possible to even frame a naturalistic science of man, warned his readers that the naturalists busy developing a "natural science of man" would be "as forward as any to propound rules of living to which they conceive that . . . man *ought* to conform." To such naturalists, Green continued, the natural science of man was "the basis of a practical art," one that aimed at installing a new moral and social order.[20]

How did philosophers react to these threats? If we are to trust James, many philosophers, "though broken-hearted at the desecration of the gar-den of philosophy," had "submitted, in a sort of pessimistic despair, to the barbarian invaders." Others were "uncertain what to do." Meanwhile, "the victors . . . intoxicated with success" assumed "that Philosophy herself" was "dead," or that "she [would] be shamed to silence, as now one, now another, of the conquering ragged regiment stands forth to face her down."[21]

One of the issues underlying these struggles was who was entitled to take up the question of human nature and, in this sense, to philosophize. To be sure, in the 1870s the issue did not yet carry with it some of the implications that it would in the early twentieth century, when philosophers would much more systematically invoke the language of specialization and develop much stronger corporate identities. Nevertheless, by the 1870s there was a growing sense that scientifically trained thinkers differed in their mental habits and capacities from more literary and philosophically trained thinkers, and seekers of philosophy wondered whether scientific training—upon the signal importance of which scientists never failed to insist—could prepare the mind for philosophical activity.

At this juncture the debate on the intellectual merits of the philosophical versus the naturalistic inquiry into human nature intersected with a debate about education and the training of the mind, and with the broader mid- and late nineteenth-century debate about the reform of college education. In the eighteenth century and through most of the nineteenth, the goal of college education was largely thought to be a disciplining, rather than a furnishing, of the mind. At American colleges students received training in the natural sciences, but moral philosophy (together with Latin, Greek, and Euclidean mathematics) was thought to provide the best training of the mind, leading to the harmonious development of its faculties, with the proper subordination of will and passion to intellect and faith. That pedagogical ideal was periodically challenged, and by the middle of the century educators felt they could not postpone the question of whether the established course of study should be replaced by a more "modern education."[22] Should the classical studies give way to deeper study of the natural sciences? Could a scientific training of mind achieve the goals of, perhaps even replace, the classical and philosophical training? Or would the more utilitarian bent of scientific education hamper the gentlemanly and liberal qualities of mind that were thought to be the goal of college education?

German, French, English, and American educators had long engaged with those or similar questions. In the English-speaking world, the controversy entered into a particularly bitter stage in the 1860s and '70s, with Herbert Spencer's, John Tyndall's, and Huxley's aggressive interventions in defense of scientific education.[23] In America the calls for a scientific reform of college education were even stronger than in Britain, as Matthew Arnold (one of the most visible participants in the English debates on college education) observed.[24] That became all the more true after the Civil War, when concerns about the national economy and swings of the business cycle raised support for technological and scientific education.[25] Many felt that the traditional classical curriculum was no longer adequate for a rapidly industrializing nation, and a widespread sense of the practical utility of science backed up the arguments of scientific reformers. Echoing and oversimplifying arguments made by British reformers (especially Huxley and Spencer), the most radical reformers on this side of the ocean argued that only the study of the natural sciences would enable students to cope successfully with their future professions and to function appropriately as members of the social body.[26] Thus, for example, drawing on mid-nineteenth-century physiological theories of "character" and "habit," and building on the popular psychological/physiological theories of William Benjamin Carpenter, Youmans contended that scientific education was the only way to instill good habits of thinking, inference, judgment, and observation—habits necessary to cope with the needs of modern life.[27]

At the turn of the century, disciplinary claims concerning the merits of the philosophical versus the naturalistic inquiry into human nature, and pedagogical claims concerning the merits of classical (and philosophical) versus scientific education were often intertwined. Reformers could easily slide from aggressive proposals for a scientific reform of college education to the discussion of the advantages of a scientific versus a philosophical study of human nature. Similarly, their opponents, especially moral philosophers, could in one breath defend the formative power of philosophical study as a unique source of mental discipline, and claim the investigation of human nature as the exclusive right of philosophers. Noah Porter, the president of Yale University and one of America's leading moral philosophers (his textbooks were frequently used in mental and moral philosophy courses across the country), is a good example. In his attacks against the naturalistic science of man, a science that strove to reduce man to a "machine" or a "voltaic battery," Porter contended that scientific education led to a "one-sided" cultivation of the mind that could not be regarded as "culture in the large and generous sense of the term."[28] Not so with the philosophical "study of man": this would lead both to a better understanding of human nature and provide a corrective to a faulty "system of education which [was] bent upon training specialists."[29] Porter also stressed that far from replacing the philosophical study of man, the natural sciences ultimately rested on it. The inductive reasoning used in the natural sciences, he argued, could find a foundation only within psychology, a philosophical discipline that Porter understood as the study of the human soul; without a foundation in (philosophical) psychology, that "mother of the sciences," the natural sciences (and the naturalistic science of man) would collapse.[30]

By the end of the century moral philosophers had largely lost both the intellectual and the pedagogical battles. Moral and mental philosophy courses, although still offered at some public colleges and religious denominational schools, had been discontinued at many American institutions of higher education.[31] The academic field of moral philosophy had almost disappeared, and so had the goal of a unified philosophical discourse of the human subject. The philosophical inquiry of human nature had been replaced by a multiplicity of increasingly autonomous and often rival disciplines, including subfields of philosophy and the human, social, and behavioral sciences.

### The Man of the "Two Disciplines"

In the 1870s William James repeatedly intervened in the disciplinary and pedagogical debates that pitted the defenders of the traditional philosophical study of human nature against the proponents of the naturalistic science of man. James knew he had to do so if he was to advance his career in the direction he wished it to go.

An instructor in anatomy and physiology at Harvard, James was increasingly drawn toward topics that, according to many, clearly lay beyond the range of physiological investigators' purview. He was interested in the study of the human mind and in psychology, a field that, according to many of his contemporaries (including Harvard philosophers), fell squarely within the field of philosophy and "literary studies." James, however, had begun exploring a naturalistic approach to mental life, one that would bring brain and sense physiology to bear on that philosophical subject. Not only did he read extensively in physiology, mental physiology, mental pathology, natural history, biology, and anthropology, but he had also begun to study psychic phenomena and to conduct experiments on himself, tracking the effects of mind-altering chemicals on his own mind.[32]

In December 1875 James began negotiating with Harvard's president Charles Eliot to be allowed to teach an undergraduate elective in "physiological psychology," a study of the physiology of the brain and the organs of sense and their relationships to the mind.[33] James probably guessed that Eliot would look favorably at such a proposal. After all, Eliot had opened his inaugural address at Harvard by declaring that the antagonism between the scientific culture and the humanities belonged to the past.[34] Nevertheless, James felt he had to proceed with caution. He was aware that his proposed physiological psychology course raised the specter of medical materialism and ran directly against the traditional approach of mental science. The textbook that James intended to adopt (Spencer's *Principles of Psychology*) smelled of religious unorthodoxy, if not antireligionism. In the 1870s, at Harvard and elsewhere, mental science, a broad field that, in James's own classification, included history of philosophy, metaphysics, logic, and the study of the mind, was still largely placed in the hands of religiously and morally concerned teachers—teachers who, James suspected, often owed their position more to their irreproachable character and religious principles than to any other qualifications.[35] James anticipated that Harvard philosophers would be opposed to his course. And, indeed, they were, the strongest opposition coming from Francis Bowen. A declared enemy of Darwin's theory of evolution, which he once described as a "fanciful theor[y] of cosmogony," and a philosopher who built the social order on a Christian framework, Bowen worried about the atheistic conclusions that he believed would follow from evolutionary doctrines. As Bruce Kuklick has shown, for Bowen evolution assumed that man derived from the brutes, and "the brutes from simpler creatures," in a regression process that, Bowen contended, ended in a "swirl of material particles."[36] The study of the human mind fell within the domain of philosophy and should not be approached by improvising scientific speculators.[37]

In his negotiations with Eliot, James was aware that, aside from the issue of the acceptability of a physiological psychology elective, he would have to address a more personal issue: Was he the right man to teach such a course? In 1872–73 James had lectured on physiology whereas in 1874–75 he held lectures on the "Comparative Anatomy and Physiology of Vertebrates." The 1874–75 lectures were delivered in the Museum of Comparative Zoology, where James also held a separate position.[38] In 1875–76 in the rooms of the zoological museum, under James's supervision, students "dissected, quite assiduously, fishes, fowls, rabbits, cats, turtles, frogs."[39] James was also a member of the Harvard Natural History Society, where during the same month as his proposal to Eliot, December 1875, he was speaking on topics such as the structure of the ear in a three-month-old fetus he had dissected and an Austrian graphing machine that recorded the "speed of nerve-force."[40] Thus to the members of the Harvard faculty, possibly even to Eliot, James was first and foremost a naturalist, perhaps even an "irresponsible doctor, who was not even a college graduate, a crude empiricist, and a vivisector of frogs," as George Santayana would put it years later.[41] A man with an "unphilosophic past," as James's friend Ferdinand Canning Scott Schiller would describe him, James was a suspicious choice for a course on psychology.[42]

Writing to Eliot on December 2, 1875, James subtly addressed these issues, seeking to assuage Eliot's and, possibly, the philosophical faculty's

anxieties about his proposal. James expressed right away the need for cooperation between the sciences and the humanities, deploying to his advantage
the conciliatory position that president Eliot had expressed in his inaugural
address. He also capitalized on Eliot's aversion to the "dogmatic," doctrinaire teaching of philosophy, as well as on Eliot's conviction that students
should be exposed to both sides of philosophical controversies.[43] In his letter
James presented himself as the man who could introduce Harvard students
to the physiological approach to the mind without harming their moral
and religious beliefs. Harvard students, James predicted, were bound to be
exposed anyway to the new naturalistic science of man, which figured so
prominently in the pages of many periodicals and magazines. Should they be
left to their own devices when engaging that potentially dangerous material?
Was it not preferable, instead, to hire somebody who could introduce them
in a safe way to that literature? But what type of man was fit for the task?
Professors with an exclusively literary and philosophical training were inadequate to it, because they could not really grasp the scientific arguments.
Narrow-minded scientists were equally inadequate to teach the naturalistic
science of man, because they were bound to fall back on "extremely crude
and pretentious psychological speculations." Obviously, the perfect (if not
the most traditional) choice for the college would be to hire a man whose
"scientific training" would enable him to "fully . . . realize the force of all the
natural history arguments," and whose "concomitant familiarity with writers of a more introspective kind" would preserve him "from certain crudities
of reasoning which are extremely common in men of the laboratory pure &
simple."[44]

Transforming his scientific and medical training from a potential weakness into a strength, James presented himself to President Eliot as just that
type of man: a man of "the two disciplines." Because of his scientific training
and his mastery of the physiology of the nervous system, James could deal
with the "natural history arguments" and their implications for the definition
of the nature of man. At the same time, distancing himself from "the mere
physiologist," James implied that, because of his introspective and serious
philosophical interests, he would be able to avoid the philosophical naiveté
often displayed by men of science.[45]

### The "Disputed Boundary" in the Field of Knowledge: The Johns Hopkins and the Lowell Institute Lectures

Over the next four or five years James continued harping on that theme,
painting himself both as a man who combined the "two disciplines" and as

an impartial, responsible teacher who could help students carefully sift through the philosophical implications of physiology and the new naturalistic science of man. That was the underlying goal of two series of public lectures that James delivered at Johns Hopkins University and at the Lowell Institute of Boston in 1878 (as well as, most likely, a widely attended lecture that he delivered at Harvard's Sanders Theater). He delivered them in the midst of parallel negotiations with Charles Eliot at Harvard and with Daniel Coit Gilman, the president of Johns Hopkins, for a possible philosophical appointment, for by then a philosophy position had become James's next career goal.

The common theme of these lectures ("the senses and the brain" at Johns Hopkins, and "the brain and the mind" at the Lowell Institute) required, again, special caution. Thus James took pains to distance himself from those men of science who indulged in uninformed philosophical speculations, and especially from those who arrogantly contended that physiology had replaced, once and for all, the old philosophical inquiry into the human mind.[46] At the same time James was critical of those "professed philosophers" who, "having hardly opened a treatise of physiology," felt authorized to dismiss physiology with the charge of medical materialism.[47] James's strategy for dealing with these delicate issues is summed up in a metaphor that he deployed both at Johns Hopkins and at the Lowell Institute, a metaphor that indicates that in those lectures James engaged consciously in what modern scholars call "boundary work." James compared physiology and psychology to two adjacent lots in the field of knowledge, and likened the philosopher and the physiologist to the two owners of those lots, each busy fighting over the dividing line and striving to "shove back the fence and reduce the size of his neighbor's lot for the benefit of his own."[48] Condemning such "jealousies" over the "disputed boundary,"[49] James invited his audience to adopt, if only for that night, "the attitude of one who should own both lots."[50] Such a man, he told them, "does not care where the fence stands and being master of all the land tries to cultivate every sq. ft. of it impartially."[51] In other words, "no mode of thinking is *against* any other. . . . If we think clearly & consistently in Theology or Philosophy we are good men of Science too. If we think logically in Science we are good theologians & philosophers. . . . It must be that truth is one & thought woven in one piece."[52]

Shifting his discussion from disciplinary issues to pedagogical issues, James noticed in passing that, thanks to his multiple teaching duties, he had felt most keenly the difficulties of understanding the brain without the mind, or the mind without the brain.[53] Being "obliged" to teach "a little" anatomy, "a little" physiology, and "a little" psychology, he had been forced to develop informed opinions about both introspective philosophy and physiology. James then implicitly compared himself as a teacher to a "judge" who, better than the "lawyers" representing the two opposite sides, feels the need to develop impartial views. He thus implied that he not only was in a position to grasp the results of both physiology and psychology, but that he was genuinely a man *super partes*, immune from the contentious attitudes and jealousies over the boundaries that he found so despicable.

How could one gain sound knowledge of the mind? Could the "subjective" methods of introspection, without the help of laboratory techniques, provide an adequate understanding of the mind? Did the experimental investigation of the brain, the nervous system, and sense organs alone suffice to unravel the mysteries of consciousness? As James told his audience, the attitudes of the professed philosopher and the invidious psychologist both ran against the intuition of the human being, who, being a "proprietor [at once] of a body and a mind," felt the inadequacy of explanations of the mind that left the brain out of the picture, as well as those accounts of the brain that left no room for the mind.[54] The attitude of the man who "owned both plots in the field of knowledge," and thus could "ignore the fence," was the only one that made sense. And, indeed, James's main thrust in the Johns Hopkins and Lowell Institute lectures was to illustrate by means of concrete examples the fruitfulness of an attitude of cooperation between the two approaches, and to expose the weaknesses of the man entrenched on one side of the fence.

A first example was provided by James's discussion of sensations. Are sensations the constituent elements of our thought? Such was the opinion prevailing among "certain philosophers" (mostly of the associationist school), who "analyzing our finished thought about *things* come to the conclusion that it is wholly made up of sensations." James contrasted the introspective speculations of these philosophers with the more humble attitude of physiologists, who, studying the sensations themselves, "have come to the exactly reverse conclusion." In particular, Hermann von Helmholtz's experiments on vision (especially on double vision) had revealed that, far from being the bricks of our thoughts, sensations are "the most fluctuating and indefinite of mental occurrences," and "are wholly plastic in the hands of our thoughts of things."[55] Thus specific laboratory work performed by this investigator, whom James portrayed as "absolutely incapable of sentimental bias" either on the side of materialism or on the side of spiritualism, happened to have broad philosophical implications and challenged the psychological theories authoritatively asserted by philosophers.

If this example indicated some of the ways in which physiology could contribute to the solution of philosophical problems, the bulk of the lectures sought to convey the reverse message: physiology, especially brain physiology, was highly indebted to introspective philosophy. According to James, that was markedly the case with the cutting edge of brain physiology, the theory of localization, a theory according to which elementary sensory and motor images were stored in particular groups of cells located in the cortex of the hemispheres.[56] This claim was meant to chastise physiologists for their aggressive attitude vis-à-vis literary and philosophical studies of the mind, and to alleviate the anxieties of his audience—specifically those of the Johns Hopkins educated public that, most likely, was still trying to recover from the scandal created two years earlier by Huxley's lecture on the campus.[57]

In order to explain action among these groups of cells, localizationists often assumed the existence of a minute network of "fibers" joining those areas of the brain. These supposed "anatomical bridges" made possible the physiological action of one group of cells on the other. But, James told his audience, those fibers (which had thus far escaped every attempt to observe them) most likely did not exist. They were fictitious anatomical constructions, the existence of which had been inferred from associationist psychology, on the basis of an analogy between groups of cells and groups of ideas.[58] The whole theory of brain localization and the belief in putative physiological actions of groups of cells on each other resulted from the main assumption of associationism, according to which the mind functions in terms of ideas and their associations.[59]

James used this example neither to attack localization theories nor to criticize associationist psychology (as he would do later in his career), but to condemn the aggressive and uncooperative attitude of those brain physiologists who baldly contended that brain physiology had exploded introspective philosophy and had replaced it with a new psychology exclusively founded on physiology. James depicted Henry Maudsley, the British brain physiologist, as a black sheep. Maudsley attacked vehemently the application of the metaphysical and theological point of view to the study of mental phenomena and critiqued the "psychological method of interrogating self-consciousness"[60] as palpably inadequate, implying, as James put it, that "the only sound psychological science [was] that founded on Physiology."[61] James told his audience that such claims were "ludicrously false." Things stood the other way around: the "subjective method" of introspective philosophy, which Maudsley and the jealous physiologists attacked, provided, for good or for worse, the foundation of brain physiology.

Localizationists, James went on, relied on introspective psychology not only in their overall assumptions but also in order to interpret their experimental findings. Even the pathological findings about aphasia that localizationists inevitably used to prove their point would be "utterly inexplicable" without the guidance provided by introspective psychology.[62] Indeed, as

James had recently argued in his review of Maudsley's *Physiology of Mind*, in his *"odium anti-theologicum"* Maudsley failed to realize that his own theory of aphasia "could not have been conceived without a preliminary psychological analysis of language."[63]

In order to show further the extent to which brain physiology depended on psychology, James described certain brain physiology experiments performed by the British physiologist David Ferrier. The manuscript notes for the Lowell Lectures are extremely sketchy, but can be deciphered on the basis of a joint review of brain physiology works by Maudsley, Ferrier, and Jules Luys that James had published the previous year.[64] Like other physiologists, with his experiments Ferrier had sought to provide evidence for the localization theory and to map out motor and sensory functions by associating them with specific cortical points. To that end, physiologists would usually supplement occasional postmortem observations of individuals affected by aphasia, paralysis, or epilepsy with two main types of experiments. The first consisted in stimulating areas of the cortex with electric currents ("galvanic" or "faradic" currents) and observing the resulting movements. The second consisted in destroying a particular cortical area (by ablation, excision, cauterization, injections of chromic acid, and other means) and observing the consequences. Rabbits, cats, dogs, and monkeys were the most usual test subjects.

According to most researchers, experiments of the first type indicated that electric stimulation of certain areas of the brain cortex *induced* muscular contractions in certain parts of the body, thus making it possible to establish homologous areas in the brain cortex of different animals. However, in order to conclude that those cortical areas were genuine "motor" areas, it was also necessary to prove that "destruction" of those areas *inhibited* the movements that had previously been reproduced by electrically stimulating the same areas. Ferrier's experiments on monkeys pointed directly to that conclusion. They showed how destruction of the convolutions surrounding the "fissure of Rolando" on one of the two hemispheres of the monkey's brain induced paralysis of the limbs on the opposite side of the body. However, experiments on dogs produced facts that appeared irreconcilable with the cortex localization view. The experiments showed that, while irritation of certain areas of the cortex of a dog had given rise to certain movements, ablation of the same areas of the cortex did not always induce paralysis of the parts that previously moved.[65] In fact, it often happened that in dogs that had been subjected to ablation of those cortical areas, the power and control of the limbs improved, and if the dog did not die of complications, "ultimate recovery" would occur "sooner or later."[66] How to account for the difference in the cases of the monkey and the dog? Ferrier argued that, while in the monkeys and in higher animals certain movements (for example, movements of locomotion) could only be performed under the direct control of the will, in the lower animals, such as the dog, the same movements could also occur in an automatic way, without the supervision of the will.

Reviewing Ferrier's book, James was quick to point out Ferrier's answer to that question ultimately rested on a claim about the differences between the *psychology* of the dog and that of the monkey: the dog could be supposed to be largely an "unconscious automaton," whereas the monkey's psychology was far more complex and more similar to that of human beings. "In the monkey, on the contrary, as in man, we are to suppose that such innately co-ordinated responses to outward stimuli are extremely few."[67] Like human beings, monkeys act from motives, or "ideas," and most of their motor reactions take place only when the stimulus is "understood," that is, "interpreted in the light of past experience or education." Thus one could account for Ferrier's experimental findings by assuming that in the monkey destruction of certain cortical areas damaged the animal's memory, rendering the monkey unable to understand a certain stimulus (to recognize *"what* it is he sees") and respond to it. Paralysis, James continued, could be explained as due to the fact that "the storehouse of the ideas" had been destroyed. Being unable to recognize the stimulus, the monkey could not initiate the appropriate motor response. But of course, James asked, how could we talk about ideas "without what introspection tells us of their formation from coalesced residua of motor and sensory feelings"?[68] Thus, even if one attempted a largely physiological explanation, ultimately it was only by means of an appeal to introspective *philosophy* that the difference observed in the cases of the dog and the monkey could be explained away.

The conclusion James drew from this example, as from his discussion of Maudsley, was that "brain-physiologists would be still groping in Cimmerian darkness without the torch which psychology proper puts into their hands. The entire recent growth of their science may, in fact, be said to be a mere hypothetical schematization in material terms of the laws which introspection long ago laid bare. Ideas and their association in the mind, cells and their linking fibres in the brain—such are the elements. But, whereas we directly see their process of combination in the mind [by means of introspection], we only guess in the brain what it *may* be from fancied analogies with the mental phenomena."[69] James reiterated the point in a public lecture that he delivered to an audience of one thousand people at Harvard's Sanders Theater on March 1, 1877, concluding that "without the aid of introspective psychology physiology [could] do nothing in the way of comprehending the nervous centres."[70] James concluded that nobody could be a physiologist without also being an (introspective) psychologist: "If it be true, as Johannes

Müller used to say, that *nemo psychologus nisi physiologus*, it is doubly true that, so far as the nerve-centres go, *nemo physiologus nisi psychologus.*" The "bad temper" that cerebral physiologists displayed about "introspection" was "not simply wrong, but monstrous."[71]

James was thoroughly familiar with brain physiology. In 1873–74, in collaboration with his friends J. J. Putnam and H. Bowditch, as well as alone, he conducted innovative work on brain localization in dogs, through experiments designed to address a technical problem posed by previous experiments in which the electrical stimulation could not be confined to the specific cortex area to be tested.[72] Despite this work, in the Johns Hopkins and the Lowell lectures, as in other papers he published at the time of his negotiations with Eliot and Gilman, James carefully avoided siding with the partisans of brain physiology. Instead, he laid great emphasis on the ways in which physiology depended on psychology, the "antechamber to metaphysics," and hence on philosophy, thus seeking to portray himself as a person whose attitude was genuinely above the opposed factions.[73] This would have assuaged the anxieties of those who might oppose his appointment in a philosophy department at Harvard or Johns Hopkins.

When, years later in his *Principles of Psychology* (1890), James went back to a discussion of the theory of localization, aphasia, mental blindness, and Hermann Munk's and Ferrier's experiments on ablation of cortical motor centers, he no longer needed (or desired) to emphasize the dependence of physiology on psychology. It is not surprising that in the early 1890s, when James was seeking to establish psychology as a natural science autonomous from philosophy, he avoided discussing the ways in which pathological findings about aphasia and experiments on brain localization depended on introspective psychology for their interpretation.

### Philosophy As an Inductive Enterprise

In his negotiations for a philosophy position, and in his early discussions of the emerging disciplinary boundaries between philosophy and the naturalistic science of man, James often referred to a group of German scholars who pressed "hard in the direction of metaphysical problems," feeling free to combine philosophy and physiology, and to investigate psychological and metaphysical questions from a physiological, naturalistic, and medical point of view.[74]

Among this group were Wilhelm Wundt, the German physiologist who was appointed professor of philosophy at Zurich in 1875, and at Leipzig the following year, and Hermann Lotze, the spiritualist philosopher who could "write physiology" as successfully as the ultramaterialist Jacob Moleschott and the philosophically oriented physiologist Hermann von Helmholtz. These figures provided James with role models that he could point to in his attempts to forge a new intellectual image. James praised them for their restraint from metaphysical biases (of either a materialistic or a spiritualistic kind) and their lack of "militant consciousness." He portrayed them as embodiments of the impartial attitude that he preached in the late 1870s, and contrasted them with their French and English-speaking counterparts, philosophers and scientists who were often prey to prejudice and partiality and tended to side either with materialism (Huxley) or with spiritualism (for example, Princeton president James McCosh or Yale president Noah Porter).[75] Lotze, in particular, as Sergio Franzese argues, had shown how the latest scientific results of physiology and medical psychology could be incorporated in a philosophical anthropology that did not issue in materialism. He provided James with a model of the integrated approach he wished to bring to the investigation of human nature and man's relationship to the universe.[76]

These investigators also offered persuasive examples of scientifically trained men who could successfully teach "mental science" and philosophy. Each of them, as James was quick to point out, owed his philosophical stature to his scientific training. Thus, in the context of his negotiations with Eliot in 1875, James backed up his proposal to introduce an elective in physiological psychology with the successful examples of Wundt and Lotze. Playing on Eliot's respect for the German university system, he implied that, as Skrupskelis notes, Harvard trailed behind prestigious institutions such as Göttingen, Heidelberg, and Zurich.[77] James had made the same point in a short, untitled, and unsigned note he had published in *Nation* in June 1875, where he stressed the "importance to philosophy" of the appointment of Wundt and E. Hitzig (a "medical practitioner and lecturer on electrotherapeutics") respectively to the chairs of philosophy and psychology at Zurich. James had some reservations about Wundt's work in psycho-physics and Wundt's failure to fully "combine" physiology and psychology.[78] Nevertheless, he contended that Wundt's work on the senses, the nervous system, and the "territory common to mind and matter," and his long scientific training as an assistant in Helmholtz's physiology laboratory at Heidelberg made him "an eminently well-qualified teacher of mental science." Playing on a wider American desire to catch up with German science, James continued: "In this country such appointments [as Wundt's and Hitzig's] would probably provoke a good deal of orthodox alarm. But in Germany not only is thought more fearless of consequences, but 'camps' in opinion are much less clearly defined, and materialistic and spiritualistic tendencies keep house together most amicably in the same professional brains."[79]

James hailed the appointment of Wundt and Hitzig as "hopeful tokens of *a new era in philosophical studies*—an era in which the old jealousy between the subjective and the objective methods shall have disappeared and in which it shall be admitted that the only hope of reaching *general* truths that all may accept is through the co-operation of all in the minute investigation of special mental processes."[80] This incipient "era of philosophical study" was thus to be characterized by the cooperation of philosophers and natural scientists. Indeed, James declared somewhat optimistically that a "reconciliation" between the two camps was close at hand.[81] The product of such a cooperation would be an inductive type of philosophy, one within which "solid philosophical conclusions" would gradually emerge "from the mass of details," just as it happened in the inductive sciences.[82] The era of grand philosophical systems, which sought to characterize "the entire universe in one single effort" on the basis of some a priori intuition, was over. Like the inductive sciences, philosophy should seek to achieve its syntheses by "taking things piecemeal," on the basis of the patient and methodical collection of facts. These should be provided both by the introspective philosopher and by the natural scientist.

### Philosophy As Mental Discipline

Although in 1875, in order to further his career goals, James presented a few scientific thinkers as good philosophers and good philosophy teachers, James's call in the mid-1870s for a new inductive philosophy and a more collaborative stance was by no means meant to blur the differences between the philosopher and the man of science. Nor was it meant to open the doors of philosophy indiscriminately to all men of science. While James struggled to legitimate the transmission of knowledge claims across the boundaries separating philosophy (and its subfield of psychology) from natural science, he also posited specific requirements that investigators ought to satisfy so as to be considered philosophers.

This was the message that James conveyed in a review of William Benjamin Carpenter's *Principles of Mental Physiology* (1874).[83] While praising Carpenter's previous biological work on human physiology, James declared himself less satisfied with the more philosophical and psychological speculations that Carpenter had offered in his book.[84] An essentially "psychological" work, James argued, *Mental Physiology* "place[d] itself in competition with the writings of professed philosophers rather than of naturalists."[85] However, owing to Carpenter's mainly biological training, the tone of the book appeared to be "slack" when compared "with the writings of philosophers."[86] James took the inadequacy of Carpenter's philosophical speculations as an "example of the inferiority of the natural sciences to the logical and philosophical in producing a certain quality of 'mental texture.'"[87] He used this example to draw a "passing moral," useful at a time when the scientific disciplines "threaten[ed] . . . to be triumphant all along the line."

Intervening in the pedagogical debate in which Carpenter himself had been thoroughly involved, James argued that the philosopher's type of mind differed intrinsically from that of the man of science. Carpenter had famously stressed the unique mental benefits conferred by early "scientific training," arguing that only that kind of training enabled the mind to properly observe and make correct judgments.[88] In response, James contended here that only philosophically trained minds were capable of properly performing operations of abstractions, and inferring in a sound, analytical way, rather than by means of imprecise leaps of mind, general principles from a multitude of data. Disciples of the natural sciences, trained "to adore facts as the mystic and unfathomable wells from which all truth is to be drawn," turned out to be lazy when it came to extracting abstract and general principles from the empirical data. Philosophy instead encouraged clear abstraction, reduction, and distillation. It made the mind "ever athletic and sharp-edged, and, more than all, far-looking."[89] James suspected that those qualities, when "not well furnished with facts to operate upon," might lead to "terrible imbecility." Nevertheless, he stressed that philosophical studies developed a *texture* of mind that was both "harder and subtler than other disciplines can produce." Thus he concluded that philosophical studies should never take a subordinate place in any education "where the subjective quality of the pupil is the aim."[90]

James's review of Carpenter's *Mental Physiology* provides yet another instance of the ease with which James could slide from discussion of disciplinary boundaries to a discussion of pedagogical questions, as well as the other way around. The reason why, here as elsewhere, James was able to slide from the disciplinary to the pedagogical is that he operated on the basis of a twofold notion of philosophy. In the 1870s, by "philosophy" James meant both a field of investigation and a mental discipline.[91] This double notion of philosophy emerges quite clearly in his article "The Teaching of Philosophy at our Colleges" (1876). There James at once described philosophy as a field of investigation (namely, the field concerned with the investigation of human nature and man's relations to the environment) and as a pedagogical technique, a unique tool for the cultivation of the mind.[92] Intervening in the debate over the "Greek question"—whether Greek should still be a requirement for college education—James argued that every discipline was appropriate for college education, provided it was taught "in the philosophical manner": "Teach all sciences in a liberal and philosophic manner, and Greek ceases to be indispensable. Teach Greek in a dry, grammatic fashion, and it becomes no better than Gothic. So that the decision of the Greek question depends on the decision of the philosophic question. . . . All branches must be taught from the first in a philosophic manner, must be saturated with the liberal spirit, for any good to be effected."[93] James believed that philosophical training alone endowed the mind with the qualities of openness, "flexibility," "mental perspective," and the capability of rethinking the familiar in unfamiliar terms. These were the virtues of the "liberal" character, a type of character that could be formed through a liberal arts training, and could never result from professional or scientific schools (such as the schools James himself had attended).[94] That is why philosophy should be the framework of thinking within which all the "branches" of knowledge should be taught.

Thus in the 1870s we find James developing at the same time two notions of philosophy as he intervened in the debates concerning the relationship between the philosophical and the naturalistic science of man. The disciplinary and the pedagogical conceptions of philosophy were closely correlated and intersected in several ways. James's understanding of philosophy as mental training underlay his understanding of philosophy as a collaborative type of endeavor, one in which general conclusions about the human subject would be inductively reached on the basis of empirical data. The data would be provided by the special sciences, but the extraction of general principles from empirical data, the kernel of philosophical activity, could be better performed by people whose minds had been trained philosophically.

In both the disciplinary and the pedagogical settings, philosophy remained distinct from the special sciences. As the investigation of human nature, philosophy was a synthetic activity that borrowed from the special sciences results and methods, but operated on a different level. Likewise, as a mental discipline, philosophy stood higher than the natural sciences in two ways: it provided a superior type of mental texture, and it provided the framework that should encompass education in all fields, including the sciences. Buying into a widespread metaphor that depicted philosophy as the "architectonic" science, James concluded that, whereas those trained in the natural sciences were only "brick-makers,"[95] those trained in philosophy were "architects."

Thus, in the late 1870s, through his boundary work James fashioned a conception of philosophy that allowed for cooperation between philosophers and natural scientists, but still implied a superiority of philosophy vis-à-vis the sciences, and a difference between the philosopher and the scientist. That difference could be overcome only by certain especially gifted investigators, like himself "men of the two disciplines," but was not to be erased.

## 1890–92: Psychology and Philosophy

In November 1890 James's *The Principles of Psychology*, "the enormous *rat* which . . . ten years' gestation has brought forth," was published. "No one could be more disgusted than I at the sight of the book," James confessed to his editor; *The Principles* appeared to him a prolix, "loath-some, distended, tumefied, bloated, dropsical mass, testifying to nothing but two facts: *1st*, that there is no such thing as a *science* of psychology, and *2nd*, that W. J. is an incapable."[96]

As historian of psychology Rand B. Evans emphasizes, James's "rat" had "no parallel in American psychology." Many praised the originality of the book, and James's British colleague James Sully even suggested that James had been magically able to "recreat[e] his subject," psychology.[97] In the early decades of the twentieth century, it was presented retrospectively as the beginning of a new era in psychology; some even described it as the beginning of American "psychology" *tout court*.[98] However, *Principles* also reaped criticism from all sides of the very heterogeneous emerging psychological community. It annoyed many of those still committed to the insights of rational psychologies premised on the idea of a soul or of a transcendental ego. It bugged both the psychologists who wanted to develop a metaphysics-free psychology, and those who were comfortable bringing in metaphysical assumptions and arguments. It hardly satisfied experimentalists and other proponents of the "new," "scientific" psychology, even as it disturbed philosophers who wanted to return to philosophy its lost dignity. How so?

Although some commentators had qualms about specific theories that James proposed, and others lamented the general "lack of unity" and unsystematic nature of the work, many criticisms were directed at the new way in which James framed the discipline of "psychology." Rather than associating psychology with "introspective *philosophy*" as he had done in the 1870s, in *Principles* James famously aimed to treat psychology as a "natural science." Psychology was to be the field of research whose task was to inquire into "the empirical correlation of the various sorts of thought or feeling with definite conditions of the brain," including both "conditions antecedent" to mental states and their "resultant [physiological] consequences" (for example, "changes in the caliber of blood-vessels" and "alteration in the heart-beats").[99] A psychology so understood would rely on the "postulate" that the brain was "the one immediate bodily condition of the mental operations."[100]

Hence it would include "a certain amount of brain physiology." James also
announced that he would "feel free to make any sallies into zoology or into
pure nerve-physiology which may seem instructive for our purposes."[101]

Although the psychologist could make incursions into those other special
sciences, one thing was forbidden by the rules of the game that James set out
in the preface to *Principles*: to think things philosophically. Psychology as a
natural science, James wrote, "assumes certain data uncritically," namely:
"(1) thoughts and feelings, (2) a physical world in time and space with which
they coexist and which (3) they know." Of course those data, like the data
assumed by any other natural science, were "discussable," and cried out for
explanation. However, James continued, "the discussion of them . . . is called
metaphysics and falls outside the province of this book." Psychologists, then,
should stick to the natural science point of view and avoid philosophizing.
They should expressly seek to avoid the kind of "fragmentary, irresponsible,
half-awake" metaphysics that, as James emphasized, crept into the work of
mental physiologists like Maudsley, and the work of more philosophically
oriented psychologists, especially the spiritualist writers (who based their
psychologies on the assumption of the existence of a "Soul") and associationists or "atomistic philosophers" (who, instead, denied the existence of
the soul, and described mental life as resulting from the spontaneous combination of elementary sensations). James found that when metaphysics,
"unconscious that she [was] metaphysical," "inject[ed] herself into a natural
science," it "spoil[ed] two good things."[102] This "strictly positivistic point of
view" was the only feature of *Principles of Psychology* for which James "was
tempted to claim . . . originality."[103] How did this separatist agenda fit with
the efforts that James had made in the late 1870s to promote cooperation
between men of science and philosophers? How did it fit with James's own
self-portrait as a "man of the two disciplines"? And what were its implications
for James's understanding of psychology and of philosophy?

As the author of *The Principles of Psychology*, James, as is well known,
was chastised both for his proposed "separatist" stance, and for his failure
to abide by it. George Trumbull Ladd, a philosopher and psychologist who
combined a physiological approach with metaphysical explanations premised
on the idea of the soul, complained that James's definition of "psychology as a
natural science, without metaphysics" was too narrow. According to James's
definition, Ladd charged, the only kinds of explanations admissible in psychology were physiological explanations that related mental states to brain
states. This restriction excluded from the field a good deal of what was considered to be explanatory "scientific" psychology, including, for example,
"all introspective psychology as such, almost the entire domain of . . .
physiological psychology," the psycho-physics tradition, and nearly all the

experimental work in "psychometry"—including those reaction-time experiments that constituted a significant part of German experimental work, and which many had come to consider as the emblem of the scientific experiment.[104] Ladd also observed that, despite James's intention to ban metaphysical inquiries from psychology, *Principles* amply brought metaphysics back into psychology (a point that James himself had conceded).[105] Philosophers who wanted to separate psychology from philosophy also found much to complain about concerning the kind of boundary work that James carried on in *Principles*. James's British friend Shadworth Hodgson, the leader of the Aristotelian Society that aimed to establish philosophy as an independent pursuit conceptually and methodologically distinct from all the special sciences, objected to James's strategy, finding that it implied a derogatory and humiliating conception of philosophy. In a letter to James, a concerned Hodgson insisted that James subscribed to "the old traditional view of the relations between philosophical & psychological inquiry." According to that view, the "positive sciences" came first, and discovered "all the laws, & truths of every sort they can"; "metaphysics" came later, and consisted of a depository of all the insoluble questions and false starts.[106] James, Hodgson suspected, was trying to do to philosophy what the atheists had done to God.

*Principles* also discontented some experimental psychologists, especially those who received their training in Wundt's Leipzig laboratory. German-trained psychologists perceived that James's approach to experimentation was very different from the experimental practices of German physiology, psycho-physics tradition, and physiological psychology, which they had come to identify with proper experimental psychology.[107] Like Wundt, who allegedly located James's *Principles* outside of psychology in the domain of literature, some of his American students had a hard time considering *Principles* as a contribution to a scientific psychology.[108] For example, E. W. Scripture—who received his PhD from Wundt in 1891 after a rigorous novitiate in reaction-time experiments, and then became director of the Yale laboratory—found that James's approach to psychology failed to comply with the methodological requirements that characterized the scientific experiment: precision, accuracy, and exact measurement.

In an 1893 speech delivered to the American Psychological Association, Scripture complained that American psychologists made experiments "without an idea of the first laws of experiment," and conducted measurements without even realizing that there was such a thing as a science of measurement.[109] For Scripture, James's work hardly reflected the mental attitudes of precision, exactitude, and patience, that "exact sensibility," as Kathryn Olesko calls it, that most German and German-trained experimenters considered as the true mark of the scientist.[110] Scripture remembered well (and quoted sarcastically) the famous passage of *Principles* in which James provocatively described the method used by German experimental psychologists as a method that "taxe[d] patience to the utmost," and that "could hardly have arisen in a country whose natives could be *bored*." He was outraged when James scorned the (self-described) patient, detailed, analytical work conducted by G. T. Ladd and many other "eminent psychologists" as "tedious." The genuine experimentalist, Scripture argued, just could "not get bored." Calibrating one's apparatus, and devising experimental arrangements that would reduce median errors below the lowest margin (Wundt praised an arrangement of instruments that reduced the error in reaction-time experiments to about 0.03 milliseconds) and produce "records of any desirable accuracy" were essential to proper experimental work.[111] James's boredom, in contrast, revealed an "amateurism" and a "carelessness" that was entirely foreign to science. Scripture implied that, far from banning metaphysics from psychology, James's psychological work stemmed precisely from the kind of "arm-chair" philosophical work ("vague observation, endless speculation, and flimsy guesswork") that he sought to banish from scientific psychology.[112]

In the next couple of years, James tried to clarify his position, speaking to the concerns of different audiences, some of which worked at cross-purposes. As is well known, in 1892 two different publications—James's *Briefer Course*, a shorter (and modified) version of *Principles*, and his article "A Plea for Psychology as a Natural Science"—reiterated his views: psychology, like any other natural science (or aspiring natural science), necessarily made metaphysical assumptions and raised all sorts of metaphysical puzzles; however, the discussion of those did not belong to psychology, and neither did any attempt to provide a "metaphysical" explanation of psychological data and assumptions.[113] Such puzzles would have to be handed to the "specialists in philosophy."[114] Philosophers, in turn, should be careful to conduct their metaphysical work outside the framework of psychology understood as "a natural science," and avoid burdening the infant science with metaphysical questions that did not pertain to it.

Meanwhile, psychology as a natural science was to be assigned to men of science: naturalists, laboratory workers, biologists. James was confident that such men, working together and abiding by the requirement that they avoid metaphysical explanations and discussions, could produce a sophisticated science of psychology. The natural science of psychology was to be part of a broader biological "science of human nature," a science of man that, like the "naturalistic science of man" that James discussed in the 1870s, studied humans in their relationships to their changing environment. James believed that both psychology and the biological study of human nature would be rich in practical applications and would yield rules useful to people with urgent practical concerns, from asylum superintendents to jail wardens and educators.[115]

In these works James reiterated the point he had made in *Principles*: that the separation between metaphysics and psychology would only be "provisional." The special sciences were just "beginnings of knowledge," parts of a larger "body of truth" cut off from the rest for practical reasons.[116] The provisional separation of philosophy from psychology was essential for the development of the new science itself, as well as for the development of a unified psychological community. Metaphysical explanations and discussions, James argued, were a source of division and contention. By excluding them "peace" would be secured, and, hopefully, "an enormous booty of natural laws" would be "harvested in with comparatively no time or energy lost in recrimination and dispute about first principles."[117] The provisional separation, moreover, was also in the best interest of philosophy. The task of philosophy, James reiterated, would be to clarify and address the metaphysical puzzles raised, and the assumptions made, by the sciences, a conception shared by several contemporary philosophers. However, other philosophers contended that philosophy should analyze the notions on the basis of which the special sciences operated before the sciences could start to carry on their work, and that philosophers' work was thus to lay the foundations for further scientific activity. James instead perceived that a proper philosophy, one aware of its scope and its duties, could take off only after the sciences had carried out their job of empirical inquiry and reached a kind of maturity. At present that point had not yet been attained. Mature sciences alone, with the "definite 'laws' of sequence" that they found to hold between the "things" that they assumed, could "furnish general Philosophy with materials properly shaped and simplified for her ulterior tasks."[118] Philosophy would then work with those empirical materials and relations to clarify them. The separation between philosophy and psychology that James had proposed thus was only a necessary prelude to a more fruitful and intensive future interaction.

This conception of philosophy differed in some ways from the architectonic conception of philosophy that James had promoted in the middle and late 1870s, but in other ways it carried forward his previous plans for redefining the field. While in the mid-1870s James presented himself as a man of two disciplines, here he more clearly separated the role of the philosopher from that of the man of science. Philosophy was no longer characterized as an inductive science—an activity that consisted in inducing general principles from the data provided by the sciences; nevertheless, it remained strictly tied to the sciences, especially psychology. In order to elucidate the new ties

James suggested, we need to take a close look at another occasion in which James brought up the issue of the relationship of philosophy to psychology, one in which he renegotiated the boundaries that were increasingly separating them.

## Psychology, Philosophy, and the "Compounding" of Feelings

In December 1894 the members of the American Psychological Association convened for the society's third annual meeting were treated—or in some cases subjected—to James's presidential address, "The Knowing of Things Together." The address consisted of a long and difficult discussion of what to some of the people in the audience must have appeared an abstruse problem: the problem of the "synthetic unity of consciousness." Those who patiently followed James through his thorny analysis of that abstract topic were rewarded with a succulent bit of "gossip."[119] Coming to the end of his discussion, James made a sudden declaration: since publishing *Principles of Psychology*, he confessed, "I have become convinced . . . that no conventional restrictions *can* keep metaphysical and so-called epistemological inquiries out of the psychology-books."[120] Coming from a psychologist who had preached the necessity of separating psychology from metaphysics, this seemed a startling and unexpected turn. What did James mean? Why did he change his mind? And what implications did "The Knowing of Things Together" have for James's understanding of psychology and philosophy?

In order to answer this question, we need to take a step back and look at a criticism of *Principles* that was made in 1893 by G. S. Fullerton, a philosopher and psychologist at the University of Pennsylvania. Fullerton, who used James's *Principles* as a textbook in some of his psychology courses, advocated, like James, a clear-cut separation of psychology from philosophy. Yet, like others, he complained that James had violated his own requirements, and that in *Principles* he shifted into epistemology and metaphysics. His criticism, however, in contrast to criticisms voiced by such as Ladd, struck a vital nerve, one that persuaded James to suspend his separatist agenda.[121]

Fullerton's critique, as Ignas Skrupskelis has shown, involved a technical point of James's psychology, but one central to James's *Principles of Psychology* and to his strategy of reframing psychology as a natural science, namely James's assumption that all mental states, including feelings of complex objects, or states that "knew" many things at once, were, in fact, integral, undivided wholes.[122] That assumption radically differentiated James from spiritual psychologists, modern-day associationists, and laboratory psychologists who were busy identifying elementary sensations and tracing the modalities of their "integration" into higher mental states. (Wundt and his followers, for example, belonged to this tradition.)

James's decision in *Principles* to regard mental states that "knew" complex objects as undivided represented James's take on the question of how to understand, and talk about, the synthetic unity of consciousness: that is, the question of how different simple, elementary feelings (for example, the feeling of lemon and the feeling of sugar) could "combine together," or "be combined together," into a complex feeling (the feeling of lemonade). That simple mental states could constitute the "parts" of higher mental states, ones "composite in nature," was an unproblematic assumption for many philosophers, especially idealists, as well as for many psychologists.[123] Experimentalists working in certain traditions of physiology and physiological psychology influenced by Fechner's psycho-physics, for example, had performed experiments that they took to demonstrate that simple feelings (the feelings of green and red) could combine into complex feelings (the feeling of "yellow" perceived by an experimental subject whose retina was hit simultaneously with green light and red light in an experiment on stereoscopic vision).[124] Many psychologists working in the spiritualist or transcendentalist tradition, likewise, did not doubt that elementary feelings could be combined into complex feelings. However, whereas the associationists explained the "self-compounding" of feelings in terms of the laws of some "mental chemistry," spiritualists and idealists assigned the task of composing separate feelings to an external agent, the soul—an individual soul according to spiritualistic writers, a transcendental one according to the idealists.

To James's mind, the claim that parts of consciousness can combine was one "of the obscurest" assumptions that could be made in psychology.[125] In *Principles* he argued that there was no experimental proof that "simple" feelings (or elementary feelings, the existence of which he strongly doubted) could "combine themselves" into a higher feeling. Assuming that they could do so was "logically unintelligible" and "flies in the face of physical analogy." Talking about "mental states . . . made up of smaller states conjoined" made as much sense as talking about "the public opinion" or "the spirit of the age."[126] Smaller feelings did not add up to a more complex feeling, in the same way that the "private," "ejective" minds of the individuals constituting "the public" did not "agglomerate into a higher compound mind."[127] The explanation offered by spiritualists and idealists was equally unacceptable to James. It involved an appeal to a metaphysical entity, the soul, that would "compound" the feelings; however, since the soul was underdetermined and lacked any clearly describable properties, this metaphysical explanation simply amounted to restating the facts, and failed to explain anything. In short, neither the associationist explanation nor the spiritualist explanation could belong to a scientific psychology, since neither of them stuck to the domain of the empirically verifiable.

The strategy that James adopted in *Principles* to avoid those problems consisted simply in eschewing the question of how simple mental states could be compounded (or could compound themselves). Instead, as we saw, James assumed that mental states, including those that "knew" many things at once, were *simple* and undivided. This assumption avoided both of the logical difficulties he had identified, that is, the appeal to "metempiric" assumptions (be they souls or "psychic atoms") and the need for metaphysical explanations. Thus James felt that it could provide "a useful basis for united action in psychology," a platform on which everyone could agree, and which would allow investigators to get on with the business of finding regularities.[128]

James's postulate, however, annoyed more psychologists than he had anticipated, and brought more discord than peace in psychology. Fullerton simply echoed the discontent of many when he complained that James's own postulate was highly metaphysical. James's assumption of the undivided nature of complex mental states, Fullerton argued, violated the intuitions of the "common man" and the scientific psychologist. Moreover, by characterizing feelings on the basis of their "cognitive function," of what they could "know," Fullerton continued, James shifted from psychology to epistemology and, contrary to his own restrictions, opened the doors of psychology to a "confused" epistemology that was entirely inappropriate to a scientific psychology. When "the psychologist . . . tries to be psychologist and epistemologist at the same time," he warned, psychology ends up "a loser."[129]

James's 1894 American Psychological Association presidential address was his response to Fullerton's criticism. James felt that his "intention" to treat mental states as "unique in entirety" was a "good one," and was "surprised that "this child" of his "genius should not be more admired by others—should, in fact, have been generally either misunderstood or despised." However, he reassured his audience that he would not defend or explain once again his assumption. Instead, he would simply "give it up."[130] From now on he would feel free, like others in the field, to talk of mental feelings as "compounds" "in [which] . . . one could call parts, parts."[131] By giving up his assumption that mental states were integers James also gave up his injunction that philosophy should not be mixed with psychology. In fact, it was precisely at this point that he confessed that "metaphysics could not be kept out of psychology."

Why did James think that by abandoning his postulate psychologists would have to take metaphysics on board? Because, he explained, one could not assume that the contents of simple feelings could "compound" into complex ones without asking oneself "how" that happened; and the project of answering that question, namely "the whole business of ascertaining *how* we come to know things together," fell "to metaphysics."[132] In other words, psychologists who took the compounding of mental states (their psychic integration) as a matter of fact should be aware that, in doing so, they brought on themselves the task of carrying on an "inquiry" into the "conditions" of that phenomenon.[133] Thus the boundaries between a "natural science" of psychology and philosophy, which James had traced in *Principles*, were to be lifted, at least in this particular case. Perhaps, as Skrupskelis argues, metaphysical explanations were still to be excluded from a natural science of psychology, but metaphysical inquiry could not be completely avoided; it was now a duty that, to some extent, scientific psychologists ought to fulfill.

James presented his decision to drop his assumption of the indivisibility and integrity of mental states as a gesture of reconciliation with his adversaries. However, he needed to give up that assumption for other reasons as well. As Skrupskelis has shown, the assumption that James had made in *Principles* could not be squared with his nascent metaphysics of radical empiricism, and especially could not be squared with the solution that James had offered to a related, and to him all-important, problem: the problem of how "two minds could know one thing."[134] Within the framework of radical empiricism, Skrupskelis notes, James considered "pure experience as neither mental nor physical but capable of being both." This enabled him to say that the same bit of pure experience could, in one "set of relations," be mental and, in another, be physical; therefore the same bit of experience could enter into the mental streams of consciousness of different people (me and you, for example). However, since the mental states of the two knowers were "fields," and not atomic feelings, the only way the bit of pure experience could be "known" by both knowers was if it was "present in two different mental fields." This explanation required an ability to talk about parts of mental states, to talk about "mental atoms which could pass unchanged from one mental state into another."[135] That, of course, could not be done if one continued to maintain that mental states were undivided wholes. Thus James here addressed a metaphysical conundrum that was central to his radically empiricist metaphysical agenda, and it was toward furthering that end that in 1894 he renegotiated the boundaries separating philosophy from psychology.

Like *Principles* and the mid- and late 1870s psychological-physiological works discussed above, James's 1894 American Psychological Association address was a piece of boundary work, part of a series of efforts to rearrange both boundary areas of exchange and common work, and reconfigure the relationships among increasingly separated groups of researchers. Indeed, "The Knowing of Things Together" carried on boundary work on multiple levels. As David Lamberth notes, it provided a "public presentation" of many of the theses that entered into James's (then still unnamed) radically empiricist metaphysics.[136] By using his presidential address to psychologists as a forum for launching a new metaphysical vision, James not only publicly announced his decision to lift the ban of metaphysics from psychology, but also demonstrated how to allow for "the waters of metaphysics" to flow into psychology in a productive fashion.

What are the implications of "The Knowing of Things Together" for James's conception of psychology and for his conception of philosophy? Skrupskelis contends that James did not abandon the conception of psychology as a natural science. James, he argues, continued to construe psychology as a natural science and to believe that, by avoiding metaphysical discussions and explanations, psychologists could produce a psychology "infinitely more complete" and useful than "the psychologies we now possess."[137] Nevertheless, with his American Psychological Association address, James also cleared the ground for a different kind of psychology, one that would allow metaphysical inquiries while still continuing to ban appeals to mysterious metaphysical entities, such as souls and substantive selves. For example, as Lamberth notes, in his 1895–96 seminar "Philosophy 20b: Psychology Seminary—The Feelings," James addressed the question of how a person can have feelings; since this, according to James, was a metaphysical question of causality, the seminar turned out to be all "about metaphysics and epistemology."[138] At the same time, in this seminar and in his "Philosophy 20b: The Philosophical Problems of Psychology" (1897), James's inquiry blended into metaphysics. Not only did he address some of the philosophical puzzles raised by psychology, he also resorted to his previous psychological discussions of consciousness in order to provide a metaphysical analysis of experience. For example, he deployed his psychological description of the "fringe" surrounding a cross-section of the "stream of consciousness" as a ground for an analysis of the "margins" of a cross-section of the "stream of experience."[139] In the same spirit, James had suggested that the fundamental psychological fact of the "knowing of things together" could represent not only the structure of "feelings" but also the "ultimate essence of all experience."[140] The result, fully articulated in the essays of 1904 and 1905, was a kind of philosophy in which, as a philosophical commentator of the time complained, James tried to "make psychology do the work of logic and metaphysics."[141] Considered by some as a form of "philosophical psychology," and by others as a form of "psychological philosophy," to many contemporaries James's work in the second half of the 1890s seemed to blur all the boundaries that previously he had endeavored to erect between a natural science of psychology and philosophy.

If after 1894 James framed a distinction between two different ways of understanding psychology—namely, the "natural science of psychology" and a philosophical psychology—he also seemed to operate with a twin understanding of the function of philosophy. On one side, philosophy consisted, as we saw, in thinking thoroughly through the data assumed by the special sciences and "scrutinizing their significance and truth."[142] By solving the metaphysical puzzles raised by the special sciences, philosophy would work to attain "a maximum of possible insight into the world as a whole."[143] It strove to provide an "ultimate critical review of all the elements of the world," and "some day" it would "help us" attain "rational conceptions of the world."[144] In addition to this critical function, James emphasized a synthetic function. Looked at from this side, philosophy embraced the "total body of truth" and stood for a possible "Science of all things." "Most thinkers have a faith that at bottom there is but one Science of all things," James noted.[145] Such a science did not yet exist: "instead of it, we have a lot of [provisional] beginnings of knowledge made in different places, and kept separate from each other merely for practical convenience' sake, . . . the 'Sciences' in the plural."[146] However, James believed that, with "later growth," the separate sciences might, at some future time, "run into one body of Truth"; that science of all things, if "realized," "would be Philosophy."[147] In this sense philosophy would comprise that "whole mass" of truth from which the special sciences had been cut off at the "early stages of their development," and in which they could in the future find unity. It was the philosopher's job to facilitate that emerging unity.

In the 1890s these two conceptions ran together: metaphysics was "the forum" where the special sciences could "hold discussion," and open up their assumptions and results to revision "in the light of each other's needs."[148] In the last decade of his life, however, James drew a clear-cut distinction between those two functions of philosophy. Philosophy in the first clarificatory sense, for which he now reserved the term "metaphysics," was a technical philosophical discipline whose task consisted in addressing problems such as the problem of being and becoming. Philosophy in the second, integrative, sense, which James would refer to as "general Philosophy," instead would unify knowledge and bring together the scattered special sciences. Although James retained the idea, one shared by many "architectonic" philosophers, that philosophy alone could unify the sciences, he gave up entirely the idea of philosophy as an architectonic science. As argued in chapter 7, philosophy for James would unify the special sciences not conceptually by solving their disputes over assumptions and data, nor by inducing general principles from which the principles of the special sciences could be deduced. Instead, it would bring forth what I call a "social" unification of the "body of Truth." It would counter the disintegration of knowledge into myriad specialized professional knowledges, redressing the fragmented early twentieth-century "social" geography of knowledge in which inquiry was carried out by investigators operating in "different," separate "places." General philosophy, properly conducted, would facilitate new modes of social relationships and new ways of collaboration and communication among the diverse kinds of inquirers. These new modes of engagement would bring together philosophers, psychologists, practitioners of the other sciences, and people with practical concerns, making them into a community that would engage in cross-disciplinary and cross-divisional discussion.

In the first half of the 1890s, however, the conception of a "general Philosophy"—adumbrated in James's metaphor of metaphysics as "the forum" where the special sciences could "hold discussion"—was not yet there. In order to develop it, James needed to carefully think through the concepts of society, the "public," "public opinion," and community—concepts that in the 1890s he had a hard time (and, apparently, little interest) in making sense of. The path by which James reenvisioned the relationships between individuals and society was the same as the path by which he reenvisioned the relationships among different disciplines, especially those between philosophy and the sciences. It ran through political events that made James sensitive to the big social questions of the day. It also ran through James's investigation of pathological, mystical, and "psychical" phenomena, and through his developing a practical sense of what a desirable community of inquirers should and should not look like. All these threads, as we will see in chapter 6, came together at the end of James's life, when James made one last successful, if profoundly disconcerting, effort to solve the problem of the "compounding of consciousnesses"—the problem that, in 1894, prompted him to lift the ban of metaphysics from psychology. James's solution to that problem embodied both his vision of society and his sense of desirable communities of inquirers; these, in turn, reflected the "self-compounding" of consciousnesses in higher mental states, revealing the deep relationships between James's metaphysical work and the social concerns that became increasingly prominent in the late 1890s. James's solution was also reflected in his understanding, in the last decade of his life, of the ways in which philosophy could mediate between the special sciences and other disciplines, promote the ideal of cross-disciplinary cooperation, and, in that way, give unity to knowledge.

# 3

# James and the (Im)moral Economy of Science

In "science," as a whole, no man is expert, no man an authority; in other words, there is no such thing as an abstract "Scientist"—fearful word! . . . By all means let every man who has a stomach for the fray be admitted to the speculative arena. But let it be on an equal footing with all comers, all to wear the speculative colors, no odds given, and no favors shown.

WILLIAM JAMES

Your scientific barbarian is a bad kind because he is always so arrogant.

WILLIAM JAMES

## The Ethos of Science and Scientific Personae

In a letter dated September 19, 1880, to William James, Josiah Royce observed: "Science expresses a particular kind of activity, especially distinguished by the ethical qualities of patience, self-possession, doubt, and universality of aim, coupled with much definiteness of construction. The difference between science and fanaticism is ethical."[1] James and many late nineteenth-century scientific investigators would have agreed with the principle underlying Royce's statement, even though they would have identified differently the specific virtues required for a correct practice of science. Right at the time when many scientific fields emerged as autonomous, organized disciplines, a variety of social actors underscored that scientific activity was predicated on the practice of a well-defined personal discipline, both mental and bodily, one requiring the cultivation of certain virtues and dispositions. Scientists and other observers variously identified the mental traits, feelings, emotions, sensibilities, and habits typical of the "scientific temper," and constitutive of what, following David Hollinger, Lorraine Daston, and others, I call the "ethos" or "moral economy" of science. The "scientific temper" was thought to regulate the cognitive behavior of scientific inquirers, and was considered necessary to the production of scientific knowledge and to differentiate it from other types of knowledge. Thus moral economies of science were often tightly linked to scientists' epistemological commitments and research practices.

In the late nineteenth century and early twentieth century, discourse about the proper scientific ethos was often cast in social or political terms, and could also convey indications about the most desirable forms of government, and the role and position that scientists should occupy in society. That is, science practitioners and other social actors occasionally presented the scientific temper as a key ingredient for the correct functioning of society and used formulations of the ethos of science as a means to support and publicize political and social visions.[2] On another level, moral economies of science also functioned as social economies; they conveyed social norms, even political values, which dictated modes of social interaction within a scientific community as well as between scientists and other constituencies. As Steven Shapin and others have noted, these norms and the modes of governance characterizing scientific groups also reflected and promoted their approach to epistemological questions.[3]

Sometimes moral economies of science, scientists' epistemological commitments, and the patterns of social interaction they sought to promote came to be fused in normative, collective images of the scientist or, as I use the term here, "scientific personae." Scientists often mobilized these collective images in order to maintain barriers separating science from other types of knowledge, to reinforce disciplinary divides, or to promote forms of sociability within the academy or the broader society. In fields where investigators could not yet agree on the goals they should pursue, the methodology they should adopt, or the site where research should be conducted—such as psychology—the persona of the man of science was used in attempts to redirect the field, redefine its boundaries, and promote particular research programs.

This chapter describes James's reaction to the epistemological regimes and the moral and social economies of science advertised by many self-described scientists at the turn of the twentieth century, and traces James's proposals for the creation of an alternative ethos of science. The chapter builds on the work of other James scholars, especially David Hollinger and Deborah J. Coon, who have located the moral and social/political attitudes that James increasingly associated with the practice of science starting in the 1890s. However, it offers a new perspective on James's image of the scientific investigator, by looking at it from the angle of James's approach to questions of evidence and testimony. As Simon Schaffer, Steven Shapin, Peter Lipton, and C. A. J. Coady, among others, remind us, notions of evidence and testimony involve important questions of trust and, because of that, lie at the intersection of epistemology and morals. An examination of these notions offers ideal terrain in which to study how moral economies of science intersect with scientific epistemologies as well as with modes of interaction and social exchange or, as Jill Morawski calls them, interpersonal styles. We will discuss James's approaches to questions of evidence and testimony both in philosophical settings and particularly in the context of his inquiries into psychic phenomena. Our goal here is to explore the interconnection between the epistemological, the moral, and the social in James's scientific research practices.[4]

In the second half of the nineteenth century and at the turn of the twentieth, reports of communications by the dead, apparitions of faraway loved ones as they died, and cases of clairvoyance and of what would later be called "telepathy" quickly propagated throughout the West, creating huge waves of interest in spiritualism—that is, the belief that the human personality survived the death of the body, and that the spirit of the dead could communicate by various means with the living.[5] The phenomena were considered by some to be supernatural and by others to be entirely natural (that is, explicable within the known laws of nature and/or trickery). Still others regarded them to be "supernormal"; that is, according to F. W. H. Myers, the person who invented the word, they went "beyond what usually happens—*beyond* in the sense of suggesting unknown psychical laws." Coined after the term "abnormal," the adjective "supernormal" by no means implied that supernormal phenomena "contravened natural laws." It only acknowledged that such phenomena simply "exhibited" the laws of nature "in an unusual or inexplicable form."[6]

Supernormal phenomena raised pressing questions of evidence and testimony. Debates surrounding these issues involved not only disagreements regarding the reliability of specific pieces of evidence but also an ultimately unresolved conflict among rival "conceptions," or "regimes of evidence," and the moral and social economies of science embodied in those conceptions.[7] Because participants in these debates did not share the same conception of evidence, even when they accepted the same "rules" of evidence, they applied those rules in very different ways and with different results. For example, depending on their social assumptions and on assumptions regarding the nature of the human mind, participants inflected in widely different ways the established rule requiring the admission only of "competent" witnesses and the exclusion of "incompetent" ones. Due to the resulting discrepancies and ensuing contests over the validity of testimony, some investigators found it necessary to open up to public scrutiny their conceptions of evidence and worked to pry open those of their opponents. In this way, debates regarding the evidence supporting the existence of a particular psychical phenomenon shifted to a broader epistemological, moral, and social discussion centered on the nature of evidence. William James was among the participants in this wider debate. Placing James in this broader context will enable us to trace the links between the epistemological, moral, and social aspects of the scientific persona that he offered.

Psychical researchers laid much emphasis on their "scientific methods," and were careful to draw a clear distinction between their activities and those of the spiritualists.[8] To most scientists of the time, however, the "supernormal" phenomena investigated by psychical research occupied an ambiguous borderland of science. Despite their proclaimed scientific methods and despite the solid scientific reputation of quite a few psychical researchers, these investigators of the supernormal hardly excited the sympathies of most men of science. A few scientists conceded that "enough of the spirit of true science [had] oozed over the boundary" that they believed separated science from the investigation of the supernormal; nevertheless, even they contended that psychical researchers and their methods remained largely unscientific, "pseudo-scientific," and even "anti-scientific."[9]

For James, investigating supernormal phenomena and discussing the regimes of evidence that he found prevailing in scientific and philosophical discussions were chiefly a means of exploring the complications of the human mind and, hopefully, in some distant future answering the question of human immortality. However, these investigations and discussions fulfilled for him a second, no less important, goal. They were a means to challenge boundaries that many of his scientific and philosophical colleagues posited as natural and absolute: the boundaries separating "orthodox science" from "superstition" and "heterodoxy," those separating the "academic" scientist, the "specialist," and the "expert" from the amateur and the "average man," as well as the chasm that many argued divided cultivated gentlemen and elite women from uneducated working-class or lower middle-class constituencies. In short, James's activities in psychical research functioned importantly also as a form of boundary work.

Through this boundary work James identified the features characterizing the bad "scientist," and framed a new image of the "genuinely scientific inquirer."[10] This image endowed the scientific investigators, and more broadly the "intellectuals," with the kind of sensibilities that would enable them not only to open the boundaries of science to marginal phenomena but also to transgress social divides and to create a new type of community of inquirers. This kind of community, as we will see more fully in chapter 6, resonated with the social vision that James began framing in the late 1890s. In this way, boundary work along epistemological, methodological, and social lines was central to James's efforts to shape the practice of good science.

## Scientists, Graphic Instruments, and the Ethos of Self-Restraint

Late nineteenth-century science saw the emergence of a new epistemological ideal, that of scientific objectivity, and the consolidation of an ethos whose features were linked with the practices and instruments of objectivity. As Lorraine Daston and Peter Galison have argued, the epistemological ideal of mechanical objectivity, the type of objectivity that strove to eliminate emotional, aesthetic, and volitional factors in the observation and interpretation of nature, was associated with an ethos that demanded of the scientist self-restraint, emotional detachment, and disinterestedness, as well as patience and humility.[11] Although hardly the only type of scientific ethos available at the time, by the mid- and late nineteenth century those attitudes gained wide currency, and became for many researchers the distinguishing mark of the "scientific temper." The ethos of self-restraint reinforced the diffusion of certain instruments, including the kymograph and other self-recording instruments, that promised to record natural phenomena without the mediation of human beings. In turn, that particular brand of scientific ethos was reinforced by the increasing popularity of those instruments, which many argued provided the very model of the perfect scientist.[12] Disciplined, patient, ever-alert, humble, indefatigable, devoid of will, free from the deceitful influence of emotions, passions, and desires, these machines appeared to many to embody the virtues deemed indispensable to the scientist and for the production of objective knowledge.[13]

In the late nineteenth century and early twentieth century, the instruments of mechanical objectivity and the associated scientific ethos informed scientific practice in a variety of fields. As experimental psychologists strove to imbue the profession with such ideals as objectivity and disinterestedness, these machines were emblematic of the standing that the field sought to acquire among the rapidly expanding natural sciences, and quickly became essential hardware in many psychological laboratories. At Harvard such machines were introduced by the physiologist Henry Bowditch. A close friend of William James's, Bowditch had spent a year in Leipzig working in the laboratory of the German physiologist Carl Ludwig, whose kymograph was the prototype and symbol of self-recording instruments.[14] In 1871 Bowditch returned to Harvard, bringing back from Germany not only the self-recording instruments but also a scientific ethos that extolled the "patient, methodical, and faithful way" of studying physiology: the German ethos of "pure research."[15] James's former student G. Stanley Hall had also systematically used graphical instruments in experiments he conducted together with German investigators in the laboratory of du Bois-Reymond in Berlin, and in those of Ludwig and Wilhelm Wundt in Leipzig.[16] Hall was exposed to Ludwig's belief that "the whole philosophy of life" could be translated into curves and graphs registered by self-recording instruments, and embraced (if only half-heartedly and provisionally) a scientific ethos that extolled the virtues of self-control, discipline, hard labor, and self-sacrifice.[17]

James was familiar with the graphic method and occasionally deployed self-recording machines. For example, as Eugene Taylor has shown, in the early 1870s in a series of experiments on brain physiology, he and James Jackson Putnam used a self-recording instrument attached to a manometer in order to record changes in blood pressure.[18] At a meeting of the Harvard Natural History Society held in December 1875, James presented another graphing machine and gave illustrations of its practical applications.[19] In a study of differences induced by hypnosis in subjects' "simple reaction-time," James deployed a Baltzar kymograph in conjunction with a telegraphic key, a tuning fork, and a galvanic circuit.[20] Later on, in *Principles of Psychology*, he described a number of such instruments, including the kymograph, chronograph, psychodometer, plethysmograph, and sphygmograph, and referred to the classical work of French physiologist Etienne-Jules Marey on automatic inscription devices, *La Méthode graphique*.[21]

James was profoundly aware of the ethos that associated impersonality, disinterestedness, and self-restraint with the production of objective knowledge.[22] While James did not reject that ethos of science, he had some reservations about it. In "The Will to Believe" (1896), the first essay in the collection of the same name, he described it in ironic words: "When one turns to the magnificent edifice of the physical sciences, and sees how it was reared; what thousands of disinterested moral lives of men are buried in its mere foundations; what patience and postponement, what choking down of preference, what submission to the icy laws of outer fact are wrought into its very stones and mortar; how absolutely impersonal it stands in its vast augustness—then how besotted and contemptible seems every little sentimentalist who comes blowing his voluntary smoke-wreaths, and pretending to decide things from out of his private dream! Can we wonder if those bred in the rugged and manly school of science should feel like spewing such subjectivism out of their mouths?"[23] James also captured the tight relation between the ethos of self-restraint, the ideal of scientific objectivity, and the use of self-recording instruments as emblems of the genuine scientific investigator: "[According to the men who cultivate science and the "positivists,"] to reject a conclusion on the sole ground that it clashes with our innermost feelings & with our desires is to make use of the subjective method, and the subjective method according to their creed is the original sin of science, the root of all scientific mistakes. According to them, so far from going whither his desires lead him, the man who is searching for truth ought to reduce himself to a mere recording machine, to make his *scientific consciousness* ["conscience de savant"] a sort of blank page—a dead surface, on which external reality will write itself without change or deflection."[24]

As a psychologist James was also aware of the goal of "mechanization" and standardization of the scientist that informed psychological research in many laboratories. Deeply concerned with the subjectivity of introspection, as well as with the goal of attaining the kind of "precision" promised by the physical sciences, many experimental psychologists strove to "bring themselves . . . under a regime of control and calibration" similar to that they enforced over their machinery.[25] As Deborah Coon writes, they "yoked" introspection to self-recording instruments and other "brass" apparatus in an attempt to "mechanize the subject" himself.[26]

James rejected both the image of the man of science as a machine and the image of the scientific investigator as a "passive, reactionless sheet of white paper, on which reality will simply come and register its own philosophic definition, as the pen registers the curve on the sheet of a chronograph." James made fun of "chronograph philosophers," that is, of experimental psychologists who lay siege to the human mind by applying with "deadly tenacity and diabolical cunning" the experimental method, the method of "patience, starving out, and harassing to death." He described their approach as quintessentially "German."[27]

The image of the scientist as a self-recording instrument was not attractive to James because it was at odds with his conviction that knowledge, as a moment of a larger reflex-arc motor phenomenon and as part of an irreducibly human reaction to, and interaction with, the environment, was oriented to action and was intrinsically governed by interests, purposes, and ends set by the knower.[28] The image of automatic, mechanical inscription of knowledge also conflicted with the evolutionary account of consciousness that James had developed since the mid-1870s.[29] For James, consciousness was neither the blank surface described by Spencer and other British empiricists nor a causally inert epiphenomenon of the brain, as the naturalist Thomas Henry Huxley and the mathematician, cosmologist, and metaphysician William Kingdon Clifford would have it. To the contrary, it was a "fighter for ends," a problem solver, an organ that had emerged in the course of evolutionary processes and had been preserved by natural selection because it helped the organism to better adjust to the changing environment.[30] Cognition was not a purely theoretical activity; it was a spontaneous activity heavily oriented by the knowers' purposes, practical values, and by (in part evolutionarily evolved) "emotional and active propensities" and interests—including both aesthetic and practical interests. These factors constantly selected among various aspects of experience and structured it. Together with the physiology of sense organs, a person's aesthetic and practical interests were responsible for shaping a person's perceptions, and in producing what individuals considered to be the transparent "objectivity of things."[31] For James, far from threatening the objectivity of things, a person's choices (of an emotional, practical, or aesthetic kind) would even seem to constitute it.

James later developed that position in the framework of his pragmatist accounts of knowledge and truth, one that made room for practical interests and active or emotional propensities in the production and assessment of truth. Although James never fully articulated a conception of objectivity and instead at times used the term in hardly compatible ways, he clearly rescinded the link between objectivity and the attitudes of impersonality, self-restraint, and disinterestedness. Instead, within the pragmatist context, James seems to have operated on the basis of what one of his critics called an entirely "negative sort of objectivity." According to that view, a minimal kind of (epistemic) objectivity was guaranteed by the existence, within "reality," of external, unverbalized sensory elements that forced themselves upon human beings quite independently of their will—although human will intervened as soon as sensations were identified and named. That dumb reality exerted an objective control of a negative kind only: it precluded some interpretations of reality, without suggesting others. This conception satisfied what at the turn of the century was often considered as the minimal requirement for objectivity—suspension of the will. Nevertheless, it left ample room for the play of those subjective factors that were central to James's epistemology, simply by reducing the guarantee of epistemological objectivity to that part of reality that impinges on the human being before knowledge processes of any sort take place, without setting any requirements on the control of subjectivity.[32]

Long before James fully articulated his pragmatism, he had concluded that nobody, not even the scientist, could do away with certain subjective psychological factors in pursuing knowledge. A purely intellectual, disinterested, and unimpassioned vision of natural phenomena was fundamentally beyond human reach.[33] In short, to James the knower was the very opposite of a passive machine. Thus the representation of the scientist as a self-recording instrument and the attendant ethos of self-restraint were intrinsically misguided.

The distance separating James from the ethos of self-restraint can best be seen perhaps in *Principles of Psychology*, a work in which, as we saw in chapter 2, James strove to frame psychology as "a natural science" and to maintain a "positivistic" standpoint. Faithful to that goal, in *Principles* he denounced the "psychologist's fallacy"—the "snare" that consisted in "the *confusion of* [the psychologist's] *own standpoint with that of the mental fact about which he is making his report*" (James's emphasis). In *Principles* James also devised what Jill Morawski has described as a "(social-epistemological) hierarchy of knowers," the goal of which was to establish a distance between the psychologist qua psychologist and "his (her) lesser self."[34] Although this move has been seen as a bow to the ideals of self-restraint and impersonal objectivity, it was not. Not even in this programmatic scientific work could James force himself to practice the scientific virtue of self-effacement. *Principles*, as G. Stanley Hall complained, was "saturated with the author's personality"; it had "at least in places" a "Rousseau like . . . confessional flavor." "The important works of art and science," Hall continued, attacking his former mentor, "have usually been done by men who sink their personality in their work. These traditions of self-effacement are effete for our author, who tells us incidentally of his age, of his early school life, his daily habits, tastes, etc." Hall clearly could not stand the work's "personal frankness," which he disapprovingly asserted was "unequalled in the history of the subject."[35]

Like his psychology, so the alternative moral and social economy of science that James developed radically questioned the tenability, even the desirability, of an unqualified scientific ethos of disinterestedness, self-restraint, impersonality, and emotional detachment.

## Psychic Phenomena and Regimes of Evidence

James's classic philosophical discussion of evidence is found in "The Will to Believe," where he discussed the legitimacy of religious and moral beliefs unsupported by "sufficient" evidence. Yet as a scientist active in various fields, James also had to deal with questions of evidence in very practical ways. His investigations of supernormal phenomena provide a case in point. James's approach to issues of evidence in the course of these investigations may help shed light on the philosophical analysis of notions of evidence in "The Will to Believe," and, vice versa, the position James took in his classic essay illuminates his approach to questions of evidence in various concrete cases. As we will see below, in both contexts James's discussion of questions of evidence unfolded at once on an epistemological and on a moral plane. By engaging such questions he formulated a novel type of scientific ethos and challenged the scientific persona promoted by many self-described "scientists" and votaries of "Science"—all terms that James used, prevalently, in a derogatory sense.[36]

James's interest in supernormal phenomena has been traced back to the late 1860s, but he started more actively investigating psychic phenomena in the 1880s. In 1884 he joined the Society for Psychical Research of London, serving as president of the society in 1894–95 and as vice president for the subsequent decade. He was also one of the founders of the American Society for Psychical Research (ASPR), established in 1884 and absorbed by the British Society in 1889. He served as chair of the ASPR committees for the investigation of hypnotism and of "mediumistic phenomena," and as honorary vice-president. After 1907 he was an active member of "Section B" of the American Scientific Institute, a similar institution created in 1907 by Columbia University philosopher and psychologist James Hyslop. James continued to investigate the subject, with the exception of a few lapses, almost until his death in 1910.[37]

In 1885 James met a Boston trance medium, Leonora Evelina Piper.[38] James was so impressed with the kind of information that surfaced through her trances that he painted Mrs. Piper as his "white crow"—the counterexample that disproved for him the rule that all crows are black and that all mediums are cheats.[39] Despite some convincing experiences with Mrs. Piper, James remained profoundly aware of the problems of evidence that psychic phenomena raised, and wrestled with the standard objections leveled against evidence and testimony for psychic occurrences. James was also aware that participants in those debates approached questions of evidence in different ways. To provide a context for James's own approach, this section maps some of the conceptions of evidence that structured debates over psychical phenomena in late nineteenth-century circles.

## Four Standard Objections against Psychical Researchers

Late nineteenth-century opponents of psychical research endlessly complained about the poor quality of the evidence, the rarity of the phenomena, and the difficulty of reproducing them under controlled conditions. However, their attacks were more targeted at those who investigated psychical

phenomena than at the evidence itself. Critics mobilized many objections
to demonstrate that psychical researchers and, a fortiori, spiritualists lacked
the mental qualities necessary in order to witness psychical phenomena and
to evaluate the evidence for their existence in a scientific way. Here I will
consider only four such objections among the many that were lodged.

The first regarded the honesty and competence of witnesses and investigators. Retrieving David Hume's famous argument against miracles, many
contended that the improbability of the phenomena was higher than the improbability that otherwise honest and competent witnesses might behave in
dishonest and incompetent ways. Thus, from early on, psychical researchers
felt forced to confront and publicly address the issue of "trust." Henry Sidgwick, the first president of the Society for Psychical Research, found that
the objection was unavoidable. Yet he pointed out that relationships of trust
were involved in "all the reasoning of experimental science," since readers
of experimental reports had to trust those who witnessed the experimental
procedures and the results of the experiments. In the sciences, Sidgwick continued, "the honesty of scientific witnesses [was] so completely assumed that
the assumption [went] unnoticed"; testimony and a moral factor were ubiq-
uitous, but they had become invisible. Not so in psychical research, Sidgwick
complained. Here, due to the improbability of the phenomena reported, the
assumption of the witnesses' honesty was constantly called into question.
For that reason, psychical researchers felt continually compelled to persuade
their audience of their own honesty and good character as well as the honesty and good character of other witnesses on whose reports they relied, or
of the people who had been instrumental in bringing about the phenomena
discussed.[40]

Sidgwick and the members of the Society for Psychical Research no doubt
had in mind the awkward case of William Crookes.[41] A renowned chemist
and self-trained physicist, best known for the discovery and accurate measurement of the atomic weight of a new chemical element (thallium) in
1861 and for his experiments on cathode rays, Crookes was a member of
the Royal Society. In 1871 he conducted a series of "careful scientific testing
experiments" that, in his opinion, proved the existence of a "new" force
that was connected "in some unknown manner . . . with the human organization." In the presence of a medium, this "psychic force" appeared to be
able to alter the weight of bodies, and to "play tunes upon musical instruments . . . without direct human intervention."[42] Crookes's reports were met
by many criticisms, most of these "perfectly fair and courteous," and Crookes
endeavored to "meet [them] in the fullest possible manner." Other critics,
however, challenged Crookes's competence, "trustworthiness," and "veracity," demanding that the experiments be repeated in front of more reliable

witnesses.[43] Crookes, a chemist "accustomed . . . to have [his] word believed
without witnesses," was very much offended. He assumed that the word of
a man of science, like that of a gentleman, should not be questioned. Indeed,
to question a man's factual claims was to question his honor and honesty,
and to transgress the codes of politeness and mutual trust regulating scientific discussions and dissent. Thus he refused to address objections brought
up in "uncourteous commentaries based on unjust misrepresentations."[44]
(In the mid-1870s, Crookes's reliability as a witness was questioned on other
grounds as well. Apparently, he was fond of "touching and embracing" the
fleshy spirit of an attractive young lady, Katie King, who materialized during his séances with a fascinating young medium, Florence Cook. It was
rumored that he was romantically involved with the medium; although nobody could actually prove that the rumors were true, Crookes's reputation
remained tarnished.)[45]

However, even when the honesty, good character, and reputation of the
investigators were beyond suspicion, critics found many reasons to question their ability to function as reliable witnesses and/or to scrutinize the
evidence. At a time that witnessed a proliferation of conceptions of the "unconscious" mind and techniques to probe it, many contended that psychical
researchers' testimony for psychic phenomena was warped by unconscious
factors. A locus classicus of this second objection is found in William Benjamin Carpenter's attacks against mesmerism and spiritualism, and, later on,
against psychical research. A British forensic expert who conducted psychological and physiological research, and a self-appointed guardian of science,
Carpenter had punctiliously argued that all the phenomena of spiritualism
(such as, for example, table-turning and the trance of a medium) could be explained away in terms of "unconscious cerebration": that is, they were simply
the products of automatic reflex-arc processes occurring in the brain.[46]

Carpenter mobilized his theory of unconscious cerebration to disqualify
the testimony of spiritualists and to question their competence in assessing
evidence. Automatisms of the sensory ganglia and of the cerebrum, he theorized, always played an important role in shaping a person's perceptions
and judgments. Only "early scientific training" could produce the "habits of
thought" and exact observation that could make a person into a trustworthy
witness and endow him with the ability to think correctly "in regard to *every*
subject."[47] Bad habits of thinking—which tended to develop in the absence
of such training—could warp the judgment or observational powers of the
most virtuous witnesses. Carpenter emphasized that specialized training in
a discipline was *not* the same as proper "scientific training," and urged that
physicists and other "specialists" (including especially a "specialist of specialists" such as William Crookes) lacked the proper kind of scientific disciplining

of the mind that would make them constantly reliable witnesses.[48] Honesty and good intentions had little or no impact on their already improperly formed mental habits, and could offer no guarantee of reliable testimony.

A third objection leveled against testimony for psychic phenomena held that otherwise perfectly reliable investigators could be subject to influences emanating from the medium. These influences were theorized in different ways. Balfour Stewart, for example, wondered if Crookes would consider the possibility of having been "electro-biologized" by the mediums with whom he worked. Perhaps, Stewart speculated, Crookes had been induced to believe that he had seen phenomena that in fact were *"subjective* rather than *objective,* the result of an action upon the [witness's] brain rather than an outstanding reality."[49] For other critics, particularly later in the century, "suggestibility" and "hypnosis," variously understood, did the same trick. Experiments on the influence of hypnosis and suggestion on memories, perception, and the emotions indicated that in those states the experimental subject could fail to perceive or recognize visual stimuli that entered his or her field of vision or could be induced to experience "false" sensations.[50] Critics reasoned that similar effects might be induced by the medium in witnesses who believed that they had witnessed genuine psychical phenomena.

Finally, especially in the late nineteenth century, many critics worried that psychical researchers lacked the necessary emotional detachment and the ability to keep their hopes, desires, and expectations in check. As a result they fared poorly as witnesses of psychical phenomena and were too "subjective"—and unscientific—in their evaluations of the evidence.

On the basis of these objections, which sometimes appeared in various combinations, in the late nineteenth century not a few self-described "scientists" argued that psychical researchers needed to be guided by better-trained colleagues or needed to be replaced by scientific experts. The neurologist George Beard and the psychologists Joseph Jastrow and Hugo Münsterberg were among those who, in the name of disciplinary expertise, theorized the need of such patronizing tutelage, even as they worked to exclude psychical researchers from the investigation of occult phenomena. Beard, one of the first American "neurologists" and the "inventor" of neurasthenia, contended that the study of psychic phenomena pertained to the domain of "cerebro-physiology." Part of a group of neurologists who endeavored to colonize the domain of mental illness, formerly the province of asylum superintendents, in an attempt to expand the territory of the new discipline, Beard charged in the late 1870s that only neurologists had the expertise necessary to understand psychic phenomena for what they really were, clinical symptoms, and to evaluate testimony regarding them. Beard even articulated the basics of a new "science of human testimony," which, he contended, was to be a branch of neurology. It would provide the much-needed guidelines that people dealing with situations involving human subjects (for example, lawyers, psychologists, and psychiatrists) ought to follow to produce correct judgments regarding evidence and testimony.[51] Following Carpenter, Beard reinterpreted trance, automatic writing, and other alleged occult phenomena in neurological terms: as products of the automatisms of the nervous system.[52] Given the true nature of such phenomena, psychical researchers who could not present neurological credentials were to be excluded from the study of them.[53]

Joseph Jastrow insisted that psychic phenomena always involved "deception" or "self-deception." The "average man" was unqualified to deal with either form of deception. He was "accustomed to implicitly trust the evidence of his senses," and was as easily duped by a medium as by stage magicians. Furthermore, the emotional appeal of the phenomena of mediumship, which seemed to disclose the possibility of immortality, set up the "best possible conditions for *self*-deception." The "layman's" inability to keep emotions in check, added to "mal-observation and mal-description," made his inquiries into supernormal phenomena "an interesting chapter [in] the natural history of error, showing how readily the emotions carry away the reason, and what a child the layman is before the professional expert in sense-deception." Jastrow concluded that the investigation of psychic phenomena belonged to the "specialist," and that "dilettanti" should be precluded from toying with such occurrences.[54]

Hugo Münsterberg agreed that most people were bound to fare poorly as witnesses of psychic phenomena but extended that predicament to men of science like himself. Scientists, he argued, were trained throughout their lives to trust nature and to place "an instinctive confidence in the honesty of their cooperators." Because of that, they were utterly unable to imagine that the experimental subjects instrumental to the production of psychical phenomena, or other people who observed the experiments, could act in a deceitful way. Detectives or stage magicians were much more adept at assessing the evidence and should be "substituted" for scientists in such experiments.[55] Münsterberg used that claim to explain why he, and by implication all serious scientists, should not even bother to attend "telepathic experiments or psychical séances." His admission that scientists were no better qualified than others in the task of witnessing psychic phenomena and evaluating their evidential value, however, did not imply that scientists were ill-equipped to "explain" psychic phenomena, which he squarely placed within his own scientific domain, psychology.

In the end, as Sidgwick had complained in his inaugural address before the Society for Psychical Research, whether they were considered honest or not, those who investigated psychic phenomena found that their testimony was always questioned. Not all of them, of course, were straightforwardly accused of fraud or gullibility, but they found that they were often depicted, even ridiculed, as incompetent, naïve, and "unscientific" folks.

## "Faggots" of Evidence and Mechanical Witnesses

To address the objections raised against them by their opponents, psychical researchers resorted to a variety of strategies. Here I examine two that are particularly relevant to a discussion of William James's approach to questions of evidence. The objection that the probability of the witnesses' dishonesty or incompetence was higher than the probability of any psychic occurrence was met by a statistical argument known in the psychical research literature as the "cumulative evidence" or "faggot" argument. The term "faggot"—a bunch of sticks woven together—lent itself to the argument formulated by James's British friend Edmund Gurney, who applied it to the study of the psychic phenomenon known as "phantasm of the living" or "veridical hallucination," that is, the apparition of a distant person who, at the time of the vision, was dying or going through a serious crisis. In each individual case, the faggot argument conceded, the improbability of the phenomenon was higher than the improbability that a witness otherwise known as honest and reputable might act in dishonest or careless ways. However, as the number of witnesses increased, so did the improbability that they all acted in dishonest ways. By adding more and more witnesses, the improbability that they all cheated would, in the end, become greater than the improbability that the phenomena witnessed had occurred. In other words, while each stick might not be "free from flaw," "the multiplication" of sticks would "make a faggot of ever-increasing solidity." Thus, Gurney argued, an increase in the number of witnesses, and in the "quantity" of testimony of this kind, improved its quality and its convincing power.[56]

However, the faggot argument was powerless against the second and the third objections, that is, the claims that, due to uncontrollable factors of an "unconscious" or "subconscious" kind, witnesses could make mistakes of judgment or observation, or could unwittingly be induced by the medium to perceive things that were not there. It was also powerless against the fourth objection, according to which, due to the lures of subjectivity, even the most honest witnesses could produce warped testimony. In response to these objections, at the turn of the twentieth century a few psychical researchers strove to reduce or even ban human testimony from their investigations. They replaced human witnesses as far as possible with mechanical ones—self-recording instruments and other pieces of apparatus that were intrinsically "honest" and "objective," because they were immune from un-conscious, hypnotic, and "subjective" influences. The psychical researchers who replaced human witnesses with such apparatus did it in the hopes, as an Italian investigator put it, that cameras and self-recording instruments would provide "objective irrefutable evidence."[57] At the same time, these investiga-tors endeavored to adhere, at least provisionally, to the ethos of self-restraint that accompanied the rhetoric and practices of mechanical objectivity.

This was exactly the strategy deployed by a group of French psychical researchers based at the Paris Institut Général Psychologique, a research institute founded in 1901 and presided over by the renowned French psy-chopathologist Pierre Janet. In 1902 a committee for the investigation of psy-chic phenomena was established, including a few physiologists, physicians, physicists, instrument makers, and philosophers, such as E.-J. Marey, by then an international expert on self-recording devices; Jacques D'Arsonval, the inventor of the D'Arsonval galvanometer; Pierre and Marie Curie; and Henri Bergson, whom James met in 1905.[58] Between 1905 and 1908, members of the group performed three series of experiments on Eusapia Palladino, one of the most famous mediums of the time (fig. 3.1).[59] Eusapia was known to cheat whenever she had a chance, and had been caught at it in 1895 in Cambridge, an episode that James feared "might prove a blow" to the "pros-perity" of the Society for Psychical Research.[60] (In 1910 Eusapia was caught again, this time by Hugo Münsterberg, who turned out to be an excellent "detective" after all [see fig. 3.2].)

Some of the experiments were conducted at the institute, others in a private apartment.[61] Packed with instruments—including cameras, self-recording machines, electrometers, galvanometers, and a battery of other tools some of which were designed or specifically adapted by members of the group—these spaces contained a rectangular table and a medium's "cabinet," separated from the rest of the room by a curtain. The graphing instruments were located in the same room or in an adjacent one (fig. 3.3).[62] They were used for a variety of purposes, including recording the medium's respiration, pulse, and other vital parameters (fig. 3.4). However, the chief goal of these instruments was to mechanically record the phenomena that took place during the séances, including the levitation of a table (fig. 3.5). During the séances Palladino would sit on one side of the table (side 1–2 in fig. 3.6), while two investigators (the "controllers") monitored her hands, feet, and knees. In one group of experiments each leg of the table was out-fitted with a contact key, a spring-loaded device, shown in figure 3.7, which completed an electric circuit when the leg of the table lifted off the ground (but did not complete it if a foot was used to lift the table). The status of

Fig. 3.1. Eusapia Palladino. From Jules Courtier, "Rapport sur les séances d'Eusapia Palladino à l'Institut général psychologique en 1905, 1906, 1907 et 1908." *Bulletin de l'Institut général psychologique* (1908). In the public domain.

each of the four legs was recorded at the same time on a graphing device (a Duprez indicator), which jumped upward when a table leg lifted. This electric outfit enabled the investigators to check whether or not the table was lifted by a pressure exerted from underneath its legs (for example, by a foot). During the experiments Eusapia's weight was recorded along with the position of the legs by means of a self-recording Marey scale placed underneath her chair. Whenever a leg of the table lifted, the self-recording system would graphically record the event.

The movement or elevation of the table's legs was a frequent occurrence. For example, on June 22, 1905, all four legs levitated simultaneously (fig. 3.8).

Fig. 3.2. Newspaper report of Hugo Münsterberg conspiring to catch Palladino.

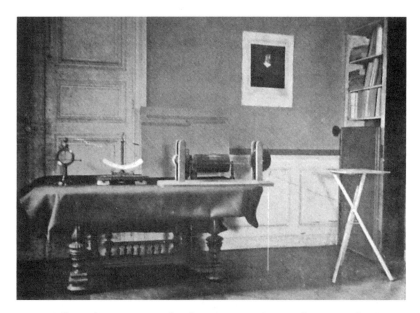

Fig. 3.3. Self-recording instruments placed in room. From Courtier, "Rapport sur les séances d'Eusapia Palladino à l'Institut général psychologique en 1905, 1906, 1907 et 1908," *Bulletin de l'Institut général psychologique* (1908). In the public domain.

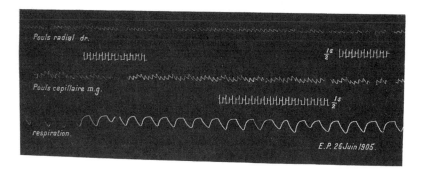

Fig. 3.4. Graphs recording Eusapia's vital parameters at the end of the séance held on June 26, 1905. From Courtier, "Rapport sur les séances d'Eusapia Palladino à l'Institut général psychologique en 1905, 1906, 1907 et 1908," *Bulletin de l'Institut général psychologique* (1908). In the public domain.

Four days later, legs 1 and 2 of the table levitated simultaneously (fig. 3.9). The graphs of leg elevations recorded in parallel with the results of the Marey scale indicated that, at exactly the same instant, Eusapia's weight on the chair had increased. Later, during the same séance, legs 3 and 4 of the table levitated, while the Marey scale recorded that the medium's weight had decreased (see fig. 3.10).

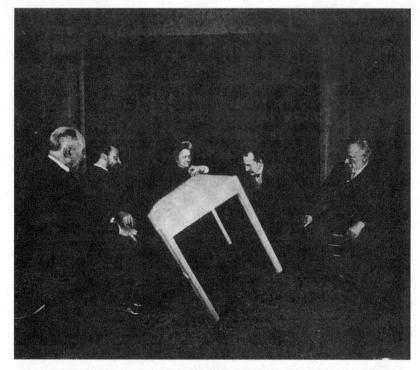

Fig. 3.5. Instant photograph of a levitation of the table. From Courtier, "Rapport sur les séances d'Eusapia Palladino à l'Institut général psychologique en 1905, 1906, 1907 et 1908," *Bulletin de l'Institut général psychologique* (1908). In the public domain.

How to interpret that evidence? The graphs recorded by the Marey scale were compatible with the laws of mechanics, that is, in the first case Eusapia could have been pushing up side 1–2 of the table with her knees, in the second she could have exerted pressure with her elbows on side 1–2, causing the opposite side to lift.[63] However, the graphs could not rule out the intervention of supernatural causes. All the while, the controllers felt confident that Eusapia had not played any tricks.

What conclusions could be drawn? In order to rule out trickery with absolute confidence as the cause of the levitations and alterations in Eusapia's weight, the investigators would have to rely on the sense impressions of the controllers. However, they were not willing to do that. They felt that in a dark setting the eyes cannot appreciate distances and directions; reduced to touch and muscular sense, the control of the movements of the medium lost "rigor."[64] Furthermore, they stressed that, to properly observe the phenomena, the controllers would have to be in a constant state of "divided attention" between the phenomena that occurred and the movements of Eusapia's limbs.[65] In addition, the investigators could not exclude the possibility that suggestibility could have led the controllers to "complete" uncritically their sense perceptions, seriously spoiling their observations. They also suspected that their emotions, of which the observers could retain only a vague memory at the end of the séance, might place them in a state of "over-excitement" that would detract from their ability to produce reliable judgments.[66]

On the other hand, the graphs did not eliminate a supernatural explanation of the facts, and the investigators felt that ascribing the phenomena to trickery would have been a subjective act of interpretation. In the end, these psychical researchers preferred to leave the question of the cause and authenticity of the phenomena unanswered, and limited their claims strictly to what was shown by the graphs traced by the self-recording instruments. That way they satisfied what appeared to be a widespread and enduring

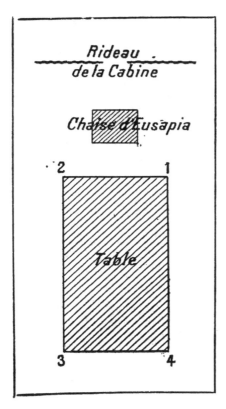

Fig. 3.6. Layout of one of the two experimental rooms. Palladino would sit at side 1-2 of the table, in front of the medium's cabinet. From Courtier, "Rapport sur les séances d'Eusapia Palladino à l'Institut général psychologique en 1905, 1906, 1907 et 1908," *Bulletin de l'Institut général psychologique* (1908). In the public domain.

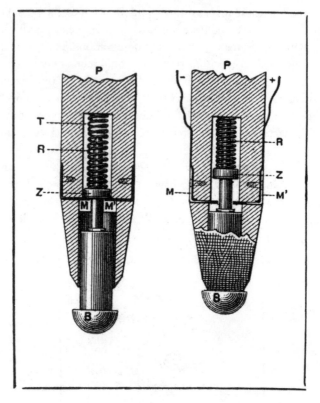

Fig. 3.7. Spring-loaded device for detecting motions of the table legs. From Courtier, "Rapport sur les séances d'Eusapia Palladino à l'Institut général psychologique en 1905, 1906, 1907 et 1908," *Bulletin de l'Institut général psychologique* (1908). In the public domain.

demand from the public, as exemplified by a 1910 letter to the editor of the *New York Times*. The writer, commenting on experiments recently made in New York City with Eusapia Palladino, stated:

> It . . . seems to the writer that expert psychologists are as likely to pass under such a controlling influence as other, untrained, persons and that their testimony . . . cannot be considered as trustworthy, for [their] mind must ever be open to the charge of having been subconsciously controlled. The witnesses, on the other hand, must be such as are naturally beyond the reach of the act of suggestion; that is to say, they must be possessed of sight and touch but wholly without mind, they must be mechanical automata capable of observing and recording the phenomena of a seance and the action of all its participants. Nothing short of such an automatically recorded seance is worthy, nowadays, of being

classified as scientific experiment. . . . Indeed it must be set down as an axiom that in determining psychological phenomena, the mind as a witness is itself wholly incompetent, and, if progress is to be made, must be replaced by mechanical instruments.[67]

Abiding by a similar requirement, the French psychical researchers transferred to their readers the responsibility for drawing conclusions concerning the events. In fact, their authorial voices disappeared from the published record, which, they hoped, would provide only purely objective evidence. Perhaps to avoid ridicule or charges of lacking the proper scientific attitude, and unable to trust their senses, these researchers did the only thing that, as a group, they found they could do: shut up, and let the instruments speak.[68]

Fig. 3.8. Graph recorded during the June 22 séance by self-recording instruments attached to the four legs of the table. The graph shows a complete levitation of the table. From Courtier, "Rapport sur les séances d'Eusapia Palladino à l'Institut général psychologique en 1905, 1906, 1907 et 1908," *Bulletin de l'Institut général psychologique* (1908). In the public domain.

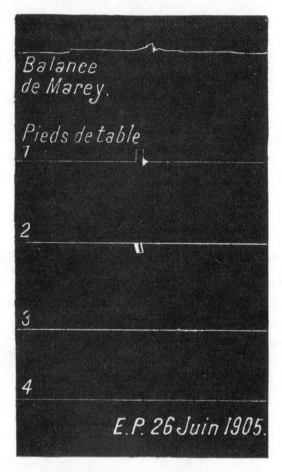

Fig. 3.9. Graph recorded by the self-recording instruments attached to the four legs of the table and to the Marey scale on which Palladino was seated. The levitation of legs 1 and 2 of the table was concomitant with an increase in Eusapia's weight. From Courtier, "Rapport sur les séances d'Eusapia Palladino à l'Institut général psychologique en 1905, 1906, 1907 et 1908," *Bulletin de l'Institut général psychologique* (1908). In the public domain.

### The Doctrine of "Absolute Evidence" and Dogmatism

James was familiar with all the approaches to evidence that underlay the objections and responses examined above. He was no stranger to Carpenter's conception of "unconscious cerebration" and his reinterpretation of mesmeric and spiritualistic phenomena in terms of automatic processes. James was acquainted with Beard, who had treated him for seasickness and, apparently, for nerve ailments. James considered him somewhat of a "con-ceited man," and was clearly aware of Beard's "war" against "spiritualism."[69] James was familiar with Jastrow's attacks against psychical researchers, amateurs, and laymen, and read everything that Münsterberg wrote about psychical research. James appreciated the cumulative nature of "faggots"

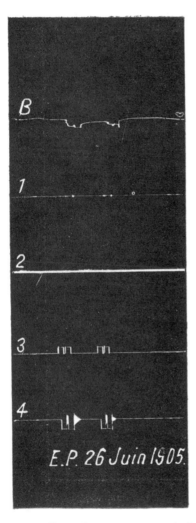

Fig. 3.10. Graph recorded by the self-recording instruments attached to the four legs of the table and to the Marey scale on which Palladino was seated. The levitation of legs 3 and 4 of the table was concomitant with a decrease in Eusapia's weight. From Courtier, "Rapport sur les séances d'Eusapia Palladino à l'Institut général psychologique en 1905, 1906, 1907 et 1908," *Bulletin de l'Institut général psychologique* (1908). In the public domain.

of evidence, and was never seduced by mechanical approaches to the production of evidence. He also resented all attempts to dismiss evidence or disqualify witnesses on a priori grounds.

Some of these issues came together in his discussion of evidence in "The Will to Believe," which James composed one year after his turn as president of the Society for Psychical Research. In that seminal work James examined a conception of evidence that he believed underlay many debates about the legitimacy of religious beliefs: a conception that he named "absolute" or "scientific" evidence. As is well known, James defended the legitimacy, under certain conditions, of religious belief unsupported by "sufficient" evidence. The essay had two main targets: first, the view that the evaluation and interpretation of evidence for religious and moral doctrines should be grounded in purely logical processes, ones in which intellect was divorced from emotions and the will; second, the position that it was immoral to believe something in the absence of sufficient evidence. James ascribed both positions to Huxley, whom he had briefly met in London in 1883, and to William Kingdon Clifford.[70] He quoted a passage from each of them. "My only consolation," Huxley had written, "lies in the reflection that, however bad our posterity may become, so long as they hold by the plain rule of not pretending to believe what they have no reason to believe because it may be to their advantage so to pretend, they will not have reached the lowest depths of immorality."[71] Clifford had made a similar point: "Belief is desecrated when given to unproved and unquestioned statements, for the solace and private pleasure of the believer. . . . If a belief has been accepted on insufficient evidence (even though the belief be true), the pleasure is a stolen one. . . . It is sinful, because it is stolen in defiance of our duty to mankind. That duty is to guard ourselves from such beliefs as from a pestilence, which may shortly master our own body, and then spread to the rest of the town. . . . It is wrong always, everywhere, and for anyone, to believe anything upon insufficient evidence."[72]

In James's view, the request that the evaluation of evidence should be carried out in a dispassionate, purely intellectual way was perfectly legitimate. Indeed, one ought, for the most part, behave that way. Yet in some cases that request turned out to be impossible, even undesirable. In his words, "the absurd abstraction of an intellect verbally formulating all its evidence and carefully estimating the probability thereof . . . is ideally as inept as it is actually impossible."[73] James further suggested that in that regard Clifford and Huxley contradicted themselves.[74] They demanded the exclusion of passions and volitions from the evaluation of evidence and the assent to belief; yet, in making that demand, they actually acted upon a hidden principle of a passional kind. Their reservation concerning assenting to a theory in the absence of sufficient evidence derived ultimately from a powerful *passion*—the "horror of becoming a dupe."[75]

James also criticized Huxley's and Clifford's precept that one should never believe anything in the absence of sufficient evidence. This principle, like the demand for purely intellectual evaluations of evidence, was untenable both on practical and moral grounds. Despite good intentions, knowers (including scientists) frequently disregarded that precept and embraced hypotheses as truths, even though they lacked the requisite evidence. Such actions nevertheless advanced knowledge. Furthermore, at least for certain moral theories, the decision to hold them to be true even in the absence of compelling evidence, and to act on the assumption of their truth, was perfectly legitimate. In the long run, James continued, such a decision might even make those theories true. As a consequence in certain such cases it would simply be immoral to suspend belief from horror of error alone. James famously concluded that under certain conditions, when one felt impelled to make a choice between two mutually exclusive theories, and the choice could not be based on pure logic and intellect, it was permissible to accept a belief in the absence of sufficient evidence.[76]

James found that Huxley's and Clifford's demands for sufficient evidence ultimately rested on the assumption of the existence of a brand of evidence that would instantly and unambiguously reveal the truth of the hypothesis it supported—that is, an evidence capable of "extorqu[ere] certum assensum," of forcing assent.[77] He observed that this idea of "objective evidence" had largely prevailed in philosophy, especially in "scholastic philosophy." Implicitly referring to the perceptual processes he had reframed in *Principles*, James argued that, although the belief in the existence of such a kind of evidence was natural, even instinctual, nevertheless it represented a weakness of human nature—one to be gotten rid of. James's criticism of this "absolutist" conception of evidence had, again, descriptive and normative sides. In the first place, absolute or objective evidence simply did not exist: it was never to be found on this "moonlit and dream-visited planet."[78] Second, the doctrine of objective evidence and objective certitude was bound to lead to what James deemed to be unacceptable intellectual attitudes: dogmatism and absolutism. As James saw it, belief in the existence of objective evidence led people to dogmatically assume that theories supported by that kind of evidence were absolutely true. He suspected that, left to their instincts, even the greatest empiricists would tend, like anyone else, to "dogmatize like infallible Popes."[79] The "chief and prior" aim of "The Will to Believe," as James wrote to his friend Leslie Stephen, was "to establish the logical rules of the game against all dogmatists, rationalists, scholasticizing 'scientists,' and their mirmidons."[80]

James denounced the dogmatism of "scientists" as early as the mid-1870s.[81] However, in the years that followed the publication of The Will to Believe, his attacks against it became more intense. A better understanding of the vehemence of James's rebuttal on these issues is facilitated by a consideration of the political and social context of those years. As Deborah Coon has shown, in the second half of the 1890s James's frequent attacks on scientists' dogmatism and "scientific authoritarianism" went hand in hand with a deep transformation that, beginning in the second half of the 1890s, rescued him from a "relative political complacency," and general lack of interest in political and social events, making him into a politically and socially committed intellectual.[82] Interferences by the United States in Venezuela and Cuba under the Cleveland and McKinley administrations, and the invasion of the Philippines in 1898, which the United States had helped to gain independence from Spain just months before, roused James's anger.[83] He denounced politicians' claims that it was necessary for the United States to make the Filipinos fit for democracy and self-government. Instead, he exposed the claims for what they were: thinly veiled justifications for an emerging American imperialism. He rejected all attempts to dogmatically impose on the Filipinos or anyone else certain ways of life and modes of social order, without taking into account the desires and visions of the affected social groups.

Meanwhile, as Coon has also shown, James followed closely, and with no less anger, the Dreyfus affair in France. Issues of evidence abounded throughout this case in which a Jewish army officer Dreyfus, accused of spying for the Germans, was court-martialed and convicted on the basis of a handwritten note found in a wastebasket in the German Embassy in Paris. Two years later, a second piece of evidence was found, apparently incriminating another French army officer, Esterhazy, and simultaneously indicating Dreyfus's innocence. Esterhazy was court-martialed but was acquitted, chiefly due to the court's decision not to consider the first piece of evidence, on the ground that it had already been used in the Dreyfus trial and thus stood as already judged, a "res judicata." Meanwhile, it turned out that one of the main pieces of evidence had been forged. Dreyfus was court-martialed a second time and, despite the fact that other forgeries had been disclosed, was condemned once again—to the great dismay of those trusting in the justice of the French Republic.[84] Here was an especially maddening example of a case in which the decision to exclude a priori certain pieces of evidence, combined with an implicit belief in absolute, and thus irrevocable, judgments about matters of fact (res judicata), served transparent political purposes and subverted justice.

No doubt in response to these events, James intensified his efforts against dogmatism, and its epistemological counterpart, the conception of "absolute evidence." He painted dogmatism as the distinguishing trait of the "unscientific" mind, as well as "the great disease of philosophical thought" and the philosopher's "besetting sin."[85] James would have agreed with what his friend Josiah Royce had written him some fifteen years earlier: the difference between science and fanaticism is ethical. Dogmatism—always allied with fanaticism—was at once unethical and unscientific.

Against dogmatism and the dogmatic doctrine of absolute evidence James proposed a new epistemological regime that, in the preface to The Will to Believe, he labeled "radical empiricism." Radical empiricism (a term that James would later use with a different meaning) was a method and a "philosophic attitude." Its main tenet was that philosophers and scientists ought to adopt a fallibilistic stance, and to "regard [even their] most assured conclusions . . . as hypotheses liable to modification [by] the course of future experience."[86] Radical empiricists, James emphasized, were no skeptics. They believed that truth was attainable. However, they also believed that one could never know for sure whether one had attained it. Accordingly, they rejected the idea of absolute, "objective" evidence or reinterpreted it as an "infinitely remote ideal of our thinking life."[87] Consequently, they held their theories to be eternally revisable and remained attentive to new evidence, including possibly contradictory evidence. They would keep an open mind and be tolerant of alternative opinions.

Radical empiricists embodied the attitudes and sensibilities that by the late 1890s James took to be essential to the practice of good science and good philosophy: open-mindedness, modesty, tolerance, respect for other people, and pluralism. As James put it, "No one of us ought to issue vetoes to the other, nor should we bandy words of abuse. We ought, on the contrary, delicately and profoundly to respect one another's mental freedom—then only shall we bring about the intellectual republic; then only shall we have that spirit of inner tolerance without which all our outer tolerance is soulless, and which is empiricism's glory; then only shall we live and let live, in speculative as well as in practical things."[88]

As the passage makes clear, the attitudes that James perceived to be essential to maintenance of the "intellectual republic" were at once epistemological and social. James's essay "The Will to Believe" created (or attempted to create) the epistemological conditions that would ensure "self respect and mutual respect."[89] I will come back later to the ways in which James's approach to evidence embodied a "social" economy of science. Before doing that, however, it will be helpful to examine in detail some of James's attacks against scientific and academic dogmatism.

## "Veridical Hallucinations" and Telepathy

Psychical research, marginal sciences, and fields of inquiry that the scientific establishment located outside of the boundaries of "orthodox" science became the intellectual terrain on which James fought his epistemological, moral, and social battle against dogmatism.

The census of "veridical hallucinations," in which James was involved, provides a good example. This investigation was a continuation of the study of "phantasms of the living" (apparitions of people at the point of death or recently dead) previously conducted by the British psychical researchers Edmund Gurney, Frederic Myers, and Frank Podmore, who regarded those phenomena as (most likely) "centrally" (or brain-) initiated "sensory hallucinations," that is, "percept[s] which lack, but which can only by distinct reflection be recognized as lacking, the objective basis which [they] suggest."[90] In other words, a "phantasm of the living" was a hallucination of sight, hearing, or touch that was not produced by any stimulus acting on the sensory organs. It was "veridical" if it represented a scene identical to, or similar to, another scene that happened in a different place, roughly around the same time. In 1886 Gurney, Myers, and Podmore had concluded that the large number of veridical hallucinations meant that they could not all be due to chance. Their statistical analyses indicated that the odds against the chance occurrence of veridical auditory and visual hallucinations were, respectively, "more than a trillion to 1" and "nearly a trillion of trillions of trillions to 1."[91] Their results were sharply criticized. For example, James's friend Charles Sanders Peirce, a champion of "accuracy" and "stern logic," found the statistical argument seriously defective. He contended that, in several cases, the testimony standing behind "Gurney's ghost-stories" was produced by "loose or inaccurate witnesses," or stemmed from errors of memory, lies, and pathological conditions such as anxiety, drunkenness, delirium, and intoxication caused by opium and chloral hydrate. Peirce argued that if the cases due to abnormal or pathological mental or bodily conditions and inaccurate witnesses were sifted out, the remaining numbers could be reinterpreted as the result of chance.[92]

In 1889, during the International Congress of Experimental Psychology held in Paris (a congress that gave ample room for a discussion of hypnosis and psychic phenomena), a new census of veridical hallucinations was launched. The census was to be carried out internationally, through the cooperation of several national committees. James, who attended the congress, chaired the U.S. committee. As in Gurney, Myers, and Podmore's study, the goal was to establish by means of statistical analysis whether veridical hallucinations could be explained away as chance coincidences. The various national committees circulated a questionnaire asking whether people had ever experienced hallucinations "while believing [themselves] to be completely awake." If the answer to that question was yes, the reader was asked to answer other questions aimed at determining whether the hallucination had been "veridical."[93] Both the report of the British committee (1894) and that of the American committee (1896) showed that, after the subtraction of all suspect cases, the veridical cases were "too numerous to be due to chance"; the frequency of veridical hallucinations exceeded 440 times the most probable number. On that basis, Henry Sidgwick, his wife Eleanor, and other British psychical researchers suggested that veridical apparitions could not be due to chance. James described the report of the British investigators as a "master-piece of intelligent and thorough scientific work," though emphasizing "I use my words advisedly."[94] He also called for caution: the British census, covering only 17,000 cases, was too small to warrant the conclusion some were drawing. Nevertheless, resorting to the "faggot" conception of evidence dear to Gurney and Sidgwick, James contended that, even though individually weak, when cumulatively considered the pieces of evidence collected by the Society for Psychical Research "carried heavy weight."[95] Thus the results of the report implied that "the telepathic theory, or whatever other occult theories may offer themselves, have fairly conquered the right to a patient and respectful hearing before the scientific bar."[96]

James's hopes, however, were disappointed. Few scientists seemed to give the census of hallucinations the patience and respect that James believed it deserved. Even those who paid some attention to the report, like the experimental psychologist James McKeen Cattell, dismissed the results because of the "weakness of the evidence."[97] Cattell, who had received his PhD with Wundt in 1886 and directed the Columbia University psychology laboratory, pointed out that the "'faggot' argument" could be turned "both ways": "when we have an enormous number of cases, and cannot find among them all a single one that is quite conclusive, the very number of cases may be interpreted as an index of the weakness of the evidence. The discovery of a great many gray crows would not prove that any crows are white."[98] Faggots, in other words, would hardly affect the "'rigorously scientific' disbeliever."[99]

James found that Cattell had entirely missed his "argument." He admitted that the inquiry remained "baffling," and that in innumerable cases the evidence was inconclusive. It could have been the product of "malobservation, illusion, fraud" or of "good and true report[s]."[100] He also conceded that evidence of this kind had to be supplemented by a working hypothesis, a "presumption," on the basis of which a verdict could be drawn. This was a concept that James borrowed from "scientific logic," and that was also crucial to the established laws of evidence. In judicial proceedings often a presumption would be made (for example, that "a child under seven years of age cannot commit a felony"). According to legal terminology of the time, the effect of the presumption was to "throw upon the party against whom it works the duty of bringing forward evidence to meet it," or the "burden of proof."[101] James promptly admitted his own presumptions: telepathy could be a genuine phenomenon, and the quality of the evidence in many cases was good.[102] For him, the poor quality of the evidence could not simply be assumed but had to be proved. James also insisted that Cattell's and other scientists' decision that the reports were false was *equally* based on a presumption: they assumed that the known physical laws that stood in opposition to supernormal phenomena were "superabundantly proven," and that it was certain that they would never allow a contravention.[103] By refusing to admit that this was a presumption, and by refusing to revise it in the light of new experience, Cattell showed his commitment to the dogmatic conception of evidence. Since he took the laws of nature to be absolutely true, he could conclude, with the best of scientific consciences, that the evidence for psychic phenomena was bad. This was thus a case in which a conclusion over the quality of evidence was reached without even examining the evidence.[104]

To James, such objections were "shallow," and the "dogmatic assurance" of scientists was inexcusable, even immoral. James emphasized just that point in his defense of mind curers and other healers against a proposed medical licensing law that, if accepted, would prevent people who had not passed a state examination (especially Christian Scientists and mind curers) from practicing medicine in Massachusetts. James, whose only academic and professional title was, in fact, that of "M.D.," chastised his dogmatic colleagues because they refused to even take a look at the "startling medical cures" and therapeutic facts produced by mind curers and their like. He contended that, by doing so, regular physicians might actually impede possible therapeutic progress.[105]

However, it was the social consequences of the dogmatic regime of evidence that concerned James most. James rebelled against many scientists' dogmatic contention that only certain classes of people—professionals, men of science, experts—were entitled to investigate certain phenomena and to produce knowledge about them. He suspected that that contention protected class and professional interests, and concealed the desire of some groups to prevail over and dominate others. James believed this to be exactly the case with the proposed medical licensing bill. He criticized the medical "profession" for attempting to make medical "experimentation the exclusive right of a certain class," and emphasized that the physicians campaigning for the bill considered the public unable to make responsible decisions about their health. As Coon has shown, such physicians worked to deprive the public of the right to choose the therapies that they considered best. James condemned this attitude as paternalistic, maternalistic, and even "grandmotherly."[106] Dogmatism, James insisted on many occasions, bred haughtiness, self-right-eousness, authoritarianism, intolerance, and "contemptuous . . . disregard" for other people and their proposed modes of life.

James found these attitudes amply instantiated in the many dogmatic men of "academic-scientific mind" who liberally ascribed to psychical researchers the mental qualities of "soft-headedness" and "idiotic credulity," and closed the door of science to the large constituency of people who, in the present as in the past, had vouched for occult phenomena. It certainly did not escape him that critics who painted psychic phenomena as "bosh," "rubbish," and "trash" made those who investigated such phenomena into unwitting collectors of garbage.[107]

In 1899 James approvingly read an article in which his friend F. C. S. Schiller, who a few years later would become the leader of British pragmatism, attacked Münsterberg and other self-proclaimed scientists who dismissed psychic phenomena and passed judgment on psychical researchers. To Schiller, such condemnation represented a close reenactment of the Dreyfus trial. "Psychic phenomena," Schiller wrote, were indeed "the Dreyfus Case of Science." Like the court-martial that sentenced Dreyfus, Schiller continued, the "General Staff of the Army of Science" did not give the case a fair hearing. Adopting the "policy of the *chose jugée* [the already judged thing]," they did not publicize the evidence on the basis of which they condemned the whole business as superstition. Indeed, like their French counterparts, they did not even make any "serious pretence of examining the evidence."[108] To Schiller as to James, the "scientist's" "hold[ing] utterly aloof from the beliefs of the vulgar" was a "stupid and dangerous . . . practice." However, whereas for Schiller the danger of that aloofness lay ultimately in the scientists' inability to dispel what they viewed as the superstitions of the "ignorant masses," for James the danger lay in the disrespectful *social attitudes* that underlay such attempts, and in the tendency to stop a conversation before it even started. That, for James, represented the core of the problem.

This point appears clearly in a famous attack that James carried out in 1898–99 against the British-born psychologist Edward Bradford Titchener. An experimental psychologist who had received his doctorate from Wundt in Leipzig and who taught at Cornell, Titchener had recently stepped into a controversy regarding the evidence for the telepathic transmission of numbers. Henry Sidgwick and his wife Eleanor had conducted an experimental study on the thought transmission of two-digit numbers from 10 to 90. During the experiments an "agent" concentrated mentally on a number, and a "percipient"—who had been hypnotized by the transmitter—tried to guess

it. They had concluded that the correct guesses outnumbered the most probable number, that is, the number resulting from chance combinations.[109]
Two Danish physiologists, F. C. C. Hansen and A. G. L. Leehman, attacked
the report of their study, arguing that the British investigators had forgotten
to quantify the guessers' "number-habits," that is, their tendency to mention
certain numbers more frequently than others.[110] James retorted that Hansen
and Leehman's article was not as conclusive as they claimed: the trial series were too short, and one needed larger numbers before arriving at any
conclusion. Titchener entered the debate taking the side of the Danish critics, and bluntly stated that "no scientifically-minded psychologist believed
in telepathy."[111] In other words, those who believed in telepathy were not
"scientifically-minded" and, as such, deserved no attention—James obviously among them. James's "credulity and his appeals to emotions," Titchener wrote to James McKeen Cattell, were "surely the reverse of scientific."[112]

To James, Titchener's attitude prevented genuine interaction and exchange. He painted Titchener not only as an authoritative, dogmatic, "soi-
disant" (self-styled) scientist but also as a *lonely* investigator. In a letter to the
journal *Science*, James wrote that Titchener wandered "in isolation" "upon
what he calls 'the straight scientific path.'"[113] Titchener replied that "if the
alternatives before [him] are scientific isolation and companionship" based
on what he considered poor "logical terms," he preferred "isolation."[114] Con-
firming James's diagnosis, Titchener turned out to be a solitary investigator,
as well as a master in the art of excluding others. He notoriously omitted
women from the exclusive society of "Experimentalists," which he founded
in 1904.[115] In the first two decades of the twentieth century, Titchener painted
the "scientist" as an individual who was withdrawn from everyday contacts.
The Titchenerian "man of science" was endowed with a "frame of mind"
entirely different from that of the common man and the "technologist."[116]
"His" life was wholly impersonal, and he inhabited a separate universe (one
symbolized by the secluded, exclusive space of the laboratory) that was
impossibly distant from theirs.[117]

Pursuing his exclusionary tactics, in the second decade of the twentieth
century Titchener mobilized that scientific persona in order to bar from the
domain of science all the psychological programs that differed from his own
"structuralism." He dismissed applied psychology on the grounds that its
practitioners were endowed with the "sort of conscience" that pertained "to
the dilettante rather than to the man of science," and disposed of behaviorism as a technology, logically irrelevant to psychology.[118] He shut out the
proponents of "act psychology" (intentionalism) because they failed to practice the scientific virtue of impersonality, and portrayed functionalists, the
psychologists associated with pragmatism and the scientific persona that

James envisioned, as unscientific men. He also sketched a geography of the sciences that placed the functionalists at the "periphery of science," more precisely at a comfortable distance from experimentalists like himself, who, in contrast, occupied "the innermost heart of science."[119] Meanwhile, Titchener, who, according to some contemporaries, had always been "authoritarian" and whose students found often inaccessible, had become more and more secluded and, in the last years of his life, avoided as far as possible contacts with everyone, including his Cornell colleagues and other psychologists.[120] James did not live to see the full incarnation of Titchener's scientific persona, but he clearly realized that Titchener embodied a pattern: one in which experts and self-described "men of science" felt entitled to exclude whole categories of actors from the domain of science and the production of reputable knowledge. James's scientific ethos and epistemology aimed to eradicate these social attitudes and to promote a different type of interaction, a kinder way of dealing with others.

## The Hodgson-Control Report

"Central to James's new beginning in philosophy," Charlene Seigfried writes, was "the assumption that there is such a thing as neutral, pure description of phenomena," something entirely different from the interpretation of phenomena. This position, however, as Seigfried also observes, was ultimately at odds with James's "contrary thesis that individuals contribute creatively to the object known," and with his contention that what we consider as the "objectivity of things" always stems, in part, from our emotional, practical, or aesthetic interests.[121] Questions of evidence raised by psychical phenomena brought the epistemological tension to the fore.

James described the Society for Psychical Research as a "weather bureau for accumulating reports" of psychic phenomena. In 1892 he praised the society's *Proceedings* for stamping each narrative of psychic occurrences "with its precise coefficient of evidential worth," and praised British psychical researchers for their uniquely "systematic attempt to *weigh* the evidence for the supernatural."[122] On many occasions he invited his fellow psychical researchers to stick to the task of collecting evidence and to refrain from offering premature interpretations. In the context of his activities as a psychical researcher, he also drew a distinction between "facts" and "theories," and between the *"judgment"* that a person passed about the evidence for psychical phenomena and one's *"feelings"* about the evidence.[123] Yet James also became aware that not only a person's interpretation of the evidence but even one's capability to perceive certain facts in the first place depended, to some extent, on certain subjective factors, such as aesthetic preferences or selective interests. He contended that the perceptions and judgments of many scientists (and many psychologists) were influenced by a "classical" kind of aesthetics that informed their "types of mind." Scientists' classical aesthetics made them blind to the bizarre, "gothic" facts of supernormal or abnormal psychology—facts that instead amply revealed themselves to the more sympathetic gaze of psychologists endowed with a "romantic" type of imagination.[124] People endowed with a "scientific-academic mind," James also wrote, were better at interpreting the phenomena, but those endowed with a "feminine-mystical" type of mind were best at ascertaining new facts. "To no one type of mind is it given to discern the totality of truth. Something escapes the best of us," he concluded, "not accidentally, but systematically, and because we have a twist. . . . Facts are there only for those who have a mental affinity to them."[125] Thus the ability of a fact to function as evidence, the determination of the direction to which it points, and, before that, a person's very ability to perceive the fact in the first place seem to be determined, at least in part, by very subjective factors.

As Seigfried notes, James never quite solved the epistemological tension between the quest for pure phenomenal description, as well as neutral evidence, and the core claim that subjective factors played a role in shaping people's perceptions as well as their interpretations. Yet, I would suggest, "pragmatically" the tension was quite productive. It allowed James to promote the modes of social engagement and kinds of communities that, given the political events of the late 1890s and the early years of the twentieth century, he had come to view as desirable. The "Hodgson-control" case, one of the last investigations of psychic phenomena in which James was involved, will help us better appreciate the social function of the epistemological tension that marked his empiricism.

In December 1905 the Australian-born psychical researcher Richard Hodgson—the director of the American branch of the Society of Psychical Research—died suddenly. Shortly afterwards, the medium Leonora Piper started falling into states of trance during which she was allegedly "controlled" by the spirit of Hodgson. Between December 1905 and the beginning of 1908, Mrs. Piper held about seventy séances during which the spirit of Hodgson supposedly tried to prove his identity. James, who had long known both Mrs. Piper and Hodgson, witnessed only few of the "Hodgson-control" séances.[126] At the time James was enmeshed in the controversy that surrounded his pragmatist account of truth. Nevertheless, probably under pressure from the Society for Psychical Research or friends of Hodgson, James agreed to prepare a report on the case. The task was enormous, and quite boring. In order to ascertain whether the details that surfaced through the trances supported the "spirit-return" hypothesis, James collated sixty-nine reports. He found that not a few fragments of evidence were perfectly compatible with that hypothesis. However, they were also compatible with other supernatural hypotheses, as well as with naturalistic explanations.

On the supernatural side, James considered the hypothesis according to which Mrs. Piper was telepathically able to "tap" into the minds of the witnesses and extract information regarding Hodgon's life that way. He also considered the hypothesis that the bits of knowledge that Mrs. Piper revealed were stored in a cosmic depository of memories, to which Mrs. Piper's subconscious mind had access. On the naturalistic side, the evidence could be easily explained away. By the time of his death, Hodgson had spent many years studying Mrs. Piper's trances. True, Hodgson was a very reserved person, even more so in his capacity as a psychical researcher. Nevertheless, Mrs. Piper might have had many chances of acquiring by natural means the information about the episodes of Hodgson's life that emerged during the séances. If her conscious memory did not recall the information, her subliminal memory might have been able to do so. James, who was interested in "independent evidence" for the spirit-control hypothesis, could find little of it. In fact, most of the evidence was spotty, somewhat incoherent, ambiguous, irrelevant, and, in some cases, demonstrably false—at best only circumstantial.[127]

In his report James tried to stick to a clear-cut distinction between evidence and the interpretation of evidence, as well as between strong evidence and more ambiguous evidence—as if evidence arrived shaded with a "degree of evidential value" independent of alternative theoretical interpretations. These distinctions, however, appeared rather hazy in practice. Since no "coercive" evidence materialized, one was left with a set of details whose degree of evidential value was measured in different ways by different people. James found that a person's feelings might inform his or her conclusions in that regard. For those who carried on a conversation with the alleged spirit of Richard Hodgson and found their questions to the spirit "answered" and "their allusions understood," it was easy to project on the whole conversation their own feelings of "warmth," and to conclude that the conversations were genuine and sincere interactions with a spirit. In contrast, those who had never taken part in any of the séances, and thus approached the phenomena on the basis of "cold" reading of the reports, would often draw very different conclusions.[128] The lesson for James was that "exact logic" and "pure intellect" could only play a small, at best a preliminary role, in the processes through which the psychical researcher scrutinized the evidence and interpreted it.[129] The decisive "vote," "if there be one," James continued, was to be cast by the investigator's sense of what was possible in nature, by "one's general sense of dramatic probability [of nature]"; and that sense seemed to "ebb" and "flow" "from one hypothesis to another . . . in a rather illogical manner."[130] This "sense" of the "dramatic probabilities of nature" provided the further assumptions that enabled the investigator to draw conclusions from circumstantial evidence.

According to James, the sense of what nature could possibly accomplish was, to a large extent, an individual aptitude. It was part of the investigator's "personal equation," that is, of those physiological and psychological factors that astronomers and, by the late nineteenth century, experimental psychologists sought to bring under control in order to eliminate individual observational error.[131] It depended on "the forms of dramatic imagination" of which a person's mind was "capable," and these in turn were shaped by a person's past experience and other personal factors, including a person's emotional constitution. In sum, depending on one's type of imagination, one could evaluate the evidence for the Hodgson-control theory in a variety of ways, and different people could come to opposite conclusions about it.[132] Personally, James confessed that "proof [of the spirit-return hypothesis] still baffle[d]" him. Yet "whether my subjective insufficiency or the objective insufficiency of our evidence be most to blame for this, must be decided by others."[133]

The Hodgson-control case thus magnified for James the limits of the ethos of impersonality and self-effacement. It confirmed that those "scientific" virtues, while in many cases desirable, in others were unattainable.[134] However, James's analysis of the Hodgson-control séances reveals more than that. It fully highlights the *social* dimension of James's ethos of science. In concluding his discussion of the evidence, James confessed to his readers that, given his own personal presumption, namely his ability to "imagine" perfectly well "spirit-agency," and his reluctance to admit that any department of human behavior should be "wholly constituted of insincerity," he was not willing to rule out the spirit-control theory (or other supernatural explanations).[135] In the end, however, he refrained from drawing any conclusion. Waiting for more facts, he limited his task to "arranging the material" in such a way that the reader could "draw his conclusions for himself."[136] Contrary to how it may appear, this strategy was hardly an homage to the ethos of impersonality and self-effacement. While James certainly considered it more appropriate to "wait for more facts" before offering interpretations, the self-restraint he displayed in this case was very different from the self-restraint adopted, for example, by the French investigators of Eusapia Palladino or the scientific "suspension" of judgment recommended by the Cliffords and Huxleys. If James avoided drawing any conclusion, he did it not out of fear of ridicule, or because he mistrusted the witnesses, or as part of an attempt to delegate the task of evaluating testimony to machines. Instead, he refrained from offering a conclusion in order to avoid imposing his own interpretation, and his own very personal presumptions, on others. In the absence of logically compelling factors and facts that would carry a determinative degree of "evidential value" for one of the competing explanations of the phenomena, James wanted to allow people with different casts of mind to draw their own conclusions. James's self-restraint was an act of modesty, of open-mindedness, and of respect, and it reflected his horror of intolerance and of dogmatism. Thus James was "extremely pleased" that one of his friends, Mrs. Mary Chapman Goodwin Wadsworth, a woman with strong spiritualistic interests, could conclude from James's report that the "utterances" during the séances were "essentially sincere."[137]

## Uses of James's Scientific Ethos: Boundary Work

James's moral and social economy of science was part of his boundary work, as his contemporaries sensed. Hugo Münsterberg, for example, accused James of violating, especially in his activities as a psychical researcher, the scientific ethos of emotional detachment, impersonality, and mechanical thinking, and, through those transgressions, of blurring the divide that separated science from metaphysics and other types of thought. James and other "mystics," Münsterberg complained, intruded illegally into the territory of science and scientific psychology when using "emotional," "teleological," and "aesthetic" ways of thinking and categories (including the "category of personality"). These categories belonged to metaphysics, poetry, or ethics, not to science. The latter, Münsterberg argued, aimed to produce an impersonal, unemotional, mechanical, and causal ordering of the world of spatio-temporal existence. Thus proper "scientists," including "scientific psychologists" like himself, were by definition committed to an impersonal, unemotional, and mechanical way of thinking. Metaphysics, ethics, religion, and poetry dealt instead with the eternal world of values and the domain of the Absolute. Only here were teleological, ethical considerations, and arguments that involved the categories of personality and will, acceptable and appropriate.[138] To Münsterberg the "transgression of boundaries" perpetrated by psychical researchers resulted in a "miserable changeling," an unsavory mixture of science and metaphysics, which spoiled both.[139] He also complained that, by doing so, James and others like him blurred the boundaries that separated the "real world" (namely, that of the Absolute) from the "scientific world," resulting "in a new world controlled by inanity and trickery," one both "unworthy of our scientific interests and unfit for our [absolute] ethical ideals."[140] James was not tremendously impressed by Münsterberg's philosophical argument. Indeed, he privately described Münsterberg's article as a "monumental exhibition of asininity."[141]

James mobilized his moral and social economy of science in order to tear down other existing or emerging intellectual boundaries. For example, Coon and Eugene Taylor have suggested that James worked to expand the reach of what was taken to be the realm of acceptable "psychology," and to make more fluid the boundaries separating normal adult psychology from "abnormal" and "supernormal" psychology.[142] His ethos of science also enabled him to challenge the line of demarcation that others saw as the natural separation between "orthodox" science and "superstition" or "pseudo-science."[143] When G. Stanley Hall dismissed the results of Gurney, Myers, and Podmore regarding veridical hallucinations, implying that they did not satisfy scientific standards, James turned the tables against him: "To take sides as positively as you do now and on general philosophic grounds, seems to me a very dangerous and unscientific attitude. . . . I should express the difference between our two positions in the matter, by calling mine a baldly empirical one, and yours, one due to a general theoretic creed. . . . I don't think it exactly fair to make the issue what you make it—one between Science and Superstition."[144] Hall (like Titchener, Münsterberg, and many others, including some psychical researchers) assumed the existence of natural, absolute divisions separating science from superstition, and placed himself on the side of science. However, the boundary line was not set in stone, and the labels could be easily reversed; in a specular, and equally legitimate, toponymy, what Hall considered science could be redescribed as unscientific, and what Hall regarded as superstition could turn out to be an instance of good science.

James's ethos of science enabled him to switch labels, and to make "outsiders" into paragons of the good scientific investigator, while redescribing dogmatic scientists as unscientific men. James hammered the point home by arguing that psychical researchers such as F. W. H. Myers were better scientific investigators than their more intolerant antagonists. Myers, the British investigator whom James credited with the discovery of the "Subliminal" self and with the first coherent attempt to map it out, was a "psychologist" who worked "upon lines" hardly admitted by the more academic branch of the profession to be legitimate, and James knew psychologists who would "wish to read Myers out of the profession." (Among them was Münsterberg, who, determined to locate "the exact limit which the scientific psychologist is unwilling to pass," peremptorily disposed of the idea of the subliminal self as a "superstition.")[145] James observed that Myers had not been "crammed with science at college," nor had he been "trained to scientific method by any passage through a laboratory." Yet James found that Myers's mental attitudes—notably his "keenness for truth"; his attention to isolated, anomalous facts that did not fit the "pigeon-hole" classifications of science; his tolerance of a range of interpretations; and the open-mindedness that enabled him to entertain unorthodox hypotheses—made him as good a scientific investigator as those who officially styled themselves men of science.[146]

However, it was especially Myers's social attitudes that James admired. Although Myers, a man of "imperious desires" and "at bottom . . . egoistic" interests with whom James had had some fights, was "decidedly exclusive and intolerant by nature," James nevertheless lionized Myers for his love for truth, which led him to suppress his "social squeamishness," to modify his *amour propre*, and to "unclub" himself completely.[147] Myers had given up his social prejudices, becoming a model not only of "patience" but also of "tact and humility wherever investigation required it." James found that this new "temper," one at once epistemological and social, was "the only one henceforward scientifically respectable."[148]

James's scientific ethos was also a tool to which he resorted when he strove to tear down some of the "social" boundaries enforced by scientists to separate themselves from dilettantes and members of the less educated public. Scientists, James found, definitely obeyed "social prejudices."[149] "I once invited eight of my scientific colleagues . . . to come to my house at their own time, and sit with a medium [Mrs. Piper]," James narrated. "Five of them declined the adventure." Another "psychological friend" of his refused to have his wife attend a séance; he would "never consent to his wife's presence at such performances."[150] James and his wife Alice meanwhile entertained friendly and respectful feelings for Mrs. Piper, whose messages had provided some consolation after the death of their little boy Herman.[151] They even invited her to spend a week with them at their summer house in New Hampshire, not an uncommon practice among psychical researchers. "We college-bred gentlemen," James wrote as early as 1890, "who follow the stream of cosmopolitan culture exclusively, not infrequently stumble upon some old-established journal, or some . . . author, whose names are never heard of in *our* circle, but who number their readers by the quarter-million. It always gives us a little shock to find this mass of human beings not only living and ignoring us and all our gods, but actually reading and writing and cogitating without ever a thought of our canons, standards, and authorities."[152]

Startling as the fact might be to professionalizing scientists and educated gentlemen, knowledge could be produced outside of gentlemanly, scientific, academic, and even educated circles by groups of people who inhabited different social worlds. It was the boundaries framing such a social geography of knowledge, making communication between different social groups impossible, that James especially challenged through his activities as a psychical researcher. He found that dogmatic scientists, determined to ignore the opinions and "facts" of the spiritualists, had created a "hideous rift" between themselves and the rest of humanity.[153] Yet he also found that the spiritualists who were determined to "return ignorance for ignorance" were also responsible for making "intercourse" between the two groups impossible.[154] A caring "intercourse" was vital, since, James wrote, the spiritualists "know the facts," while those with a scientific training "are most competent to discuss them." Only by cooperation between the two would it be possible to seriously explore human nature, the nature of the subliminal, and the question of human immortality.

## Communities and "Faggots": Epistemology and Sociability

James's activities as a psychical researcher, and his cooperation with other investigators in projects sponsored by the Society for Psychical Research and similar societies, made him aware of the importance of cooperation if knowledge were to be attained. Just as isolated pieces of evidence proved to be of little value unless supported by many other pieces of evidence considered collectively and over the long run, so isolated investigators (despite the great contributions that some of them made) were better off when they collaborated with other people. Especially when it came to certain kinds of psychic phenomena, James found that a whole community of inquirers was necessary. This call for a community of inquirers reflected James's commitment to the argument of the "faggot of evidence." To see how the epistemological practice resulting in the production of evidential faggots reflected James's vision of the community of inquirers, we need to return briefly to the "Hodgson-control" case.

   When James was busy trying to determine the evidential value of the various fragments of evidence that surfaced during the Hodgson-control investigation, he found that he could not carry out the task alone. He needed the help of others. For example, in one séance Hodgson's alleged "spirit" confessed that during his life Hodgson had been very fond of a certain lady (he called her "Hulda"), to whom he had proposed, and who had refused to marry him.[155] Was that piece of information true? Could it function as evidence of Hodgson's survival after death? To answer these questions, it was necessary, first of all, to figure out who the lady might have been. Next it was necessary to talk with the lady. Furthermore, it was necessary to ascertain whether Hodgson had confided this to anyone at the time. Gossip about such things spread quickly, James reasoned, and might have easily reached Mrs. Piper's ears; if that was the case, the details disclosed in the séances regarding the lady had no evidential value. To settle these issues and eliminate those doubts, James "inquired of a dozen of R. H.'s most intimate friends." He contacted the sister of the lady in question, the lady herself (who corroborated part of the story, even though she turned out to have a different name), as well as a New York "lady-doctor," in whom apparently Hodgson had confided, and another psychical researcher. Putting together the pieces of information provided by these people, James concluded that Mrs. Piper might well have heard about Hodgson's emotional involvement by ordinary means (gossip), and that the evidence relative to the lady was not conclusive for the spirit-return hypothesis.

In the end the report James published regarding that episode resembled the recording of a multivocal performance. It was based on the records taken by the witnesses during the séances, and on the automatic-writing messages produced by Mrs. Piper's hand, but these were interspersed with remarks made by others, both during the séances and afterwards. To be sure, during the investigation, differences of opinion and tensions materialized among the circle of people interested in attaining clarity over the veracity of the trances. Ultimately, however, as James was relieved to notice, "since we all had fair minds and good-will and were united in our common love for Hodgson, everything got settled harmoniously, . . . the situation . . . smoothed itself out, leaving nothing but a new system of friendships among persons who before Hodgson's death had for the most part been unacquainted with one another."[156] This small group of people, some of whom were not psychical researchers at all, provides an example of the kind of community that James favored—a kind very different from the formal, bureaucratic, and distant investigators required by the new scientific objectivity. This, in marked contrast, was simply a group of people willing to cooperate out of their "love" for a common friend.

The "faggot" conception of evidence was reinforced by the idea of a community of investigators. James agreed with Gurney and Sidgwick that little "sticks" supporting a psychical phenomenon, although each individually weak, could form a strong "faggot," if collected over the long run and considered together with the evidence for other psychic phenomena. "Weak sticks make strong faggots," James reiterated a year before his death.[157] And woven faggots of evidence, as Gurney had also emphasized, worked because they stemmed not from one isolated witness but from a whole collection of people, who, so to speak, joined strengths. The number of witnesses was of the essence. Notably, there were no restrictions on *who* could provide the "sticks" composing the "faggot." For example, for the purpose of the census of veridical hallucinations, the witnesses might include anyone who responded to questionnaires published in newspapers and magazines. In the late 1890s this aspect of the questionnaire method raised many criticisms. Hugo Münsterberg, for example, attacking the use of the questionnaire method in child psychology, complained that the reliance of that method on "cheap and vulgar materials" provided by untrained and untechnical observers who obviously lacked the proper scientific attitude, resulted in a "caricature" of psychology, and made this method hardly suitable for a scientific investigation.[158] Furthermore, faggots of evidence were essentially different from "chains" of evidence. In a chain, as James stressed, each link depended on previous ones. A faggot, instead, owed its strong evidentiary value not only to the weaving together of its "sticks," but also to its being, precisely, a motley "faggot"—that is, something composed of strands that were "independent of each other and came from different quarters."[159] In order for the faggot to work, James argued, witnesses had to be "independent" of each other, and the testimony provided by each of them must not be logically or otherwise dependent on the testimony provided by others.

To see how anomalous these communities of inquirers could appear to some of James's contemporaries, it may be helpful to compare them to the more famous community of inquirers explicitly postulated by James's friend C. S. Peirce in the late 1860s. Peirce had delineated a community of self-transcending inquirers who were willing to sacrifice their individual perspectives in the quest for truth. He had painted the search for truth as a statistical process in the course of which idiosyncrasies, assumptions, and other individual factors depending on the personality of the investigator would statistically cancel out. His vision of the ideal community of inquirers thus elided personal differences and individual traits, indeed even reduced the "individual man" to a "negation," though, as Thomas Haskell has argued, it might well leave room for self-interest and self-aggrandizing tendencies.[160] Friendships had no part in the process of inquiry. Finally, as Haskell has also argued, Peirce's community of inquirers was modeled after *professional* scientific societies. It was a homogeneous, exclusive, professional community.

The communities of investigators that James envisioned—and some of the ones that he personally wove together in the course of his inquiries into psychic phenomena—differed from the Peircean community on all counts. They were to be open and heterogeneous. Within them people could be linked by deep personal feelings (quite different from Peirce's depersonalized "evolutionary love") and yet could retain independence and their individual perspectives and feelings. Chapter 6 will suggest that these communities reflected the communities and modes of social engagement that, starting in the late 1890s, James came to consider as crucial to society, in a manner that will illustrate how James's vision of the order of society went hand in hand with his vision of the epistemic and social order of knowledge.

## The Ethos of Science and Pragmatism

James's activities as a psychical researcher reveal how his epistemological approaches embodied and promoted moral and social sensibilities that, by the late 1890s, he considered essential to the practice of good science. Psychical research, however, was by no means the only field of inquiry in which James found it necessary to weave these elements together. Indeed, James's intense concern with the moral, social, and epistemological attitudes of the genuine scientific investigator (as opposed to the "dogmatic" scientist) found expression in his formulation of pragmatism. Pragmatism, in an important way, was the embodiment of the radical empiricist method that, by the late 1890s, he had come to view as central to science. Introducing pragmatism to a broad public in the fall of 1906, James told his audience that pragmatism was primarily "a method," and one that did "not stand for any special results."[161] Pragmatism had "no dogmas, and no doctrines"; it was "completely genial"; it would "entertain any hypothesis, it [would] consider any evidence."[162] Pragmatists, James also noted, might be committed to whatever theory they find it expedient to believe, provided they would regard their doctrines as *provisionally* true, or "half true," vulnerable to criticism, and potentially revisable in light of future experience—never as "absolutely true." The technical distinction between "half-truths" and "absolute truths," which James elaborated amid debates on his pragmatist theory of truth, expressed his passionate rejection of dogmatism and intolerance in favor of the dispositions and virtues that he had come to consider the defining traits of the genuine scientific person as well as of the intellectual: pluralism, open-mindedness, and tolerance. In particular, pragmatism was open both to science and religion, since it endorsed the scientists' "craving for facts" while allowing for religious doctrines to be true in the pragmatic sense.[163] In refusing to embrace science to the exclusion of religion, or the other way around, it further embodied the moral and social economy of open-mindedness, tolerance, and respect.

As we have seen, James's approach to questions of evidence raised by psychic phenomena was paralleled by his uneasiness with the extreme self-restraint often demanded by turn-of-the-century notions of scientific objectivity and objective evidence. James's concerns with the unqualified requirement of self-effacement, emotional detachment, and disinterestedness found expression in his pragmatist accounts of knowledge and truth, which he presented as processes shaped, not in small part, by psychological and physiological factors. Pragmatism was also a form of boundary work. Indeed, it continued, albeit in a different context and with different audiences, the transgressive boundary work that James carried on through his activities as a psychical researcher. The next two chapters will explore some of these aspects of James's pragmatism.

# 4

# Mental Boundaries and Pragmatic Truth

My failure in making converts to my conception of truth seems, if I may judge from what I hear in conversation, almost complete. An ordinary philosopher would feel disheartened, and a common choleric sinner would curse God and die, after such a reception. But instead of taking counsel of despair, I make bold to vary my statements, in the faint hope that repeated droppings may wear upon the stone, and that my formulas may seem less obscure.
WILLIAM JAMES, 1907

On December 28, 1907, the annual meeting of the American Philosophical Association, held that year at Cornell University, hosted a panel on the theme of "the meaning and criterion of truth." The panelists included William James; Charles Montague Bakewell, of Yale; C. A. Strong, of Columbia; the Cornell philosopher J. E. Creighton; and John G. Hibben of Princeton.[1] The session, the high point of the conference, was entirely devoted to the pragmatist conception of truth, which, as it turned out, was attacked by each of the panelists—except for James. Bakewell and Strong, both of whom had studied with James at Harvard, focused on technical problems of James's account of truth. Hibben argued that pragmatism failed to "stand its own test" (namely, that "whatever works is true") and was inadequate even as a working hypothesis.[2] Creighton declared that the pragmatist theory of truth (especially its emphasis on the "practical") had "been definitely refuted by the flood of criticism which it [had] called out." There had "not . . . been a shred left of its original form, or of its claim to supersede all the older [philosophical] systems."[3] To make things worse, the president of the Association, Harry Norman Gardiner, also delivered a substantial criticism of pragmatist accounts of truth in his address before the whole APA.[4]

James's defense of pragmatism that day sounded tired and halfhearted, and in a letter to his British pragmatist friend Ferdinand Canning Scott Schiller he described the discussion as "abortive."[5] James was not new to such situations. James's *Pragmatism*—in his own opinion the "most important thing" he had "written yet"—had been published only months before the APA panel, but James's conception of truth, like those of his philosophical allies John Dewey and Schiller, had been under attack for years.[6] Instead of "curs[ing] God" and dying "after such a reception," James chose to patiently answer all the objections. His efforts, however, appeared to be largely useless, and James continued to be appalled by what he described as his opponents' "almost pathetic inability" to understand the theories that they strove to criticize.[7]

The dispute over the pragmatist account of truth reached levels of acrimony that many felt inappropriate to refined professional (or professionalizing) philosophers. Why such disdain? What made communication unproductive, and discussion almost invariably inconclusive, with both parties sticking to their positions? What was at stake that could enrage participants to the point that they would give up their academic dignity, labeling each other as "incompetent," "stupid," "philistine," "senile," a "monkey," and even "sausage seller"?[8] In this chapter, on James's account of truth, and the next, on the pragmatist controversy, I seek to answer these questions.

Instead of underscoring the amply discussed philosophical novelty of James's pragmatist account of truth, and the technical ways in which it diverged from period conceptions of truth, I emphasize its conventionality by placing it in the context of contemporaneous theories of the mind and psychological-physiological discussions about the relationships between the three "faculties" that philosophical and psychological writers, from Kant on, had traditionally assigned to the mind: feelings, cognition, and volition.[9] By embedding James's account of truth in this context, the chapter argues that some of the features of James's conception of truth that most upset his contemporaries were actually commonplace within psychological and psychophysiological theories of cognition—a fact that some of his friends and opponents alike noted. Like psychological and physiological doctrines, James's account of truth blurred the boundaries between those three "functions" of the human mind, as James and many other psychologists now preferred to regard them. By unpacking the psychological elements of James's conception of truth, especially the notion of "satisfaction" that James most insistently presented as a criterion, even a mark, of truth, I also argue that James firmly rooted truth processes in the physiology of the human body. Thus I suggest that James offered not only a psychology of truth, as Bruce Kuklick has shown, but also a physiology of truth—a project that other psychologists and students of human nature had envisioned as worthy of being pursued. The contentions arose only when James and other pragmatists dragged psychological doctrines into the domain of philosophy, a move that was increasingly being banned in philosophical circles. Based on these conclusions, in the next chapter I reinterpret the pragmatist controversy as a war over disciplinary boundaries and ultimately as a clash between two different visions of the future of philosophy as a discipline.

## A Philosophical Analysis of James's Conception of Truth

Some years ago Hilary Putnam, a philosopher to whom we largely owe the late twentieth-century resurgence of interest in William James and pragmatism, proposed a philosophical anatomy of James's notion of truth. While the majority of philosophical analyses of James's account of truth aimed essentially to improve on it and make it into a viable philosophical conception, Putnam's goal was primarily "exegetic."[10] His analysis provides a helpful introduction that will allow a better understanding of the storm of protest that greeted James's notion of truth.

Putnam identified four central components in James's account of truth. These components, he argued, emerged at different times, and were patched together into James's pragmatist conception of truth only sometime in the early years of the twentieth century. The first was the "ultimate consensus" notion of truth, or the notion of truth in the long run. This kind of truth was the ideal limit to which knowledge would converge over time. Putnam noticed that James borrowed this notion from C. S. Peirce and adopted it as early as 1878, that is, some twenty years before he considered pragmatism as a working philosophy.[11] Later James dubbed ultimate truth "abstract truth" and "absolute truth," and, like Peirce, understood it as the "final opinion to be converged to and determined [although not exclusively so] by reality."[12] This notion was different from the more "concrete" and temporary truths or, as James also labeled them, "half-truths"—the only kinds of truths he believed to be available to human beings. As an example of a half-truth James offered "Ptolemaic astronomy," which was true for a certain time and "became" false later.[13] Putnam equated truth proper—a tenseless property—with "absolute truth," rather than with the revisable and temporary "half-truths."[14]

Second, there was an "interest component," that is, "the unpeircean idea that truth is partly shaped by our interests." As Putnam explained, the idea was "unpeircean" because Peirce insisted that ultimate consensus would be brought about by an "external permanency," which was "nothing human." In contrast, James believed that absolute truth would be shaped by the total drift of experience *and* by something human—interests as well as "more practical and more immediate aims and sentiments."[15] James emphasized that these personal factors also entered into the production of concrete truths. Putnam, however, insists that James's emphasis on interests "should not obscure the fact that James, like Peirce, declares his allegiance to a notion of truth *defined in terms of ultimate consensus.*"[16]

Third, there was a "realist strain." In contrast to other readers, Putnam underscores that James was (or in any case always claimed to be) an epistemological realist.[17] In particular, James insisted that truth was a relation between an "idea" (or, in James's varying terminology, a "belief," a "theory," and so on) and the "object" or "fact" to which the idea "referred." Both the idea and its object contributed to determining (or bringing about) the truth of the idea. This point was perceived as very controversial by James's critics and admirers alike. Many denied that James was a realist, ignoring his disclaimers to the contrary. Others contended that James could never be a realist, even if he wished to be one, as epistemological realism was incompatible with certain metaphysical claims that underlay his pragmatism. Putnam locates this strain of James's account of truth in the context of James's metaphysics of relations, and associates James's understanding of the notion of agreement with a conception of reference that James had developed earlier, while he was working on his metaphysics of radical empiricism. Neither of these conceptions, Putnam noted, involved a notion of self-transcendence of mental images and ideas.[18] Instead James understood both as processes taking place wholly within experience, and linking one part of experience (ideas, beliefs, mental states) with another (the "object" of the idea or the object's "surroundings"). This leads Putnam to the fourth component of James's conception of truth, an "empiricist component." This component is "summed up in the claim that "truth *happens* to an idea." An idea, James wrote, "*becomes* true, is *made* true by events. Its verity *is* in fact an event, a process: namely of its verifying itself, its veri-*fication.*" Putnam argues that James did not conflate truth with verification, and that "it can only be the entire process of *verification in the long run* that 'makes' an idea true."[19]

Scholars have long understood the ways in which James's account of truth differed from others. First, it radically diverged from "correspondence" theories of truth, according to which the truth of a proposition consisted of a relationship of correspondence between the proposition and the state of affairs to which the proposition referred. James's account of truth remained superficially compatible with the correspondence or "agreement" accounts of truth. However, it replaced the general and empty notion of agreement with a plurality of concrete processes, each of which was responsible for the agreement of an idea with its objects in some specific case. Moreover (except for cases of "direct-acquaintance" in which the idea coalesced with reality in a single experience), within a Jamesian framework the agreement of an idea with reality, however defined in the specific case, was always to be thought of as a dynamic relationship, one that involved processes. James illustrated this by means of a famous example, which led some to describe this aspect of his conception of truth as "ambulatory" truth: "If I am lost in the woods and starved, and find what looks like a cow-path, it is of the utmost importance that I should think of a human habitation at the end of it, for if I do so and follow it, I save myself."[20] The experiential process through which the lost individual, armed with his idea that at the end of the cow-path there should be a hut, walks along the path and, step after step, reaches the hut, is an instance of truth process, a simple example of verification. Like other pragmatists, James viewed these processes as ones of adaptation of the knower to the environment, which, in turn, James understood to include all sorts of theoretical and practical situations, as well as the body of past truth collected by humankind during its history.

Finally, James's conception of truth linked truth inextricably to action. For James and for other pragmatists, "the possession of true thoughts mean[t] everywhere the possession of invaluable instruments of action," and "our duty to gain truth" could "account for itself by excellent practical reasons." Indeed, "far from being here an end in itself," the possession of true thoughts was "only a preliminary means toward other vital satisfactions," as the example of the person lost in the woods and eventually finding a hut illustrates. As James put it, the truth of an idea consisted in its practical consequences over the long run, in its "workings," and in its ability to "lead" to a part of experience to which it was worth being led.[21]

## Truth, Volitions, and Emotions

Despite its mainly "exegetic" intent, Putnam's very helpful analysis of James's theory of truth was guided by Putnam's own philosophical agenda—making James a precursor for a type of neopragmatism immune from the relativist "menace." For that reason, it does little justice to an aspect of James's conception of truth that many of James's contemporaries instead found central: the contention that not only interests but also emotions, volitions, and purposes could play an important role in the production of truth, and perhaps even enter into its definition. This was something that James's contemporaries could not quite swallow.

Critics as well as supporters of pragmatism were especially appalled by the importance that James seemed to confer on satisfaction in his account of truth. In 1907 James famously stated that the truth of an idea was "determined by its satisfactoriness," and that "the *matter* of the true" was "absolutely identical with the matter of the satisfactory."[22] Two years later he reiterated the point: "satisfaction has to figure in the definition of truth. . . . Truth at any time is that quality in a belief which is satisfactory. Truth is the quality which satisfactory beliefs possess—truthfulness we will call it, if you like."[23] But what did James mean by satisfaction? Characteristically, James denied that the term "satisfaction" could be defined univocally, "so many are the practical ways in which it can be realized."[24] Satisfaction was a multidimensional term, and it had "to be measured by a multitude of standards."[25] He included aesthetic agreeableness, emotional satisfactions, satisfactions of taste, moral satisfactions, practical satisfactions, satisfaction associated with consistency, as well as a "theoretical satisfaction."

The idea of satisfaction pervaded James's account of truth. Satisfaction entered into the making of absolute truth (that "finally satisfactory formulation"),[26] exactly as it entered into the production of our half-truths.[27] Satisfactions were patently present in the "interest component" of truth, since for James satisfactoriness (as a mark of truth) was an interest-relative concept. James argued that point eloquently, if perhaps not persuasively, in a passage that he wrote in response to the French critic Marcel Hébert, a former Catholic priest. Imagine, Hébert argued, that someone owned a painting that he believed to be a Corot, and suppose that the authenticity of the painting was called into question. There certainly was a truth about the authenticity of the painting: either it was a Corot or not. In either case, the owner's feelings of satisfaction (in believing that it is a Corot) or uneasiness (when the authenticity of the painting is questioned) were totally irrelevant. James's provocative answer was that the *true* idea of the painting was the idea that met a person's present interest most satisfactorily. That might have been the "proprietary interest," or interest "in the picture's place, its age, its 'tone,' its subject, its dimensions, its authorship, its price, its merit."[28] Each of these interests would be matched by a different kind of satisfaction (or uneasiness). Thus there were as many truths (or falsenesses) about the Corot painting as there were interests.

Satisfactoriness also entered into what Putnam described as the "realist strain" of James's conception of truth. "Agreement" between an idea and reality yielded several practical satisfactions by putting the knower in better working touch with reality.[29] Satisfaction, finally, was also embodied in the "ambulatory" aspect of James's conception of truth, the view according to which true ideas *lead* the knower to their object through a sequence of intermediate experiential links. Each such intermediate link arouses pleasurable feelings insofar as it is "felt" to "fulfill the goal" of the previous link, and to be "continuous with it." This kind of satisfaction, James wrote, "grows *pari passu* with [the knower's] approximation" to the final terminus (namely, the hut in the cow-path example).[30]

I have spent some time emphasizing the ubiquity of satisfaction in James's notion of truth because late twentieth-century philosophical accounts of pragmatism have often tended to downplay it.[31] James's contemporaries, however, perceived that satisfaction (or satisfactoriness) was central to James's account of truth, and not a few accused him of conflating truth and satisfactoriness. Others reproached James for his maddening ambiguity over the issue. Did satisfactions follow from the fact that one held a belief to be true? Did they "accompany" truth? What did it mean to say that satisfactions were "marks" or "criteria" of truth? As Arthur Lovejoy, a Harvard graduate, observed, the claim that satisfactions were "concomitant" with truth was a "psychological truism." But the contention that satisfactions were marks or criteria of truth was an "illicit" epistemological claim. Lovejoy felt justified in requesting from all professed pragmatists "an oath . . . to cease speaking of satisfaction as a 'criterion' of validity."[32] Even well-meaning friends were awfully concerned about such questions. James Bissett Pratt, a former student of James's, stressed that the making of satisfactions into marks of truth led to paradoxes. For example, if at the same time the proposition A and its negation not-A were felt satisfactory by two different individuals, it followed that both A and not-A were true, which violated the principle of contradiction.[33] Émile Boutroux had problems "with the word 'satisfactory.'"[34] And Ralph Barton Perry, James's faithful student, rejected the "radically pragmatist" theory of truth that seemed to maintain that "the proof, mark, or guarantee of truth [was] the *satisfying* character of that moment in the process in which the cognitive interest is fulfilled." To Perry this claim appeared to be identical to the claim that truth *consists* of feelings of satisfaction, and was thereby unacceptable.[35]

Critics also took James to task for equating theoretical satisfactions with practical, emotional, and aesthetic satisfactions, and making all of them marks of truth. Lovejoy, for example, complained that at times pragmatists seemed to assume the "commensurability and equivalence" of all those types of satisfaction for epistemological purposes. But satisfactions certainly had to be arranged in an epistemic hierarchy, with theoretical satisfactions at the top. If a claim "aroused pleasant feelings" but "violated the logical notion of consistency," he enjoined, it could certainly not be held to be true. (On the other hand, if pragmatists accepted that point, then their theories would be no different from those of their adversaries.)[36]

In common parlance, "satisfaction" indicated a feeling or an emotion. Thus, for some critics, James's emphasis on satisfaction implied a reduction of truth (and falsehood) to "pleasure," "sentiments," or the "emotional expression" of ideas.[37] They called for a clear separation between knowledge, truth, and the intellect on the one hand, and emotions and feelings on the other. Those psychological processes, they charged, might indeed be concomitant with the discovery of truth, but they were logically distinct from it.

Other critics framed similar arguments in an attempt to challenge the role that pragmatist accounts of truth conferred on interests, volitions, and purposes. Some critics, for example, chastised James for equating truth with "what is useful to believe" or with "what it pays to believe," while others took issue with a passage in which James seemed to identify truth with "the expedient in the way of our thinking." To these critics, the "useful" and "the expedient" were interest-dependent concepts and had "no cognitive value."[38] One concerned critic even sent James a letter, stating that "the word expedient . . . has no other meaning than that of self-interest," and that "the pursuit of this has ended by landing a number of officers of national banks in penitentiaries," and concluding that "a philosophy that leads to such results must be unsound."[39] Others agreed that utilitarian value and cognitive value might well harmonize but contended that they should be kept logically distinct, while still others complained that pragmatists placed truth in the domain of "volition" rather than that of "cognition." They argued that, by linking truth to interests, purposes, and needs of the knowers, pragmatists were guilty of offering "teleological" theories of truth and, worse, of making truth depend on the private purposes of the knower. While often admitting that volitions could "occur simultaneously" with cognitive processes, or even "accompany them," these philosophers firmly denied that volitions could have any cognitive import. Edwin Creighton, the first president of the American Philosophical Association, and one of the panelists at the 1907 APA session on the pragmatist theory of truth, is a case in point. While he acknowledged that "ideational and volitional components of thought" could occur simultaneously, he insisted that they were "distinguishable." He also pointed out that personal factors such as will and purpose had nothing to do with the logical problem of truth, while James's colleague Josiah Royce stated that according to the pragmatists "one [could not] sunder will and intellect," and even "the most remote speculations" were, "for the man who engage[d] in them, modes of conduct."[40] These critics all urged that the cognitive, the emotional, and the volitional should be kept distinct.

## Embodied Truth

If critics accused James's conception of truth of blurring the boundaries separating the cognitive from the volitional and the emotional, some of them also perceived that James went further and blurred the divide that separated the intellect from the body—something they found unacceptable.

That James's account of truth did just that becomes apparent in a closer analysis of what he meant by "satisfaction," and what he called the "intellectual emotions," as well as by a closer discussion of the "ambulatory" component of his theory of truth. Although James never defined the term, he took satisfaction to be a kind of pleasure. In a psychological seminar he taught in the early 1890s, James examined a variety of theories about pleasure and pain (or pleasure and displeasure). These included physiological theories that sought to localize pleasure and pain by assuming the existence of (unobservable) specific pleasure and pain nerves or brain centers. James rejected those theories, but he agreed with physiological and psychological writers in viewing pleasant and unpleasant feelings as "diffused" physiological processes, usually "irradiated" throughout the organism.[41] Satisfaction, as a pleasant feeling, then could not be disembodied, since it involved some sort of physiological process. The same was true more generally of the intellectual emotions, such as, for example, the pleasure of consistency.

James, like many other psychologists of the time, drew a distinction between the "standard emotions" (for example, anger, fear) and the subtler emotions (including aesthetic, moral, and intellectual emotions). In an initial formulation of his theory, he had argued that while the former represented "the *feeling* of the reflex bodily effects of what we call [the] object] of the emotion," the latter were not accompanied by bodily changes.[42] Later, however, he revised his position, and suggested that in the aesthetic, moral, and intellectual emotions, "the bodily sounding-board" was "at work far more than we usually suppose." Those emotions, James continued, were "assuredly for the most part constituted of other incoming sensations aroused by the diffusive wave of reflex effects" set up by their objects. They were accompanied not only by brain processes, but also by bodily changes that took place outside of the brain: respiration, motions in the chest, in the heart, in the nostrils, and tension in the fingers. Thus James concluded that, "as far as [their bodily] *ingredients*" were concerned, the subtler emotions represented "no exception" to his (and Carl Georg Lange's) account of emotions "but rather an additional illustration thereof." Indeed, a "purely disembodied human emotion" was "a nonentity."[43]

The bodily dimension of intellectual satisfaction (a distinct kind of pleasure) was clear in James's statements about the "satisfaction" of "consistency" and the unsatisfactory feelings aroused by inconsistency. In 1905, describing the consistency afforded by "theoretical truth," James argued that the "pleasure" of consistency depended on "habits."[44] For James habits were both psychological and physiological processes. Physiologically, they were paths where nervous energy freely flowed.[45] Psychologically, habitual actions were characterized either by pleasure or by absence of unpleasant feelings. Because of their habitual character, consistent trains of thought were characterized by pleasure or absence of pain. Inconsistent thought processes, in contrast, were unpleasant and resulted in "intellectually unsatisfactory" situations.[46] Several years earlier James had argued that situations in which different representations "harmonize"—for example, when there is intellectual consistency—followed the "laws of composite pleasure" and obeyed the German physician Karl Spamer's "theory of pleasure," according to which pleasure "results from an addition of motion to nerve matter 'adequate' to the intrinsic motion."[47] Thus satisfaction (including the paradigmatic case of "theoretical" satisfaction, the satisfaction afforded by consistency) had a physiological component, even though that component could not be pinned down to any specific, localized process. Contemporaries of James who were familiar with his theories of habit, and with period theories of pleasure and pain, would not have failed to perceive that James's emphasis on satisfaction (as a concomitant process, a mark, or even part of the definition of truth) linked truth to the physiology of the body.[48]

James's theory of truth involved the body in other ways as well, and his critics did not fail to appreciate that point. In 1909 the Columbia realist philosopher W. P. Montague referred to the "ambulatory" theory of truth as "the motor theory of truth."[49] The expression was modeled after the phrase the "motor theory of ideas" or the theory of "ideo-motor action," a psychological-physiological theory that James ascribed to the British physiologist and psychologist William Benjamin Carpenter. In Carpenter's formulation, the theory of ideo-motor action amounted to the claim that "the strongest or the most dominant idea resident in the mind would automatically lead to a corresponding action." Ideas had a tendency to act out, to incite a sequence of psychic as well as motor associates leading to an action.[50] Carpenter and other mental physiologists and psychological and philosophical writers limited ideo-motor action to a specific class of ideas (ideas of movement or of action), and confined the application of the ideo-motor theory to those cases in which the power of the will was in abeyance (as in somnambulism or hypnotism, for example).

James, however, extended the motor view to all ideas and made it into the cornerstone of his conception of will. According to James, ideas and thoughts were "impulsive." In *Principles* he wrote: "We do not have a sensation or a thought, and then have to *add* something dynamic to it to get a movement. Every pulse of feeling which we have is the correlate of some neural activity that is already on its way to instigate a movement. Our sensations or thoughts are but cross-sections, as it were, of currents whose essential consequence is motion."[51] Underlying that claim was the physiological assumption—which, as James acknowledged, could only be vaguely stated—that each mental state, each thought, including our theoretical beliefs, was connected to some physiological motor path and, consequently, was capable of inciting a sequence of motor, as well as mental, associates.[52]

Montague's use of the phrase "the motor theory of truth" was supported by James's own terminology. Shortly after the publication of *Pragmatism*, James brought up the question of why true ideas had a tendency to work. He answered that a true idea works because it *"has associates peculiar to itself, motor as well as ideational*; it tends by its place and nature to call these into being, one after another; and *the appearance of them in succession is what we mean by the 'workings' of the idea."* The "tendencies" that true ideas had to give rise to such "chains" of motor and ideational associates were important to an epistemology of truth: "the whole chain of natural causal conditions [excited by a true idea] produces a resultant state of things in which new relations, not simply causal, can now be found . . . these new relations being epistemological relations, e.g., truth, reference, adaptation, instrumentality, etc."[53]

Years earlier, James had made similar claims in an essay in which he had investigated what enabled one part of experience (an idea or a feeling, for example) to "apprehend" another part (its object). What does it mean to say that one's idea of the tigers in India "knows" actual tigers or "points" to them, James had asked. His answer was that an idea could "be cognizant" of its object only through a procession of experiential links that led continuously to that object or to some of its surroundings: "the pointing of our thought to the tigers is known simply and solely as a *procession of mental associates and motor consequences that follow on the thought*, and that would lead harmoniously, if followed out, into some ideal or real context, or even into the immediate presence of the tigers."[54] Since, as we saw, the notion of ideo-motor action involved physiological processes taking place in the nervous system, by incorporating that notion into his account of truth (and reference) James linked in yet another way truth (or its workings) to the physiology of the knower.[55] The point was not lost on his contemporaries. Thus James's former student Charles Augustus Strong described James's theory of "cognition" as a "naturalistic" theory.[56] And McGill philosopher A. E. Taylor pointed out that when the pragmatists talked about the practical consequences of truth they deployed the word "practice" not to indicate (as other philosophers of the time did) a "self-initiated" change in some "datum of presented facts" but rather to indicate physical motor changes.[57]

Critics also perceived that, by embodying truth in the psychological and physiological workings of the knower, James's account radically changed the image of the "knower." No longer abstract minds that could pursue their quest for truth in complete disregard of sentiments, feelings, passions, interests, and utilities, pragmatic knowers were bodily beings fully engaged in action and practice. They were Darwinian organisms constantly acting in, and reacting to, the changing environment on the basis of their needs, interests and purposes.[58] That conception of the knower, to some philosophers, appeared to endanger not only the philosophical dignity of the notion of "truth," but also the dignity of philosophers, as it called into question the century-old self-representation of the philosopher as a paradigm of pure knowledge, a disembodied mind engaged in contemplation and in the dispassioned, disinterested search for truth.

### The Psychology of Truth: Feeling, Cognition, and Volition

Although critics strongly disliked the ways in which James and other pragmatists made truth "lean" on emotions, volitions, interests, purposes, and physiological processes, the pragmatists were hardly doing anything novel.[59] Most Anglo-American psychological works written in the second half of the nineteenth century associated cognition with emotions, practical goals, and volitions. The distinction between emotion, cognition, and volition, which had structured traditional faculty psychology, could still occasionally appear in systematic psychological treatises, but, more often than not, it retained a merely formal flavor.[60] The classical tripartition of the faculties of the mind appeared to be only a useful theoretical tool, and many questioned whether it reflected the actual organization of the human mind.

A few examples will illustrate this point. James Ward, a British psychologist whom we met in chapter 1, considered feeling, cognition, and volition not as independent faculties of the mind but rather as functionally and genetically related aspects of mental activity. In his widely read 1886 entry "Psychology" for the *Encyclopedia Britannica*, he stated that the three constituents of mind were connected to each other and implied each other, and it was only by "ideal analysis that they could be discriminated and considered apart."[61] Similarly, for the British psychologist James Sully each state of mind had a cognitive side, a volitional side, and a feeling or emotional side; pure feelings, volitions, and cognitions were seldom met: "Feeling, knowing and willing . . . cannot exist in perfect isolation from one another any more than the colour, form and odour of a plant."[62] The American psychologist George Trumbull Ladd described consciousness as a rope. Any portion of it was "complex with an irreducible threefold complexity," being at once a "fact of intellection," a "fact of feeling," and a "fact of conation." He added that, although one of the three aspects "could be emphasized . . . at the expense of the others . . . , no one of the three could be destroyed without destroying the psychic fact itself as an object of consciousness."[63] In his *Psychology* (1887), John Dewey stated that the three modes of mind were irreducible and distinct. Nevertheless, he emphasized that they were equally dependent on one another: "knowledge is not possible without feeling and will; and neither of those without the other two."[64] One generation earlier, Herbert Spencer had insisted "on the impossibility of dissociating the psychical states classed as Intellectuals from those seemingly most unlike psychical states classed as emotional."[65]

Among the "emotional" states, psychologists had long recognized an affinity between satisfaction, or pleasure, and truth. The Italian anthropologist Paolo Mantegazza, author of widely read psychological, ethnographic, and philosophical works (including the scandalous ones that for a while fascinated Freud's "Dora"), emphasized the pleasures engendered by the search for truth. In the first of a series of "physiological" studies of "the moral man," he declared that truth belonged not only to the domain of "mind" (intellect) but also to that of "sentiments," and coined the phrase "the sentiment of truth" in hope that someone would someday offer a full-fledged "physiology of truth."[66] The associationists too had long pointed out the existence of a link between truth or consistency and pleasurable feelings. The British psychologist Alexander Bain, for one, noticed that "the operations of the Intellect may be attended with various forms of pleasure and pain," such as "Consistency and Inconsistency in truth and falsehood."[67] The American psychologist Hiram S. Stanley observed that, in the "person of strong intellectual interests," the discovery of "disagreement" among ideas produced "acute pain."[68] Ward associated the "pleasure of consistency" with the "form of the flow of ideas" rather than their content. H. R. Marshall spoke of a distinguished "pleasure of judgment" that "stands on the same footing as do many pleasures which are called purely emotional."[69] For Dewey "every identification" (and every "clear distinction") was "accompanied by a peculiar thrill of satisfaction; a feeling which seems to be a combination of the feelings of harmony and of the broadening-out of the mind through the performed identification."[70] The Danish philosopher and psychologist Harald Höffding observed that when thought was free from external or practical concerns, there could arise "joy in agreement, sequence, and connection, and a feeling of pain at discord, contradiction, and lack of connection." Pleasure and pain, he continued, were felt "not merely because our standard of truth is maintained or disregarded, but because in harmony or discord itself there is something immediately satisfying or painful."[71]

James was quite familiar with all this literature, and personally acquainted with many of these psychological writers, including Spencer, Bain, Sully, Ward, Dewey, Ladd, and Höffding, whom he once invited to give a lecture in one of his classes.[72] He read at least one of Mantegazza's physiological studies of moral man,[73] discussed Stanley's work in his 1891–92 seminar on "Aesthetics," and wrote a positive review of H. R. Marshall's book *Pain, Pleasure, and Aesthetics*, presenting it as "almost 'epoch-making'" in the "still uncertain science of aesthetics."[74]

In his own psychological works James had not hesitated to crisscross the uncertain divides separating feeling and volition from cognition, and to offer an account of cognition that fully embedded it in the bodies of the knowers. For example, his psychological account of belief regarded belief both as a cognitive phenomenon (one that, as Gerald Myers notices, presupposed conception, judgment, or thought); as an emotion (a verbally indescribable but no less distinct "emotion of conviction" or "emotion of belief"); and as a "psychic act" psychologically—though not physiologically—indistinguishable from the will.[75] Similarly, in *Principles of Psychology* James emphasized how interests, expectations, pleasant and unpleasant feelings, as well as active, emotional, and aesthetic needs, played an important role in the processes by which the knowers came to ascribe a cognitive value to sensations and to select the sensations that they ultimately regarded as "objective." Discussing the selective activity of the mind, he claimed that the sensations whose objects people took to be real were those that were endowed with liveliness and coerciveness of our attention (a factor ultimately boiling down to personal interests of the knower), and had a stimulating effect upon the will and emotional interests (such as, for example, "love, dread, admiration, desire"), in addition to being "congruous" with certain aesthetic interests.[76]

Drawing on evolutionary theory, from early on James viewed cognition as part of a reflex-arc process.[77] Roughly until the middle of the nineteenth century, the notion of reflex action was restricted to the spine, and understood to consist of two phases, one during which external stimuli were conveyed to the spine through "afferent" sensory nerves (action), and another in which a reaction was conveyed from the spine to nervous terminations through "efferent" motor nerves (reaction). However, around 1845, the theory was extended beyond the spine to encompass also the cerebrum and brain, and later in the century many argued that it provided a model for an understanding of human action in general.[78] By the 1880s the application of the reflex-action theory to mind, intelligence, and intellectual processes, such as thinking and knowing, could be found everywhere: from textbooks in psychology and mental physiology, to articles in popular scientific and literary magazines, to the novels of George Eliot. James considered reflex action to be a physiological process articulated in *three* phases. First, incoming currents were transmitted along sense nerves to the nervous centers (the spinal cord and the brain); second, certain changes took place in the nervous centers; third, a reaction followed, in the form of an outward current discharged along motor nerves. James called the three stages "impression," "reflection," and "reaction," and associated them with the three "departments of mind": feeling (or sensation), cognition, and volition.

Investigators of "mental physiology" and other men of science often deployed the reflex-arc theory of the mind to depict human beings as machines, "conscious automata." In a widely read essay, for example, T. H. Huxley depicted consciousness as a powerless spectator of the mechanical actions taking place in the body, and denied that volitions could cause any bodily movements.[79] James instead used his modified, tripartite notion of reflex action (together with the theory of evolution) in order to justify a teleological notion of the mind and of cognition.[80] His conception of reflex action made the "conceiving or theorizing faculty" of the mind into something that functioned *"exclusively for the sake of ends."*[81] The first stage of reflex action (sensation) took place for the sake of the second (cognition), and the second took place for the sake of the third (volition), so that the final stage governed both sensation and cognition.[82] Thus cognition, James famously wrote, was "incomplete until discharged in act"; in this scheme cognition turned out to be "but a fleeting moment, a cross-section at a certain point of what in its totality is a motor phenomenon."[83] By making cognition (and "conception") into the intermediate state of a reflex action, James challenged the traditional distinction between the faculties of the mind, and made cognition functional to volition and action.[84]

In the late 1870s James offered a conception of rationality that—like his accounts of cognition, sensation, and belief—radically questioned the idea of a separation between the intellect, feelings, and volitions. He associated rational trains of thought with habits, and identified the "sentiment of rationality" (the subjective mark by which a person recognized a rational conception) with a "strong feeling of ease," "peace," and "rest," or with the absence of unpleasant, distressing feelings that arise when a habitual train of thought is interrupted or impeded by some obstacle.[85] He also observed that "the transition from a state of puzzle[ment] and perplexity to rational comprehension" was "full of lively relief and pleasure."[86] Philosophers "craved" conceptions, James wrote, but which kinds of conceptions were "the right" ones for them? Ignoring the Kantian indictment of passions, those "diseases of the soul" and enemies of rationality, James suggested that philosophers found a conception to be rational when it satisfied one of the two main "theoretical passions" or "cravings": the passion for simplicity and unity and that for clarity and distinction.[87] Thus, ultimately, the "right conception for a philosopher depend[ed] . . . on his interests."[88]

James expanded the discussion of rationality to include what he called "practical rationality." He contended that "of two conceptions" that satisfied equally "the logical demand," the conception that awoke "the active impulses, or satisfied other aesthetic demands better than the other" would be considered to be "the more rational conception."[89] Two types of conceptions violated the requirements of practical rationality, and were bound to appear irrational to the layperson and the philosopher: first, philosophical conceptions that failed to banish uncertainty from the future, thus leaving people prey to the unpleasant feeling that "something [was] impending," and, second, conceptions that painted the universe in a way that made human beings into ineffective and insignificant agents in the total scheme of things.[90] These conceptions legitimated only emotions of "fear, disgust, despair, or doubt"; consequently, they would be deemed less rational (practically) than conceptions that engendered more positive emotions.[91] Thus the appreciation of the rationality—theoretical and practical—of a philosophical conception was tightly linked to interests, passions, and practical and emotional factors, including faith.[92]

These examples illustrate how much James, in the 1870s, '80s, and '90s, was willing to admit that the nonintellectual nature—feelings, emotions, and volitions—and the body cooperated with the intellectual nature in the production of knowledge, including scientific and philosophical knowledge. In his pragmatist writings James continued this line of thought. Indeed, in this respect pragmatism was an "old way of thinking." Thus the question arises: Given that it was common practice, in psychological inquiries, to bring together feelings, emotions, cognitions, volitions, and action, why did precisely that aspect of pragmatism meet with such stubborn resistance? Nobody blinked when, in Principles and in his essays on "The Sentiment of Rationality," James associated intellectual processes with emotional and practical factors and with bodily processes. Why then were people now shocked by pragmatists' refusal to sunder cognition from the will and from the emotions, and the intellect from the body?

To answer that question, we need to take a detour through an issue that was hotly debated at the turn of the twentieth century: the issue of "psychologism." This detour will enable us to perceive that the pragmatist controversy was, to a large extent, a controversy about the boundaries of philosophy and those of psychology, the division of labor between philosophy and neighboring disciplines, and ultimately the nature of philosophy itself.

5

# Pragmatism, Psychologism, and a "Science of Man"

Some logicians presuppose *psychological* principles in logic, but to bring such principles into logic is as absurd as to derive morality from Life. If we took the principles from psychology, i.e., from observations of our understanding, we shall only see *how* thinking proceeds, and *what* happens under manifold subjective hindrances and conditions. . . . In logic we do not want to know how the understanding is and thinks, and how it hitherto has proceeded in thinking, but how it ought to proceed in thinking.

IMMANUEL KANT

Pragmatism begins where philosophy ends.

ALBERT SCHINTZ

In the preface to the second edition of his *Critique of Pure Reason* (1787) Immanuel Kant famously took issue with some "moderns" who "thought to enlarge" logic by "introducing *psychological* chapters on the different faculties of knowledge (imagination, wit, etc.), *metaphysical* chapters on the origin of knowledge, or . . . *anthropological* chapters on prejudices, their causes and remedies." He warned those modern philosophers that by allowing those sciences to "trespass upon one another's territory," we "do not enlarge" them but "disfigure" them.[1] The suspicions that Kant harbored against psychological and anthropological approaches to the laws of logic reflected his concerns with a broader issue: the question of the nature of philosophy and its territorial delimitation, as well as the question of the proper relationship between academic philosophy and a "philosophy for the world," rich in practical implications for education and civic virtue. As John H. Zammito has shown, in late eighteenth-century Germany these questions were subjected to intense debate. Empirical psychology—which Kant and others contrasted with "rational" psychology (a subdiscipline of philosophy) and, a fortiori, with Kant's own "transcendental" psychology—was based on observation. Anthropology, or "the science of man," was an empirical study of human beings, emerging around the mid-eighteenth century from "the confluence (or the 'con-fusion') of a number of disparate inquiries," including the "*medical* model of physiological psychology, the *biological* model of animal soul, the *pragmatic* . . . model of cultural-historical theory," the "*philosophical* model of rational psychology," and even the "*literary-psychological* model of the new novel."[2]

In the second half of the eighteenth century both "empirical psychology" and the new "science of man" were presented by some as alternatives to philosophy. The "modern" philosophers whom Kant chastised found these sciences far more seductive than the "traditional pursuit of metaphysics."[3] They worked to reframe philosophy as an empirical, naturalistic kind of inquiry, and attempted to incorporate it, even reduce it, to the discourses of empirical psychology or anthropology. Kant reacted against those tendencies. After his turn to critical philosophy, he dismissed "empirical" psychology as an "obtuse 'naturalism.'" He contributed to anthropology, even aspiring to "become the German academic authority in anthropology."[4] Yet, as Zammito also notes, Kant "deviated" from the empirical pattern that dominated anthropology, especially from physiological anthropology (the study of what nature has made of man), which he considered "a sheer waste of time." In contrast to other leading anthropologists, he did not conceive of anthropology as an empirical, descriptive study of human beings; rather, he worked to make it into a universal study of human nature, derived from a "metaphysical—'transcendental'—argument about the 'fundamental grounds of the possibility [of] human nature.'"[5] In other words, he made anthropology dependent upon metaphysics rather than the other way around.

At the turn of the twentieth century the debate about the proper relationships between philosophy, psychology, and the human sciences was once again in full swing both in Germany and in other countries. The institutional, social, and intellectual contexts were, of course, very different, as was the meaning of the term "discipline," but in both debates ultimately what was at stake was the question of the nature of philosophy, its task, and its goals.[6] In the modern debates new concerns about philosophy's institutional conditions also figured importantly. In the course of these new debates Kant's strictures against psychological and anthropological approaches to normative philosophical disciplines found new currency.[7] And so did what many, then as later, took to be Kant's proposed radical cure against such evils: the expulsion of empirical psychology and anthropology from the domain of metaphysics.[8] Kant thus became an icon for one of the two opposed sides in a controversy that raged through academic circles: the controversy over what came to be called "psychologism."

The term "psychologism" lacked a univocal definition.[9] It was mostly used to indicate psychological approaches to all sorts of domains of inquiry, ranging from mathematics and sociology to the traditional normative philosophical disciplines of ethics, aesthetics, and logic. The term had mostly a derogatory connotation, and few willingly used it to describe their own approach. Explaining mathematical laws in terms of the empirical operations by means of which the human mind combined pebbles, seeking to derive the laws of society from the psychology of the individual, and basing moral principles and the laws of logic on the psychological makeup of the human species—all of these counted as examples of "psychologism" or its no less dangerous subspecies, "anthropologism." Both psychologism and anthropologism, enemies charged, produced confusion and paradoxes, and resulted in the ultimate philosophical horror: relativism.

This chapter places the controversy over pragmatist accounts of truth and knowledge in the context of broader debates about the appropriateness of psychological approaches to the philosophical subdisciplines of logic and of epistemology or "the theory of knowledge," as it was often called. We will see that, like the psychologism debate and the debates that had raged in Kant's days, the pragmatist controversy ultimately revolved around divisions over the proper relationships between philosophy and other disciplines. Debates surrounding pragmatism also intersected and combined with debates over competing approaches to psychology. Once placed in this larger context, the pragmatist controversy appears to have been not only a controversy over truth but also a controversy over the boundaries and the very nature of philosophy as an academic discipline. From this perspective James's account of "truth" turns out to have been a form of boundary work; it represented both an attempt to rethink truth, and an effort to redefine philosophy's task, by repositioning it vis-à-vis other disciplines.

This chapter also suggests that James's pragmatist account of truth was part and parcel of an ambitious project that guided many of James's activities: the creation of a new "science of man." This science of man, as others have argued, was to be theoretical and practical. It would study the "human being in the world," and aim to reach a better understanding of what it means to be human. It would also help humans transform experience and adjust to the environment in ways that would further human life. The science of man that James envisioned, I argue, drew from psychology, physiology, and biology, as well as from philosophical subfields, and could not be confined to any particular discipline. Instead, it could only be a cross-disciplinary endeavor, one that required a transgression of disciplinary divides and the cooperation of investigators working in all sorts of different fields. I suggest that James's engagement in this cross-disciplinary project resulted in a novel definition of the nature of philosophy and its goals—one that was hardly compatible with the projects that many contemporary American philosophers intended for their discipline.

## The Politics of Psychologism: German Debates

The German psychologism debate has been carefully analyzed by scholars such as Martin Kusch, Wolfgang Carl, and Mitchell Ash, and there is no need here for me to duplicate their work. Here I will briefly summarize a few moments of the controversy as reconstructed by these scholars, highlighting figures and issues that became terms of reference in Anglo-American debates over pragmatist theories of truth.

In the 1880s the German philosopher and psychologist Theodor Lipps painted logic as a "psychological discipline": logic was a "part of psychology"; it was a *"physics* of thinking."[10] By depicting logic in this way, Lipps took issue with Wilhelm Wundt who instead, in his *Logik*, had depicted logic as an "ethics of thinking."[11] Perhaps seeking to mediate between Lipps and Wundt, Christopher Sigwart, a professor of philosophy at Tübingen, presented logic as at once an "ethics" and a "physics" of thinking.[12] Like Wundt, Sigwart contended that psychology was different from logic. Yet in the end his complicated attempt to show just how they differed ended up blurring their boundaries. Both psychology and logic dealt with "thinking." Logic was the science of a particular kind of thinking, the type of thinking that is "distinguished by the consciousness of its objective necessity and universal validity." Sigwart described that "consciousness" of certainty in terms of subjective feelings and, in that way, reintroduced psychology into logic.[13]

Lipps, Sigwart, and Wundt were all accused of psychologism, and so were dozens of other psychologists, philosophers, and double-identity practitioners. Among the accused, two deserve consideration. The first, the psychologist Carl Stumpf, whom we met in chapter 1, found himself reluctantly involved in the psychologism debate. In the early 1890s he criticized psychologism, which he defined as "the reduction of all philosophical research in general, and all epistemological enquiry in particular, to psychology."[14] Nevertheless, his position could easily be construed as psychologistic. For example, he criticized Kant and the neo-Kantians for neglecting psychology and for failing to see that an account of judgments—including a priori judgments—could not dispense with a psychological study of the nature of judgment.[15] In his view epistemological claims should be subjected to "the test of psychology," and "no claim [could] be epistemologically true and psychologically false." Thus, although psychology and epistemology dealt with different problems, they were linked by multiple ties.[16]

The second example, the Vienna-based philosopher Wilhelm Jerusalem (1854–1923), instead frankly accepted the label of "psychologism" for his work. Psychologism wanted to turn logical truths into psychological truths—a project that he fully supported. For Jerusalem, philosophy could no longer aspire to the position of queen of the sciences; nor could it afford to stay aloof from the results of the sciences. His own "philosophy," he wrote, was characterized by "the empirical viewpoint"; by "the genetic method," that is, a method that attempted to derive the laws of logic (or ethics and epistemology) from a psychological study of the human mind; and by "the biological and social method of interpreting the human mind." Jerusalem's theory of knowledge challenged the transcendent and independent nature of truth, and linked it with empirical and psychological knowledge-processes.[17]

Among the opponents of psychologism, many of whom shortly afterward ended up involuntarily joining the crowded ranks of the accused, were the "big and beery" leader of the Southwest neo-Kantian school Wilhelm Windelband, his Freiburg colleague Heinrich Rickert, and Edmund Husserl.[18] Windelband objected to the genetic method. It was "hopeless" to use the genetic method in order to derive the laws of logic, since it was not possible to justify "by means of an empirical theory, what [was] itself the precondition of every theory."[19] Rickert worked to articulate a theory of knowledge independent of any ontological or empirical considerations and premised upon the notion of an absolute, transcendental norm of truth. To him "psychologistic" and other naturalistic approaches to the theory of knowledge were unacceptable precisely because they ended up dismissing the notion of absolute truth, not to mention the notions of the absolutely good and the absolutely beautiful. Psychologism made the true, the good, and the beautiful relative to time, place, people's psychological makeup, and empirical circumstances. That temporality and limitedness were precisely the consequences that Rickert aimed to defeat with his theory of values. Psychologism was a dangerously destabilizing doctrine; it challenged aspirations toward a moral and social order premised on transcendentally grounded moral values—something that the conservative and increasingly nationalistic philosopher took to heart.[20]

Husserl, like Rickert, associated psychologism with relativism. He made that case with great efficacy in the *Prolegomena to Pure Logic*, the first volume of his *Logische Untersuchungen* (1900). Husserl expressed great concern about the blurring of the divides that separated distinct fields of science. Quoting Kant, Husserl condemned psychological, metaphysical, and anthropological approaches to logic. He warned that approaches that ran together the boundaries of those sciences ended up subverting those disciplines and compromising their growth. The "mixture of heterogeneous things into a putative field-unity," which took place whenever disciplinary divides were not respected, could have "the most damaging consequences": "the setting up of invalid aims, the employment of methods . . . not commensurate with the discipline's true objects, the confounding of logical levels." These consequences were particularly serious for philosophy, since the philosophical sciences, unlike the natural sciences, could not immediately extract from experience the laws of their "territorial separation."[21]

The great mistake of psychologizing logicians, Husserl argued, was that of treating the pure laws of logic as empirical, psychological laws. Psychologistic thinkers typically failed to understand that the laws of logic (normative) and the laws of psychology (descriptive) belonged to two different categories. Husserl also stressed that the task of logic was quite different from that of psychology: logic did not inquire into the causal origins or consequences of intellectual activities but into their truth content.[22] Husserl contended that extreme psychologism, the doctrine that sought to deduce the laws of thought from the "functioning" of the mind or the "psycho-physical constitution of man," was bound to result in relativism.[23]

According to Husserl, isolated statements "of a psychologistic tone" could be found in the work of many thinkers who did not consciously espouse psychologism. That was not the case with Sigwart, whose work, Husserl complained, was entirely framed by psychologism. Sigwart, Husserl told his readers, "push[ed] psychology into philosophical channels."[24] His "psychologizing epistemology" made the truth of a proposition depend on the existence of a concrete intelligence thinking that proposition, with the consequence that propositions not yet conceived of by anyone could not be true.[25] Thus "Newton's gravitational formula was false before Newton's times," and became true only after Newton discovered it—a patent absurdity.[26] But Sigwart was not only guilty of making truth relative to time, thus losing sight of its eternal and ideal nature; by making truth dependent on actual thinking beings he also made truth a "species-relative" concept.[27] That way, Husserl charged, Sigwart fell straight into the trap of "anthropologism," a type of species relativism that made truth and the laws of logic dependent on the actual makeup of the human species.[28] Species relativism, Husserl continued, implied that angels, "logical supermen," and other intelligent beings who did not belong to the human species obeyed a different kind of logic—something absurd.[29] Judgments, Husserl continued, are "bound by the pure laws of logic," and are independent of "time and circumstances, or . . . individuals and species." That "being bound," he emphasized, was not to be understood "psychologically in the sense of a thought-compulsion," but rather "in the ideal sense of a norm." People might be "psychologically" compelled to take a proposition to be true, but that was irrelevant to the question of whether the proposition was true or not.[30]

To Husserl many of the problems faced by psychologistic thinkers derived from their failure to distinguish between the "act" of judging (a psychological process) and the "content" of a judgment.[31] The "content" was an objective proposition; it was either true or false, regardless of what people thought about it. The "act" of judging, in contrast, resulted from psychological causes, and was neither true nor false.[32] To lose sight of that distinction meant to blur the line that separated the field of logic from that of psychology.[33]

Why was the psychologism issue so pressing to German philosophers at the turn of the twentieth century? Two reasons appear to have been particularly relevant. The first, apparent in Rickert's arguments, had to do with social order. For some of the accusers, psychologistic approaches to logic and ethics compromised the values on which the order of society rested. At a time that witnessed growing social tensions and the failure of German liberalism, philosophical doctrines that made truth and the good relative to time and place appeared dangerous to society. But the psychologism debate hit interests that were even closer to home. As Ash and Kusch have shown, German-language psychologism discussions were tightly linked to questions concerning the distribution of power and resources at German-speaking universities. Not by chance, the issue of psychologism became urgent exactly at the time when German psychologists started to compete with "pure philosophers" for philosophy chairs and for sparse funds in the universities.[34]

In the psychologism debate the social question of the allocation of resources and the intellectual question of the goals of philosophy and psychology were inextricably combined. Wundt, Lipps, Sigwart, and Stumpf carried out psychological work, and occupied philosophy positions at various universities.[35] Wundt and Stumpf were willing to present themselves both as psychologists and as philosophers. Ash has shown that their career goals and attempts to carefully negotiate their double identity shaped their interventions on the issue of logic's relationship to psychology. Wundt successfully maneuvered in order to establish the philosophical significance of psychology, broadly understood, and to avoid threatening his more strictly philosophical colleagues. That is, he proclaimed the philosophical importance of psychology but was careful not to represent psychology as "part" of philosophy (or philosophy as part of psychology).[36] Stumpf—a psychologist with training in philosophy, physics, and physiology—pursued a similar strategy. He painted psychology as an important propaedeutic to the study of philosophy, and yet he was careful to emphasize that psychology could never "replac[e] philosophy."[37] The impressive career of this philosopher-psychologist, which culminated with his appointment to a prestigious chair in the Philosophical Faculty of Berlin, was due in part to his scrupulously avoiding presenting himself as an experimentalist or as a narrow specialist.[38]

On the antipsychologistic front, as Ash and Kusch have also shown, Husserl's, Windelband's, and Rickert's rejection of psychologistic approaches to philosophy went hand in hand with their concerns about the institutionalization of psychology and the allocation of philosophical chairs to experimentalists. The battle against psychologistic approaches to logical questions, Kusch argues, was primarily a battle against the confusion of the intellectual and institutional roles of the philosopher with those of the experimenter.[39] For example, in his classification of the natural and human sciences, Rickert made psychology into a natural science, expelling it from the domain of philosophy—a philosopher could not also be a psychologist, and a psychologist could not pass as a philosopher. Windelband famously attacked the experimentalists who occupied philosophy chairs at German universities: "For a time in Germany it was almost so, that one had already proven himself capable of ascending a philosophical pulpit when he had learned to type methodically on electrical keys and could show statistically in long experimental series carefully ordered in tables that something occurs to some people more slowly than it does to others. That was a none too satisfying page in the history of German philosophy."[40] Husserl made the same point. After rebuking the "specious philosophical literature"—which consisted of epistemological, logical, ethical, and pedagogical doctrines all "based on the natural sciences, above all on experimental psychology"—he complained that the philosophical faculties, under the pressure of the natural scientists, "zealously [gave] one chair of philosophy after another to scholars who . . . [had] no more inner sympathy for philosophy than chemists or physicists."[41]

Thus German-speaking antipsychologistic philosophers aimed to clearly draw the boundaries that would separate what they believed to be two very different *Wissenschaften*—philosophy and psychology. They also insisted that the tasks of the philosopher were different from those of the natural scientist and sought to dislodge psychologists—especially experimentalists—from the ranks of philosophers. Not surprisingly, Windelband, Rickert, and Husserl were among the 107 German, Austrian, and Swiss philosophers who in 1913 signed a petition protesting against "the filling of chairs of philosophy with representatives of experimental psychology."[42] The document made the case that psychology was developed enough to be recognized as an independent science and to find for itself an autonomous academic location—a motion designed to exclude psychologists from the field and the institutional structures of philosophy.

In short, while tensions over the control of the field and institutional resources certainly were not the only factor causing the German psychologism debate, as Kusch argues they played an important role in shaping the debate.

## Pragmatism and Psychologism

In 1896 William James's friend George Holmes Howison, chair of the Department of Philosophy at Berkeley, noted that until recently it was still customary for those who wrote on logic or the theory of knowledge (two endeavors between which many American philosophers did not draw any distinction) to do so from a psychological standpoint.[43] By the mid-1890s, however, many started challenging that custom, and simultaneously an Anglo-American version of the psychologism debates sprang up. The Berkeley Department of Philosophy produced its own antipsychologistic thinker, ironically a psychologist: George Stratton. Stratton advocated a separation between logic and psychology. The two disciplines, he argued, dealt largely with the same materials but had different goals, standards, and methods. Psychological descriptions of the mental processes that preceded or accompanied logical processes, and psychological explanations of logical processes (that is, causal accounts relating logical processes to natural causes) were equally irrelevant to logic. In particular, the "feelings" that accompanied a train of reasoning had no bearing on the logical issues of its "worth" and "validity." The task of logic was neither to describe nor to explain mental processes. Its task was to identify the norms of thought.[44] Unfortunately, Stratton wrote, psychologists and philosophers often lost sight of that "simple" truth, leading to an erosion of the "boundaries" between the two disciplines.[45] It was in the interest of all to set up "more definite bounds" and work out a clear division of labor.

Howison probably read Stratton's article with some dismay. A couple of years earlier, when the Department of Philosophy was planning to hire a psychologist, Howison, uncertain whether he should offer Stratton a philo-

sophical position, had gone to William James for advice. James supported
Stratton's candidacy. The young man, James felt, could potentially become one of the few people who could at once master laboratory skills
and metaphysical depth: "a tinker" and "a metaphysician in one," a "rare"
combination.[46] Stratton's intervention on the issue of the relationship between logic and psychology proved James wrong. Howison suspected that
Stratton's separatist strategy stemmed from his ill-advised dismissal of the
"old" philosophical psychology and his wholehearted devotion to the laboratory. Like Wundt, in whose laboratory Stratton spent one year, Stratton
was adamant that thought and logic were beyond the reach of laboratory
methods.[47] Howison strongly disagreed. For him, logic might well belong
within the realm of the traditional "rational psychology," the philosophical
psychology of the soul toward which Howison's own philosophy, a form of
"personal idealism," inclined. The logician, Howison claimed, "*must* take a
hand in psychology, willy-nilly, and *perforce* sin against his own canons of
division," because of the indissoluble ties linking the two disciplines.[48]

At the turn of the century and in the first decade of the twentieth century,
"rational" psychology was fading away. The question of the proper relationship between logic and psychology was framed in new ways both by people
who considered themselves philosophers, and by self-defined psychologists,
especially psychologists involved in the heated controversy over the goals of
psychology and its proper relationships with other disciplines. These issues
divided the two leading and rival psychological camps: structural psychology and functional psychology. Psychologists working in these different
traditions promoted different disciplinary configurations. Their differences
played a role in shaping both the American psychologism debate and the
pragmatism controversy.

In the decade that followed Howison's and Stratton's interventions, the
number of separatists had multiplied, even though they had not yet won
the day. For these separatists logic and the theory of knowledge could not
possibly "maintain friendly relations" with psychology, and any attempts
to do otherwise compromised the "territorial integrity" of both disciplines.
Logic was an "ethics" of thinking, psychology was a "physics of thinking."[49]
The U.S.-based antipsychologistic thinkers found ready weapons in the work
of their German counterparts, and directed them against what many of them
now took to be the most obvious embodiment of psychologism and the
clearest attempt to blur disciplinary boundaries: pragmatist theories of truth
and knowledge.

The philosopher A. E. Taylor, discussed in chapter 1, may serve to illustrate
their position. In 1905 Taylor took issue with "the traditional and persistent

modern error of regarding logic as somehow concerned with the subjective processes of cognition." Pragmatists were the main culprits. In his article "Truth and Practice," a long and tedious *syllabus errorum*, Taylor accused the pragmatists of failing to understand the difference between three fundamentally different questions: "(1) what is the meaning of the . . . predicates true and false; (2) to what propositions [are those] predicates correctly ascribed; (3) how have we come to make the ascription in any given case." The first question asked for a definition of truth. The second asked for a criterion that would allow one to correctly assign to any given proposition the predicates "true" or "false." The third was about the processes through which the actual knower comes to ascribe truth or falsehood to some proposition. The "obstinate psychologizing logicians," Taylor charged, failed to perceive that of those three questions only the third was psychological. The first and the second were "entirely extra-psychological," and it was inappropriate to deal with them from a psychological standpoint.[50]

Because pragmatists conflated those three questions, they became involved in a chain of errors. For example, pragmatists were incapable of understanding that "a theory of the steps by which true convictions are arrived at" could offer no answers to the question of what truth is and which propositions are true. Pragmatists' genetic approaches to logical questions about truth—whether based on the psychology of the individual or the human species—were entirely misplaced. The same applied to evolutionary approaches: the evolutionary processes through which humans have come to discriminate between truth and falsehood could provide neither a criterion nor a definition of truth, the same way as the evolutionary processes through which humans have come to distinguish the color red from the color blue provide no clue as to the essence of those colors.[51] Taylor argued that pragmatists were also mistaken in reducing truth to the emotions: emotions could compel someone to accept a belief as true, but they could never provide a logical justification for the belief.[52] The pragmatists' attempt to link the truth of a proposition to its practical consequences—including especially "motor changes"—was equally unacceptable, since practical considerations were "quite irrelevant" to logic.[53] Their effort to ground truth on human opinions, be they private opinions or the opinions to which "mankind will ultimately agree," was "perverse." As if that were not enough, pragmatist doctrines of truth had paradoxical consequences. Resorting to Husserl, Taylor stressed that, from a pragmatist angle, propositions such as the doctrine of the earth's motion had been false in the past and had become true later—something preposterous.[54] However, Taylor's central concern was with the pragmatist's failure to perceive that truths were logical imperatives, and had a "claim to our recognition" independent of whether anyone "thought" those truths, or whether mankind would ever come to take them to be true. (Taylor borrowed this claim directly from Rickert.)[55] Due to that failure, pragmatist accounts of truth so distorted the notion of truth that it became unrecognizable.

Taylor's antipragmatist objections were hardly original, and one can perhaps excuse James for dismissing him as a "regular little monkey."[56] Arguments similar to his peppered the Anglo-American pragmatist controversy. For example, W. B. Joseph (1867–1943), a British philosopher, accused pragmatists of failing to distinguish between the point of view of psychology and that of logic. Questioning James's claim that "truths should have practical consequences," he insisted on the need to distinguish between the consequences of a statement's being true and the consequences of believing in the statement—the latter being the reaction to which "the statement psychologically prompts the man." Echoing Husserl, Joseph accused James of failing to distinguish between "a *psychological compulsion* that drives you to think in a certain way, and a *logical recognition* that you ought to think in that way, and that others ought to, whether psychologically they are compelled to or not." Psychological compulsions, Joseph decreed, were irrelevant to truth. The confusion to which James and other pragmatists were prone confirmed Joseph's suspicion that "pragmatic writers" tended "to reduce logic to psychology."[57]

James's friend John Edward Russell, of Williams College, pointed out that, among many other errors, pragmatism conflated truth both with the situation in which thinking and knowing arose, and with the consequences to which truth led.[58] This confusion resulted from what Russell took to be the banner of pragmatism: its attempt to unite "the ruling conceptions and methods of psychology and logic."[59] Other friendly censors agreed that pragmatism's most serious problem lay in the issue of "the relation of Psychology to Logic."[60] James's former student Charles Augustus Strong, for example, could not swallow pragmatism's "psychologizing tendency." Pragmatism, he conceded, could be amended and could be transformed into a solid doctrine, but, to do so, pragmatists would have to correct its "one-sided practicalism and its . . . psychologism."[61] According to others, the pragmatists failed to grasp the notion of "independent thought," that is, thought that "in its actual exercise *takes no account of the psychological situation*."[62] Bertrand Russell stated a point that had become obvious to many: "The facts which fill the imaginations of pragmatists are psychical facts: where others might think of the starry heavens, pragmatists think of the perception of the starry heavens; where others might think of God, pragmatists think of the belief in

God, and so on. . . . Thus their initial question and their habitual imaginative background are both psychological."[63]

As in the German psychologism debate, some of the participants in the pragmatist controversy worried that pragmatism opened the door to relativism. Hugo Münsterberg, for example, drawing upon the authority of his former Freiburg colleague Rickert, stressed that the pragmatist theory of truth could not preserve the notion of absolute truth. As a result pragmatists could not even claim that their own account of truth was true. Moreover, by subverting the notion of a transcendental norm of truth to which human claims "ought" to conform (according to a categorical imperative), pragmatism gave up the idea of absolute and eternal ethical and cultural values. For modern-day relativists "eternal values should simply be explained psychologically like the fancies of a fairy-tale." "Yes," Münsterberg continued, "modern relativism," especially pragmatism, "glories in its nakedness. [It is] satisfied . . . as soon as it is demonstrated that the tastes and norms are different at various times and among various people."

Psychological, sociological, or anthropological inquiries, Münsterberg continued, could provide useful information about various societies, but they could offer no indication whatsoever as to which ethical values could really claim an eternal and universal status. Münsterberg's own tightly-knit idealistic "philosophical system" certainly did not suffer from that fatal flaw: neatly identifying two main types of values, life values and culture values, which were further subdivided into eight "classes," each of which was further divided into three parts, he pronounced the existence of twenty-four groups of values each separated from the other by "sharp boundary-lines." He took pride in the idea that his approach solidly grounded at once absolute truth, pure philosophy, and social order.[64]

The opponents of pragmatism did not necessarily share a common epistemological or metaphysical position, and politically they were committed to all sorts of projects. Nevertheless, many, though by no means all of them, found a common ground in their rejection of a type of thought that in their view conflated philosophy with psychology. These critics protested that, because of that confusion, pragmatist accounts of truth did not actually clarify the essence of truth. Pragmatists, critics famously argued, confused "what *is* true, and what is *thought to be* true."[65] Pragmatists addressed the psychological causes and the physiological processes that led people to believe an idea to be true, and described the practical consequences of truth (or of holding a belief to be true). However, according to their critics, such considerations were irrelevant to the question of "what" was "meant by 'truth' and 'falsehood.'" Simply put, to many of its critics, pragmatism did not offer a theory of truth.[66]

## Angell, Schiller, and "Psycho-Logic"

How did the pragmatists respond to the accusation of illicitly blurring the divide between philosophy and psychology? James and his allies did not deny the charge. The Chicago-based pragmatists (or "instrumentalists" as they preferred to call themselves) were explicit in approaching logic from a psychological and empirical point of view. The collective volume *Studies in Logical Theory* (1903), which James hailed as the birthdate of the "Chicago school of philosophy," presented logic as essentially tied to functional psychology. As John Dewey put it, from the evolutionary standpoint, which was central to "instrumentalism," psychological accounts of the genesis of logical concepts should be accepted by logicians not only as legitimate but also as indispensable: "Since the act of knowing is intimately and indissolubly connected with the like yet diverse functions of affection, appreciation, and practice, it only distorts results reached to treat knowing as a self-enclosed and self-explanatory whole—hence the intimate connections of logical theory with functional psychology."[67]

Dewey's colleague James Rowland Angell, whom we met in chapter 1, made the same point. In 1903 Angell argued that the way in which one related psychology to philosophy depended on the type of psychology one referred to. He considered two types of psychology: the functionalism dear to the Chicago pragmatists and the structuralism dear to Cornell psychologist Edward Titchener. Structuralism sought to analyze the mind into its elemental processes, and confined the study of the mind to the laboratory. Functionalism, instead, studied the operations of consciousness under actual life conditions, and looked at structures as inseparable from functions. Functionalism went all the way back to Aristotle, but Angell associated the origins of its modern variety with William James's evolutionary account of the mind, a theory that made consciousness into an active factor of evolution and an active participant in the organism's processes of adaptation to the environment. Functionalists sought to find out what "function" the operations of the mind performed in adaptive processes.

Angell argued that if psychology could "confine itself to the study of structures," as structuralists believed, then there would be no problem in "distinguishing [psychology's] field from that of logic."[68] That, however, could simply not be done—hence the "unchecked" and, to Angell, welcome "invasion" of ethics, logic, and aesthetics, not to mention epistemology and metaphysics, "by [functionalist] psychology." In practice, Angell continued, despite what the structuralists and many philosophers liked to think, the "boundaries" between psychology and the philosophical sciences had been "extensively obliterated." That was not at all a bad thing. In particular, functional psychology "if not stopped, must issue in a logic, an ethics, and an aesthetics," as well as in a metaphysics and epistemology.[69]

Functionalists, Angell explained, addressed logical, epistemological, ethical, and esthetic questions from a genetic, evolutionary, and practical point of view. For them truth and falsehood were defined in terms of what they did rather than in terms of what they were. From that angle, truth should be thought of as "ultimately synonymous with the effective," while error should be defined as "that form of inadequacy that issues in the failure of practice."[70] That reconceptualization of truth and falsehood as a matter of outcomes was associated with what Angell called a "practical" turn in logic. He located that turn in James's 1898 Berkeley address "Philosophical Conceptions and Practical Results," the occasion on which James revived C. S. Peirce's pragmatic principle of meaning and launched pragmatism on a new course. The new practical logic would approach logical notions from a genetic point of view and would by no means be separated from psychology. Within it, logical accounts of notions such as judgment and inference would consist mainly in psychological accounts of their origin and function.

Three years later, in his presidential address before the American Psychological Association assembled at Columbia University, Angell explicitly linked functionalist psychology to pragmatism and humanism, even though he stopped short of claiming that they were one and the same thing. (He confessed that he was weary of bringing upon himself, and upon functionalism, the "avalanche of metaphysical invective which has been loosened by pragmatic writers.") In his address he emphasized that, by focusing on the "utilities of consciousness" and its adaptive function, functionalists blurred at once the divides that separated cognition, feeling, and emotion, and the "boundaries" that, according to other psychologists, demarcated psychology from biology (as well as from physiology, neurology, and physics).[71] Functionalism and pragmatism, then, stemmed from a new biological framework, and were part of an endeavor to study mental activities, including cognitive activities, from a biological, adaptive point of view. Both clearly ran counter to all attempts to separate philosophy from psychology and biology.

F. C. S. Schiller, the leader of British pragmatism, or "humanism," and one of James's closest philosophical allies, was even more emphatic than Angell and other Chicago instrumentalists regarding the links between psychology and logic. Notorious for his lack of proper academic manners, Schiller dismissed many of pragmatism's opponents as hopeless cases. J. E. Creighton, an idealist and one of the first philosophers to launch the antipragmatist reaction, was an "ass."[72] J. E. Russell "stupidly" repeated his accusations, failing to realize that he needed to provide an argument for them. (James agreed; after unsuccessfully trying to make Russell understand his doctrine of relations, he accused Russell of "honest stupidity.")[73] Schiller found Taylor's attempt to "exclude" all "reference to psychology from logic" "reckless." Taylor's "shallowness," Schiller wrote to James, was quite astounding: Taylor's philosophy was "nowhere more than 5 ft 6 deep," but as Taylor "himself [was] but 5 ft 0, that [was] quite eno' to take him out of his depth." Yet, Schiller continued, Taylor was "enlightened, fair-minded, unprejudiced & receptive compared with the bulk" of pragmatism's enemies.[74] Bertrand Russell was an "abstraction-jiggler," an "extraordinarily, futilely, clever young man," whom Schiller could not quite get an insight into—and so on and so forth.[75]

When accused of approaching logic and epistemology from a psychologizing angle, Schiller frankly acknowledged the fact: pragmatism, and specifically his own "humanism," was indeed a form of "psychologism." "If 'psychologism' means a demand that the psychical facts of our cognitive functioning shall no longer be treated as irrelevant to Logic, it is clear both that Humanism is Psychologism, and that the demand itself is thoroughly legitimate."[76] Logic and psychology, he wrote, dealt with the same subject, albeit from two different angles and with different ends in view. Yet, "while perfectly distinct," the two disciplines were also "perfectly inseparable," and ought to work "together hand in glove": "the function of Logic develop[ed] continuously, rationally, and without antagonism, out of that of Psychology."[77]

Schiller defended the point publicly in 1906, in a symposium of the Aristotelian Society devoted to the question "Can logic abstract from the psychological conditions of thinking?" During the debate Schiller launched a psychologistic counterattack against "intellectualistic" approaches to logic, and especially "formal logic," a "calculus of imaginaries" that he found to be "as formal and as arbitrary as algebra, and not requiring to be essentially connected with any human function of cognition."[78] Both the "intellectualists" (pragmatism's chief enemies) and formalistic logicians insisted on regarding the human knower as a disembodied entity, living in some abstract realm. They viewed the laws of logic as "discarnate existence[s] . . . bearing as much relation to actual knowing and to human truth as the man in the moon."[79] To the contrary, Schiller contended, "pure" thought, thought not conditioned by psychological factors, was nowhere to be found. When completely severed from the empirical, psychological, and organic thinker, thought died and, with it, logic. There was only one way of saving logic: "The logician [must] become a psychologist as well, and drop the notion of a 'pure' and 'independent' logic. He must conceive his business strictly as the evaluation of actual human thinking, and dismiss as unscientific presumption the wild-goose chase of an 'absolute thought,' and as illusory trifling the construction of symbolic systems which cannot be applied."[80]

Schiller took the lead, proposing what he provocatively called a "psycho-logic," that is, an epistemological (rather than formal) kind of logic based on the "actual cognitive procedures of the human mind."[81] "Psycho-logic" did away with the abstraction of a pure, disembodied knower, and recognized that the interests, emotions, and purposes of the knower, far from being "psychological disturbances" distorting cognitive and logical thinking, were necessary conditions of thought. The goal of psycho-logic would consist chiefly in showing that "the most fundamental conceptions of Logic, like 'necessity,' 'certainty,' 'self-evidence,' 'truth' are primarily psychological facts . . . [and] are inseparably accompanied by specific psychological feelings."[82] Against formalism and intellectualism, then, Schiller aimed to merge psychology and logic into an inseparable form of inquiry.

## James: Psychology and Truth

Like Schiller, Angell, and Dewey, James was aware that the pragmatism controversy reflected the broader psychologism debate. He was well read in Anglo-American and German literature for and against psychologism, and was personally acquainted with several of the participants. On the German side, James read and annotated Sigwart's *Logik* (1873).[83] James was also acquainted with Lipps's and Wundt's work, and was on close terms with Stumpf.[84] James exchanged letters with Jerusalem, and appreciated the Viennese philosopher's attempts to develop an *Erkentnisstheorie* based on a study of the activity of the mind. Jerusalem, in turn, had no doubts that James was "the greatest Psychologist of our age,"[85] and that pragmatism was the "philosophy of the future." He also hastened to state that his own theory of truth and his conception of knowledge were "pragmatic," and volunteered to translate James's *Pragmatism* into German.[86]

James also carefully read Windelband's *Die Geschichte der neueren Philosophie* (1880) and *Präluden* (1884).[87] He had little sympathy for Rickert's neo-Kantian philosophy of values. He confessed that Rickert was one of the philosophers whom he had been "antagonizing" in his articles about truth (Taylor also figured in the list).[88] James took Rickert's indictment of relativism to be an indictment of pragmatism, and responded to Rickert in a chapter of *The Meaning of Truth*. There he coupled Rickert with Münsterberg, who much admired Rickert. In a letter to Münsterberg, James took the pleasure of describing Rickert's "chapter on 'Relativismus'—Relativismus being exactly identical with pragmatism in the discussion in question" as *"erbärmlich* [pathetic]" and "really *infantile."*[89] James was also aware of Husserl's work, as Husserl was of James's.[90] Husserl painted James as an "excellent researcher," and was quite intrigued with many aspects of James's psychology. James's attitude towards Husserl was more ambivalent. Rumors have it that James advised the publisher Mifflin against publishing an English translation of *Logical Investigations*. Certainly James realized that, while Husserl had not accused him of psychologism, Husserl's "Prolegomena" (1900) had become a formidable arsenal for the enemies of pragmatism.[91] For example, the widespread argument that pragmatism made truth dependent on time and circumstances was derived from Husserl, and so was the accusation that, in James's pragmatism, a proposition could be true for one person and false for another.[92]

James worked hard to clear away the paradoxes that, according to his critics, followed from pragmatism. In addressing some of those paradoxes, he resorted to strategies that had been crafted by German "psychologizing" logicians, thus implicitly taking a position in the psychologism debate. Critics objected that pragmatism implied that a proposition could be true today and false tomorrow, and that propositions not yet thought of by anyone were neither true nor false. In response, James drew a distinction between ideas (or, more generally, beliefs, propositions, and theories) and facts. Truth, he claimed, was a property of ideas; reality was a property of facts. Once truth and reality were distinguished, James contended, the paradox vanished. James probably borrowed this strategy from Sigwart or Jerusalem, both of whom had used it to fend off Husserl's attacks.[93]

Furthermore, while James felt it necessary to take care of the paradoxes apparently generated by his account of truth, he never quite denied what for many—as we have already seen—was clearly an important underlying issue: pragmatism trespassed the boundaries that, according to its critics, ought to separate philosophy from psychology. When his friend J. E. Russell accused pragmatism of failing to distinguish between logic and psychology, James did not deny the charge. Instead, on the margins of his copy of Russell's article, he jotted down: "Logical meaning, as true or untrue, seems to me to be an integral constituent of the psychological content here. You can't separate the body of the latter from its way of functioning."[94] Similarly, when former Harvard student James Bisset Pratt depicted pragmatism as a philosophy that "reduce[d] everything to psychology," James reinterpreted the objection and read Pratt as if he accused pragmatists of *reducing* truth to the purely psychological experiences of the knower. James could easily dismiss that charge.[95] He always insisted that his account of truth was realistic, since it viewed truth as "essentially a relation" between "an idea" and "a reality" external to "the idea."

James argued that pragmatism was far from reducing truth to psychological processes. That, however, was not to deny that pragmatism allowed for psychological approaches to logical and epistemological concepts, including that of truth. Pratt, James counterattacked, preferred to talk about the "abstract trueness" of an idea rather than its "concrete verifiability"; but what could that abstract trueness mean if not "something definite" in the idea that "determined [the idea's] tendency to work"? And what could that something be? James concluded that the answer to that question would have to come from "psychology" and from other empirical points of view such as those of "biology" and "biography." Only such approaches could figure out the ways in which the idea could start a chain of motor and ideational associates, and call them into being one after the other.[96]

One of the main objections against James's account of truth was that James described "how truth was arrived at," but failed to offer a theory of the logical notion of truth. As a result, critics charged, pragmatism said "everything about truth except what it *essentially* [was]."[97] According to James, instead pragmatism did tell what truth essentially was; it did so precisely by giving an account of the experiential processes, including the psychological processes, through which truth was arrived at.[98] Truth *was* the result of that essentially embodied process of knowing. Indeed, since pragmatism aimed to describe those processes, the pragmatist account of truth was "the only articulate attempt to say positively what truth actually *consists of.*"[99]

To further answer that objection James resorted to the distinction between the "ambulatory" account of knowledge, according to which true ideas enable the knower to walk through experience and reach the things to which those ideas refer, and the "saltatory" account of knowledge, according to which knowledge and reference require a "jump" from ideas and propositions to things. The ambulatory account, James wrote, described "knowing as it exist[ed] concretely"; the more abstract saltatory account, in contrast, did not take into consideration intermediate experiential steps, and thus "describe[d] its results abstractly taken."[100] According to critics, James observed, the "more concrete" ambulatory account "confound[ed] psychology with logic." When these critics searched for "the meaning of truth," they searched for a "logical relation, static, independent of time," not to be confused with "any concrete man's experience." James contested precisely that conclusion. "This," he thundered, "indeed, sounds profound, but I challenge the profundity." James defied "anyone to show any difference between logic and psychology" in that regard: "The logical relation stands to the psychological relation between idea and object only as saltatory abstractness stands to ambulatory concreteness. Both relations need a psychological vehicle; and the 'logical' one is simply the 'psychological' one disemboweled of its fullness, and reduced to a bare abstractional scheme."[101] Since logical processes required the mediation of a "psychological vehicle," James felt he was justified in characterizing his psychologically oriented account of truth and knowledge as "logic." Indeed, "in this case," as Bruce Kuklick argues, "there was no distinction between the psychological and the logical."[102]

James's conception of logic differed from those suggested by other pragmatists, since he worked to retain the "necessary" and "a priori" character of the fundamental laws of logic and mathematics.[103] In *Principles of Psychology*, James had argued that the logical principles regulating subsumption and predication, and the principles of arithmetic and geometry, did not arise from the "experience" of the individual thinker, or even from the experience of the human race, if, by experience, one meant, as Herbert Spencer and his followers did, the "external order" impressing itself on the human mind. Nevertheless, James assigned logic and mathematics (and classification) a "natural" origin. The principles underlying them, he claimed, had first appeared in the minds of past geniuses as "spontaneous" variations caused by yet-to-be-ascertained natural causes affecting the structure of the brain. Given their beneficial value to adaptation, these variations (including, for example, "the mind's capacity for discerning difference and likeness") had been selected and preserved, so that now they appeared to be "a priori" valid and necessary to any individual.[104] Although this approach differed from the one that James found predominant in evolutionary circles, it was still a biological, naturalistic approach to logic and mathematics. It rooted logical and mathematical principles in the constitution of the human mind understood as the product of evolutionary processes.

With his pragmatist account of truth, James went one step further, and linked what many considered to be a logical notion (truth) to psychological and physiological processes, "practical interests," even "personal reasons," of the individual mind. As Charlene Seigfried observes, James stressed that it was "pernicious to oppose epistemology and truth theories to 'psychological facts . . . relative to each thinker, and to the accidents of his life.'"[105] Thus James made clear that logic and epistemology, on the one hand, and the psychology (and physiology) of truth and knowledge, on the other, were hardly mutually exclusive. Logic and psychology did indeed ask different questions concerning truth; nevertheless, in the end, a logical account of truth could not dispense with psychology.[106]

## "Pragmatism begins where philosophy ends"

To return now to the question that was asked at the end of chapter 4, why did pragmatism create such a commotion, when James's and other investigators' previous psychological approaches to cognition and rationality had hardly chafed the sensibilities of psychologists and self-described philosophers? The answer now should appear clear: like the psychologism debate, the pragmatism controversy was in significant part a dispute about disciplinary divides.

As we saw earlier, psychological theories of cognition often linked truth to satisfactions, volitions, and conduct. James's account of truth was not fundamentally different, *prima facie*, from the many mid- and late nineteenth-century psychological doctrines that associated cognition with emotions and volition. These included James's own conception of cognition as part of a larger reflex-arc process and his account of the "sentiment of rationality." The difference between these earlier accounts and pragmatism lay both in the manner in which these conceptions were presented, and in the time at which they were presented. The accounts of belief, sensation, and perception that James had articulated in the 1880s, and included in *Principles of Psychology*, were marked as "psychological": *Principles* was a psychology textbook; "The Sentiment of Rationality" was a "psychological essay."[107] In contrast, the pragmatist theory of truth was clearly presented as a *philosophical* doctrine. Within the framework of pragmatism, as Kuklick argues, James's previous "psychological description spoke directly to the philosophical question of truth." James had "transform[ed] his old psychological position into philosophy." That was exactly what its censors could not swallow.[108]

The Chicago pragmatist Addison Webster Moore zeroed in precisely on that issue:

so long as the beginnings of the pragmatic movement were confined simply to a "psychological" discussion of the connection between thinking and willing, it attracted little attention. . . . So long as pragmatism was in its "psychological" stage and contented itself with observing that all ideas are purposive, we merely exclaimed, "How interesting!" But when some adventurous souls began to carry the implications of this psychology over into logic, and ethics, into metaphysics and theology, and to point out the discrepancies between it and the current conception of truth and error, of right and wrong . . . the plot began to thicken; alarms were sounded; signals were hoisted, cobwebbed armor and weapons were seized, and the most active campaign in philosophy since Kant was on.[109]

A few years later, Schiller made the same point. James's philosophy, he argued, was "contained already" in *Principles of Psychology*, but American philosophers (except for Dewey) had been "too dull to perceive" it. "They thought that as his philosophy appeared in a work called 'Psychology' it did not concern them. They praised it as psychology, for it was such good reading;

but that a good psychologist could also have been, nay, could for that very reason be, a great philosopher, never entered their heads. When James proceeded to draw out the implications of his new psychology and to set down its applications to their disputes, philosophers were genuinely shocked to find that a mere reform of psychology [James's attack against the atomism of associationism] portended a *revolution in philosophy.*"[110]

The battle over pragmatism was thus ignited because it breached the walls that many had worked to erect between philosophy and psychology. It challenged emerging assumptions about the questions that represented the proper domain of philosophy and those that represented the domains of psychology and biology. When opponents condemned pragmatism for conflating truth and knowledge with psychological processes, they worried not only about truth, or about objectivity, but also about the nature of their discipline and its boundaries, a point often missed by recent philosophical analyses but clear to James's contemporaries. Thus J. E. Creighton lamented that, by blurring together cognition and truth with volitions and actions, pragmatists failed to distinguish between philosophy and biology.[111] Columbia philosopher Frederick J. E. Woodbridge reproached pragmatism for "forc[ing] logic and knowledge to yield to treatment by evolutionary biology."[112] He insisted that the activity of the scientist and that of the philosopher should take place in two "totally different spheres."[113] W. P. Montague, in his radical attack against pragmatism's realist pretensions, identified four types of pragmatism, including a "biological" pragmatism (the instrumentalist theory of knowledge) and a "psychological" pragmatism (to which the "motor theory of truth" belonged).[114]

For many of these philosophers attacking pragmatism was part of their efforts to protect the boundaries of their discipline. They stressed that pragmatist theories of truth and knowledge were biological, physiological, psychological, or "scientific" theories but *not* philosophical theories; as such, they did not belong to philosophy. Thus, to some, James's pragmatist account of truth was a psychological theory masquerading as philosophy: James's pronouncements about truth were perfectly appropriate in a psychology textbook but clearly out of place in a philosophy journal.[115] For James's friend and old mentor Shadworth Hodgson, a British philosopher who, as we saw in chapter 1, strove to demarcate philosophy from the special sciences, "the names 'Pragmatism' and 'Humanism' alike" indicated the "partial character" of those doctrines, and betrayed their "total unfitness to be a *philosophy.*"[116] Münsterberg perceived that pragmatism contributed to (and drew from) "ethnology, sociology, biology, and psychology," approaches that, he stressed, had no bearing on "the ultimate problems of philosophy."[117] Taylor painted

James as a great psychologist, but a poor philosopher, and Albert Schinz declared that "pragmatism beg[an] where philosophy end[ed]."[118]

During the 1907 American Philosophical Association panel on truth that James had described as "abortive," the panelists voiced the same opinion. For Princeton philosopher John G. Hibben pragmatism was a "substitute for philosophy," despite the fact that its proponents insisted that it "merit[ed] the name and rank of true philosophy."[119] And Creighton, in a transparent attack on pragmatist accounts of truth, implied that pragmatism had no "genuine title to the name of philosophy."[120] In sum, pragmatism was to be excluded from the domain of philosophy because it did not engage the proper questions in a properly philosophical fashion.

## Disciplinary Politics

Instrumentalism, humanism, and James's pragmatism became prominent at a time when the relations between philosophy and psychology were particularly strained in the United States. As mentioned in chapter 1, in the 1890s and the early years of the twentieth century many American psychologists worked to redefine the field and to establish the institutional and professional structures that would make it independent of philosophy. Tensions surfaced amid maneuvering to shape the American Psychological Association, the psychologists' professional society established in 1892. As Michael Sokal has shown, in its first several years of existence the psychological society counted among its members many scholars with strong philosophical interests. In 1895 (and again in 1898), the schedule of the American Psychological Association annual meeting was dominated by philosophy papers. More psychologically oriented members complained about the presence of philosophers, finding that the philosophical papers dealt with "topics evidently inappropriate at a meeting of psychology." These psychologists complained about the society's loose membership policy and worked to create clearer disciplinary boundaries. They moved to confine philosophical papers to a special philosophical section of their annual meeting, and made plans to establish a separate professional association where philosophers could be dispatched.[121] A formal motion to this end, together with a request for stricter criteria for membership in the society, was first made in 1896, the same year the psychologist George Stratton launched his crusade against mixing psychology and logic.

These pressures prompted philosophers to establish their own professional society, the American Philosophical Association (APA).[122] As Daniel

Wilson has shown, one of the problems that the Association found urgent
was the perceived openness of philosophy, and especially that people wholly
"unschooled" in philosophy took the freedom to philosophize and to pass
judgment on professional philosophers. The APA was meant to put an end
to that situation. While the APA did not cut its ties with psychologists and
other scientific associations, the leadership was determined to clearly distinguish the roles and tasks of the philosopher from those of the psychologist,
and it worked to limit access to the discourse of philosophy only to those
with professional training in the subject.[123]

Throughout the first decade of the twentieth century, the issue of psychology's conceptual, methodological, and institutional relationship to philosophy was periodically raised by both philosophers and psychologists. As
we saw in chapter 1, that happened in 1905 at a joint symposium of the
APA and the American Psychological Association that followed the inauguration of Emerson Hall at Harvard University. There the panelist J. R.
Angell dismissed the suggestion that psychology be separated from philosophy as ill-advised as well as impractical.[124] His opponent, A. E. Taylor, who
only a few months earlier had launched his antipragmatist salvo, expelled
psychology from philosophy, and situated it in the neighborhood of natural sciences such as physics and chemistry. He confined psychology to the
laboratory and praised mathematical treatments of psychological facts (such
as those offered by Hermann Ebbinghaus) not because he was particularly
attracted to such approaches, but because they were as distant as possible
from philosophy.[125]

The following year the relationships between philosophers and psychologists grew more tense. At the 1906 annual meeting of the American Psychological Association, its council, of which the "separatist" George Stratton was
a member, moved to adopt a stricter interpretation of the society's constitutional requirements for membership. It also "recommended an amendment
to the Constitution," to the effect that they might "drop any member of the
Association who [had] not been engaged in the advancement of Psychology
for a period of five or more years."[126] Although the amendment was not accepted, and discussion of the motion was deferred until the next year (when
it was tabled), such proposals clearly reveal a strong separatist strain within
the ranks of the psychological society's leadership.

Amid tense relations, not a few philosophers embraced the rhetoric of
specialism. According to an observer at the time, they worked to transform
philosophy into a "modest special science, dealing with definite problems
and giving definite answers."[127] The strategy "of autonomy" that they pursued required giving up the idea of a philosophical synthesis of knowledge
and all grandiose plans for an architectonic philosophy that would embrace

and unify all the sciences. In return it promised to secure for philosophers a fully insulated, protected, and autonomous territory where they could conduct their investigations undisturbed.[128] Other APA members continued to advocate cooperation between philosophers and scientists. Still, more often than not, the kind of cooperation they envisaged was premised on a clear division of roles: scientists were welcome to cooperate with philosophers, as long as they would not claim that they were entitled to practice philosophy. G. H. Howison, as we saw in chapter 1, was quite clear about that: philosophers would not forgive the scientists who "trespassed" the "boundaries" of philosophy and invaded the "interior centre and citadel of the region."[129] Likewise, when Creighton lectured to the newborn APA on the importance of "intellectual contact and personal intercourse," he restricted the scope of such beneficial interaction to "fellow workers" in the field of philosophy.

Even some of those who cultivated a role as dual practitioners were clear that, while one could switch back and forth between philosophy and psychology, one was not allowed to practice one's trade by mixing the two. Hugo Münsterberg belonged to this group. He is sometimes portrayed as a person who, like James, blurred disciplinary divides. However, although Münsterberg made the case that the Harvard psychological laboratory must be physically located in the new philosophy building, and argued that psychologists needed philosophical training, he would not have been comfortable being described as a breaker of boundaries. As seen in chapter 3, for Münsterberg psychology, as a science, and philosophy (which he largely identified with metaphysics and a theory of values) were radically different endeavors. When one put on one's scientist's cap, one was committed to an impersonal and mechanical way of thinking that allowed no room for values and teleological considerations. When one put on one's philosophical cap, one abandoned the impersonal way of thinking, and embraced values and teleology. Both activities were completely legitimate. One could be both a philosopher and a psychologist, but not both at the same time, for the ethos of science and that of philosophy were radically different.[130]

Pragmatism, humanism, and instrumentalism endangered all of these plans to defend the separate integrity of the philosophical territory. Schiller, never one to pass up a good fight, brought the point to everyone's attention. Those who opposed pragmatism, he surmised, aimed to carry out a broader philosophical attack against psychologists' encroachment on philosophy. Pragmatism's enemies, Schiller wrote, simply aimed at "abolishing Psychology" and ejecting psychologists from the field of philosophy—a goal that, Schiller suspected, "all lovers of philosophic sport" would be "cordial[ly]" interested in.[131] "Humanism," instead, worked the other way around. While his opponents aimed to fence off a neat sporting field, Schiller made it clear that his open-armed embrace of psychologism was part of a strategy intended to disconcert philosophers and to subvert the rules of the professorial game. He specifically rejected demands for a clear division between the roles of professional philosophers versus those of nonphilosophers (including psychologists). "The study of knowing," he stated, "cannot be partitioned out between the sciences of Logic and Psychology. . . . [It] is one, whether it be called Logic or Psychology."[132] His "psycho-logic" was patently an interdisciplinary enterprise, one that grew out of psychology and required competence both in philosophy and in psychology.

James too made it clear that rigid disciplinary divides had no room in his expansive conceptions of philosophy and psychology. Twice president of the American Psychological Association (in 1894 and again in 1904), he addressed *psychologists* with papers devoted to topics rich in *philosophical* implications.[133] As we discussed in chapter 2, the 1894 presidential address ("The Knowing of Things Together") dealt with the problem of the synthetic unity of knowledge. This paper gave rise to much comment from philosophers (or practitioners of a philosophical psychology, especially James's colleague Josiah Royce and the British idealist Francis Bradley). However, psychologists of a younger generation dismissed it. Months after James's 1894 presidential address, the experimental psychologist E. W. Scripture, whom we met in chapter 2, argued that issues such as the "unity of consciousness" did not, at bottom, differ very much from the question "How many angels could dance on the point of a needle?" Suppose one proved that "$19^1/_2$ angels can dance on the aforesaid needle," Scripture wrote, "well, what of it?" Even if one could demonstrate that consciousness was a unity, or a double, or whatever one pleased, one had made no step forward in the solution of practical problems of humanity—an all-important goal for self-professed "scientific psychologists" like Scripture. To them James's 1894 address was as unfit to be part of a scientific psychology as the metaphysical or theological questions debated by the scholastic philosophers.[134]

In turn, when James put on his other hat, such as in 1906 when he acted as president of the American *Philosophical* Association, he did just the opposite. In his presidential address before the philosophical society, "The Energies of Men," he provoked his philosophical audience by inviting philosophers to concentrate their efforts on a *psychological* and *physiological* problem—the creation of a "science" of mental energy.[135] Like the problems that James had selected for his presidential addresses before the *psychological* association, this was obviously hardly the type of core problem that would help build a clear-cut divide between philosophy and psychology.[136] Indeed, as we will see in chapter 8, James's choice of topic intended to do just the opposite.

Thus from the mid-1890s on, both as a psychologist and as a philosopher, James worked to prevent the consolidation of disciplinary boundaries, and encouraged the exchange and transfer of notions and data across disciplinary divides. By transferring biological, psychological, and physiological notions into the innermost core of philosophy, into the key philosophical notion of truth, James's pragmatism contributed to those same ends. James did not hide the fact: his account of truth, he stated, was at once genetic, psychological, and strictly philosophical, even "logical."[137]

## Philosophy and the "Science of Man"

Why in developing his account of truth—a concept central to philosophy—did James so insistently trespass the boundaries that others believed should separate philosophy from psychology, biology, and physiology? My answer, expounded in chapter 7, will ultimately link James's intellectual endeavors with his vision for the proper community of inquirers and for society. I will suggest that the function of the philosopher, for James, included not only the formulation of technical philosophical theories but also the task of organizing and reordering the modes of social engagement governing interaction among knowledge producers and between intellectuals and other social groups. Here I begin exploring how James's pragmatism promoted those goals by revisiting the question of the nature of the relationships between philosophy and the "science of man."

Charlene Seigfried, whose work has been seminal to James studies, and more recently Sergio Franzese have claimed that a central goal of James's many activities was the formulation of a "natural history of man" (Seigfried) or a "science of human nature" (Franzese). This goal, according to these scholars, remained central to James's understanding of philosophy. According to Seigfried, James turned away from a notion of philosophy as the a priori construction of a purely intellectual, closed system of ideas. Instead, he radically "reconstructed philosophy" by rooting it in the concrete experience of human beings in the world. James, she argues, made philosophy into a study of the "experiential structures" that characterized "the irreducibly human modes of encountering the world." At the same time James strove to begin philosophy "anew" by making it into a reflective endeavor to "harmonize self and world in pursuit of a better future," and to offer a "synthesis and reconciliation of all human motives, interests, and desires."[138] In other words,

Seigfried argues, James redefined philosophy both as a concrete analysis of human experience and as a *practical* tool for the transformation of experience into a world that humans would find most satisfactory. Philosophy, then, would help advance human life.

Franzese locates James's manifold activities in the tradition of a "philosophical anthropology" that he traces back to Kant's "pragmatic" anthropology and to Hermann Lotze. This tradition, which continued in the twentieth century, aimed to articulate an "integrated" image of "man"—one that would encompass "all human functions, higher and lower, organic and spiritual," as "activities proper to the [human] . . . mode of existence."[139] Kant's pragmatic anthropology, he notes, concentrated on the "opposition between nature and culture," and framed the problem of human freedom by examining "the relationship between what nature makes of man [the subject matter of physiological anthropology] and what man makes of himself as a free actor [the subject matter of Kant's anthropology]." Since man was comprised of both nature and reason, freedom consisted precisely in man's ability to "transcend his natural side by means of reason" and thereby to create a culture, and "impose that way his control on nature" through a process that enables man to become a "person."[140] Like Kant, Lotze emphasized the distinction between a world of nature and a world of culture. He too strove to articulate an image of human nature that did justice both to the human intuition of (or desire for) freedom and to the sense of teleology. In addition to examining what "man can make of himself," Lotze also explored the place of humans in the "wonderful Automata of Nature."[141] In a decidedly spiritualist study of human nature, he resorted to the tools offered by "medical psychology" and "physiological anthropology"—an endeavor for which, as John H. Zammito has shown, Kant had little patience. In that way, he strove to create a science of man that would explore both what nature makes of man, in the tradition of physiological anthropology, and the world of culture.

The project of articulating a broad science of man was indeed central to many of James's endeavors, including philosophy, which James had defined in the mid-1870s as "the reflection of man on his relations with the universe."[142] Indeed, James's commitment to such a project shaped not only his understanding of the methods and subject matter of philosophy but also some of his philosophical doctrines and ultimately his conception of the nature of philosophy. This section illustrates that James's pragmatist account of truth reflected his conception of proper inquiry into human nature and, further, that James's pragmatism linked philosophy to a science of human nature in multiple ways. The result was a forceful reconceptualization of the task of philosophy and its relationships with other fields of inquiry.

## Truth and the Science of Man

As James scholars have long seen, James looked at human beings as organisms acting in and reacting to an environment. The new "science of man" that, as discussed in chapter 2, he advertised in the 1870s, and then again in the late 1880s and the 1890s, was based on an essentially biological "conception" according to which human activity belonged "at bottom to the type of reflex action."[143] "The human individual," in other words, was "an organized mass of tendencies to react mentally and muscularly on his environment."[144] James emphasized that the reflex-arc process always ended "in some activity." Because of this, the new "biological inquiry into human nature," and with it the conception of psychology as a natural science that James promoted, tended to "treat consciousness . . . as if it existed only for the sake of conduct." Both endeavors focused on the "practical utility" of mental life in the process of adjustment to the environment. And both were to contribute not only theoretical innovations but also practical applications and rules of action that would help people "predict" and "control" mental states and behavior.[145] The science of man that James envisioned was indeed meant to be a pragmatic science.

The study of human nature that James proposed was to be essentially different from the previous, all-embracing philosophical science of human nature that some British and American moral philosophers had worked to assemble earlier in the century. At the turn of the twentieth century, the goal of offering an all-encompassing philosophical "science of man" appeared no longer to be feasible. It just did not seem to fit into the framework of a now more narrowly defined philosophy, nor did it seem that such an all-embracing project could be confined within the boundaries of any other special science—despite the fact that a few investigators worked to reconfigure disciplines such as sociology and psychology to take on the function of a broadly understood science of human nature.[146] To James, disciplinary fragmentation and isolation threatened not only to produce a splintered image of the human being but also seemed to compromise the whole project of a science that would capture concrete human experience and give an account of the human mode of existence. James sensed that the human and social sciences, precisely because they dealt with a being whose essence and activities straddled the divides between nature and culture, implicated each other and needed to cooperate with the natural sciences, and that any attempt to create barriers was artificial.

I suggest that the science of man and his "relations to the universe" that James envisioned could only achieve its goal by being cross-disciplinary.

It could not be carried on by isolated investigators; instead, it required exchange and cooperation across emerging disciplinary boundaries. James had made this point as early as 1892, when he promoted the new "biological," "naturalistic," study of human nature and its closely allied discipline of psychology understood as a natural science. He had then praised the heterogeneous "band of workers"—including "men of facts," "laboratory workers," "physiologists," and "biologists," as well as "practical men" such as doctors, clergymen, educators, jail wardens, asylum superintendents, and naturalists—who, "full of enthusiasm and confidence in each other," worked at a common project of a science of human nature.[147]

James's pragmatist account of truth fits tightly into the broader project and pattern he charted for creating a science of human nature describing the relationships that humans have to the external world, and human action in the world. Reflecting that focus, as we saw, James's pragmatism looked at truth processes as involving motor discharges, even as chains of reflex-action processes finalized to action. It depicted "true thoughts" as "invaluable instruments of action," and studied truth in the context of a study of human action, linking it inextricably to human purposes, interests, and needs.[148] James's pragmatist account of truth aimed to unveil how "truth function[s] in actual situations," and how true ideas help humans better act in and re-act to the environment.[149] Furthermore, as James Conant notes, James's conception of truth was also itself "a guide for action." It was not only, or not essentially, a "theory" of truth "but rather a proposal concerning how we should lead our lives."[150] In short, James's conception of truth was part and parcel of an expansive pragmatic inquiry into human nature. And it reflected the cross-disciplinary nature of that kind of inquiry. Thus it was no accident that in the course of framing a philosophical account of truth, James stomped back and forth through the fields of biology, psychology, and physiology.

## Philosophy and Temperament

One of the aspects of James's pragmatism that scholars have found most striking, even "shocking," is his so-called temperament thesis, namely, in Hilary Putnam and Ruth Anna Putnam's words, the claim according to which "the decision we make on any metaphysical question . . . is and ought to be a matter of 'temperament.'"[151] In *Pragmatism* James famously associated a person's philosophical choices with the kind of person one is: a person's *Weltanschauung* is, in no small part, a product of one's temperament. Temperament gave the professional philosopher "a stronger bias than any of his more objective premises," and was indeed "the potentest" of all our premises in philosophy.[152] As a result, James continued, the whole history of philosophy boiled down to the history of a clash between two conflicting types of temperament: on one side stood the "tender-minded" philosophers, religiously inclined minds who were prone to draw upon absolute a priori principles and frame rationalist philosophical conceptions; on the other stood "tough-minded" investigators, who, in contrast, reveled in empiricism, privileged facts above principles, and were irreligious. James presented pragmatism as a kind of philosophy that would appeal to an intermediate kind of temperament, one that he felt was more closely associated with "healthy human understanding."[153] But what did James mean by temperament? Answering that question will reveal that James linked not only pragmatism but the whole discipline of philosophy to the science of man.

James never explicitly defined the term "temperament." Nevertheless, he used it in ways that are consistent with the features that many contemporary psychologists, clinical psychologists, "characterologists," and students of physiological aesthetics ascribed to the notion of temperament. Like them, James firmly embedded temperament in the human body: temperament was part of the "congenital human constitution," and it was rooted in the nervous system, even in "neural tissues."[154] More precisely, temperament, for James, functioned as a mediator in complex human responses that he interpreted (consistently with the general approach in *Principles*) as higher-level reflex-action processes. Temperament, in short, was a factor that contributed to shaping (and characterized) an individual's reactions to given situations. It was a cluster of emotional and passional tendencies, including aesthetic passions, all of which, James emphasized, either consisted of, or required the presence of, reflex-arc processes taking place in the body. Like other psychologists and men of science who reflected on temperament, James emphasized that the reactions mediated by temperament could sometimes be unconscious. Yet they were not rigidly predetermined but allowed for freedom and choice—a result that followed from the great variability and indeterminacy that James was able to build into his three-stage model of the reflex-arc process.[155]

Reexamining the temperament thesis in light of this conception of temperament highlights the fact that, by rooting philosophical choices in temperament, James portrayed the activity of philosophy as a personal "reaction" to the universe. Indeed, for James philosophy was the expression of a person's "total reaction against the presence of the universe." That reaction, as James stated in *Pragmatism*, was to a large extent a product of temperament, which acted as a filter, sorting out those facts and relations that an individual felt belonged to the universe. Like other temperamental reactions, those that lay at the basis of a person's philosophical choices were often of an "unverbalized," "inarticulate," even "half unconscious" kind.[156] A person's sense of the universe was responsible for his or her acceptance of certain beliefs (including many philosophical, scientific, moral, and religious ones), which only later he or she would try to justify by appealing to logic and reason.[157] Not only was philosophy a reaction to the universe; it also involved an attempt to reframe the universe in ways that appealed to a person's innermost emotional constitution. This included a person's aesthetic tendencies, and throughout his work James stressed the role that aesthetic tendencies played in shaping a person's philosophical choices. Moreover, James compared the creation of philosophy to the creation of works of art.[158] Both pursuits, as James and many turn-of-the-twentieth-century proponents of the new discipline of physiological and psychological aesthetics argued, stemmed from aesthetic passions, sometimes unconscious ones. These passions were inscribed in the emotional and physiological constitution—the temperament—of the artist and the philosopher, and were constitutive of their mental organization.[159] The comparison between the creation of philosophy and the creation of a work of art—"philosophers paint pictures," he once said—reinforced the role played by temperament in shaping a person's philosophical choices.

The temperament thesis had important implications for James's understanding of philosophy.[160] Not only did it import another physiological-psychological concept—that of the temperament—into philosophy; it also conveyed the point that, in order to understand their own distinctive activity (philosophical activity itself), philosophers needed to resort to the insights offered by biologists, physiologists, psychologists, even by students of aesthetics (especially physiological aesthetics), and all the scholars who studied the nature of temperament and its role in creative processes. That meant that philosophical self-reflection could not be accomplished by the philosopher who was holed up within a narrowly defined, "pure," discipline of philosophy. This was a point that contemporary philosophers could not have missed, and one that some of them did not appreciate. James's ally Giovanni Papini made the implications of such a position clear for all. Since philosophy constituted a "psychological document" of the individual philosopher—a "sentimental, vital, reaction" reflecting an individual philosopher's way of life—philosophy would best be replaced with a "very personal" psychological study of the constitution of individuals' minds. Papini labeled this new science "egology."[161] Its goal would be to manufacture the specific kind of philosophy that perfectly fit a particular individual's temperamental and psychological inclinations. Its practitioners would then become akin to merchants selling commodities in the marketplace, the "bazaar" of philosophy. From such a perspective, philosophy had to forsake its pretensions to being a depository of absolute insights about truth and other matters of eternal importance. Instead, it would either have to disappear from the map of knowledge (as Papini recommended) or accept that it was simply an instrument that would help different individuals cope with the universe and intensify their lives.

If philosophy was to contribute to and draw from the cross-disciplinary science of human nature, it was also meant to play a special role in making such a science possible. Philosophy, as we will see in chapter 7, should first and foremost ensure that an open cross-disciplinary discourse could take place. This task led James to redefine "general philosophy," and make it into a kind of activity that would enable the emergence of collaborative communities of inquirers. In such communities people coming from different disciplines, professions, and callings could engage in cross-divisional conversation. Just such communication would allow for the creation of an integrated science of human nature. At the same time, as we will also see, James redefined the persona of the philosopher, the truly philosophical self. As he more clearly framed this broader role for philosophy, he simultaneously ascribed to philosophers—indeed demanded of them—the sensibilities, mental, and social attitudes that would enable them to act as go-betweens, boundary breakers, enabling communication and exchange across disciplinary and other types of divides.

The last two chapters will also explore the relationship between James's understanding of the task of philosophy and his understanding of the role of the philosopher in order to illuminate the centrality of the practice of breaking boundaries to both. However, to provide a fuller picture of James's evolving views on these central issues of his mature years, it is necessary to examine the conception of the self that James offered in those years, for his understanding of the human self was intimately connected to the kinds of communities, philosophical self, and modes of social engagements that James promoted. This will be our task in the next chapter.

# 6

# Ecstasy and Community
## James and the Politics of the Self

He realized, as every hireling must, . . . that he belongs to another, whose will is his law.

W. D. HOWELLS

The opposition in human nature of the two ideas of solidarity and personality may be . . . illustrated by describing as an expression of the former the sense of the sublime, of the grand, of . . . the instinct of infinity, and on the other hand as an expression of the personality, the desire of being circumscribed, shut in, and bounded, the aversion to vague limitations, the sense of coziness . . . or what may be called the instinct of finity.

EDWARD BELLAMY

## The "Crisis" of the Self and Some Uses of It

William Dean Howells's novel *A Hazard of New Fortunes*, published in 1889, enjoyed tremendous success. Critics praised the author's social vision of "humanitarianism and co-operation."[1] After reading it, Howells's friend William James reported that he could "hardly recollect a novel that ha[d] [so] taken hold of [him]": *A Hazard of New Fortunes* was a "d-d humane book."[2] Set in New York, the novel probes the relationships among a group of people engaged

189

in the publication of a new magazine. Each of the male characters struggles to preserve a sense of selfhood and self-mastery. The owner of the magazine, a natural-gas millionaire, belongs to the category of men "who have made money and do not yet know that money has made them."[3] The literary editor, a middle-aged man who took on the job in hopes of furthering his literary aspirations, becomes increasingly aware of a loss of self-direction. As he tries to comply with the whimsies of the owner of the magazine, he quickly realizes, "as every hireling must," that he is a puppet in the hands of "another, whose will is his law."[4] The selfish and ambitious artistic editor quickly loses his self-confidence and self-respect as he comes to recognize his profound lack of authenticity. Reflected in the critical eyes of a woman who does not reciprocate his love, he perceives the splintering of his self into a multiplicity of conflicting social masks.[5]

Quickly a best-seller, the novel chronicled the weakness and divisiveness of the self and the profound erosion of the conception of selfhood that had once structured social and economic activities and individuals' self-perception in antebellum America. Howells's diagnosis was unambiguous. The crisis of the unitary and sovereign self was a byproduct of industrial capitalism, an economic order that deprived many people of the conditions that throughout the nineteenth century had been associated with citizenship and selfhood: ownership of the means of production or one's labor.[6] Only one solution was left—to relinquish at once the laissez-faire economy and the illusion of the self-directed simple self.

At the start of the twentieth century, anxiety concerning the erosion of the unitary and masterful self was widespread in America.[7] According to T. J. Jackson Lears, not only intellectuals and middle-class people, but also artisans and workers who lost their craft identities, sometimes experienced an "uncanny sense of unsubstantiality" or a disturbing sense of fragmentation.[8] In what appeared to be the absence of a strong unifying principle, the self could splinter into a cluster of contradictory social roles or into a series of inconsistent behaviors. Mental physiologists revealed that heredity, instincts, and reflex-arc automatisms controlled many acts previously thought to be controlled by consciousness.[9] Rapidly multiplying cases of pathological or artificially induced "dissociation" (split personality, hysterical symptoms, or posthypnotic states), as well as states obtained by means of occult practices (automatic writing, trance, and projection of the double), displayed a self that was split by deep fault lines and appeared to be at the mercy of powers sometimes perceived to be "alien" to the personality of the subject experiencing those conditions.[10]

According to some historians, these phenomena contributed to bringing about a crisis of the self at the turn of the new century. In response to this perceived "collective crisis of identity" scores of moralists and preachers taught others how to regain self-mastery and wholeness in a new social and economic order, which they believed, deprived people of autonomy, inner unity, and individuality.[11] To other commentators, however, as James Livingston has shown, the decline of isolated individuality appeared to open up the possibility for a full socialization of life and for new forms of cooperation.[12] Both social actors who saw industrial capitalism as an end in itself and those who perceived it as one stage in the transition to socialism gladly gave up the burden of the individuated well-bounded self and explored new forms of subjectivity. They relocated agency from the individual to the social group and spread the self over social networks, depicting it as a "permeable entity with indistinct boundaries."[13] Among these theorists of the "social self," some resorted to the language of sociology and conceptualized the self as a product of associations, even of "social institutions."[14] Others instead cast the "social" self in decidedly religious, even mystical, frameworks. They linked the overcoming of what the social visionary Edward Bellamy identified as an "instinct of finity" or of personality, and the prevailing of the opposite instinct of "infinity" and "solidarity," to experiences of ecstasy and mystical unification. In such states, those mystical writers revealed, individuals could step out of the confines of their individualities. They could then participate in the life of a larger truer self and sympathize with their fellow human beings in new ways.[15] Ultimately, both the theorists of a secular social self and the more mystical or religious writers perceived that the inner division of the self and its loose boundaries made the human being intrinsically social.

This chapter revisits the question of the social valence of William James's account of the self by locating it in these realms of discourse and practice.[16] James worried about the leaky boundaries of the self, as well as about the lack of self-mastery, general weakness, and "lack of inner harmony" of the modern self, which he diagnosed as effects of the hectic life prevalent in modern America.[17] He found troubling symptoms of those conditions in himself: in his fear of becoming insane, for example, in the antagonism among his various social selves,[18] and in his early bouts of depression and, later, neurasthenia.[19] Nevertheless, like William Dean Howells, a close friend, and Edward Bellamy, a writer whom James much admired, James perceived that the breakdown of the autonomous, well-bounded self opened up new possibilities both for the individual and for society. He also realized that the crisis of the traditional self and the new social order made it necessary to rethink the relationship between the individual and society. As did many of his contemporaries, James addressed what he perceived to be a fundamental tension: that between the claims of society and those of the individual, between a new tendency toward a full socialization of life and an individual's desire to retain autonomy and moral agency.

James scholars have fully explored the issue of James's political orientation and the linked question of his reaction vis-à-vis the shift from proprietary capitalism to "corporate" capitalism. Some find that James resolutely opposed capitalism and its institutions, whereas others conclude that James, like other pragmatists, created a "framework" for the "acceptance" of corporate capitalism.[20] Despite their widely diverging conclusions, the work of these historians is important and innovative. However, I suggest that ultimately the terrain on which James addressed the all-important issues of the relationship among individuals in society, and the autonomy of the individual vis-à-vis ever more powerful social and economic institutions, was not primarily that of politics or political economy. It was instead that comprised of psychology (normal, "abnormal," and "supernormal"), metaphysics, and mysticism. Located at the intersection between those fields of inquiry, and drawing on medical, religious, and hygienic techniques of the self, James's account of the self, I suggest, represented a contribution to his broader project of a pragmatic "science of man"—a multi- and cross-disciplinary kind of endeavor that aimed, as we saw in chapter 5, not only to explore human nature but also provide concrete tools that people could use to improve themselves and their living conditions.

After a quick review of James's political views and the topology of the self that he delineated, this chapter makes two main claims. First, the Jamesian self and the techniques of self-cultivation that James promoted, particularly techniques for the unification of the divided self, were instrumental to the creation of a strong citizenry that could participate in political action and initiate effective social change. Second, James redefined the boundaries separating the individual self from society, and those separating different individuals within society. Especially in the last, politicized decade of his life he envisioned an open self surrounded by uncertain and leaky contours. The permeable boundaries of the individual self made it possible to imagine a new type of social interaction—one essentially different both from the intersections of the isolated trajectories of the economic individuals of classical liberal thought, and from the depersonalizing relationships that James believed characterized capitalistic institutions. The social interactions that he envisioned were rooted in intimacy and solidarity. James believed that they would allow for a deeply communal life fully compatible with the claims of individualism.

## James in the Political Spectrum

James was a politically engaged thinker even though the exact nature of his political vision is difficult to capture. He sometimes described himself as a "mugwump," locating himself among those who in 1884 bolted from the Republican Party, condemning in the name of civic virtue the presidential nomination of the "corrupt" James G. Blaine.[21] The Mugwumps, Robert B. Westbrook argues, perceived themselves as individuals endowed with a superior culture, character, and moral sensibility, and believed that "they were entitled by virtue of these credentials to political leadership." With them James, who came from a family of "inherited wealth," shared a concern that democracy might turn wrong, and a desire to steer it along safer lines by placing government and the choice of political leaders in the hands of an educated elite.[22] With the Mugwumps James fought his main (according to some, his *only*) political battle: a passionate struggle against the new imperialistic turn taken by the United States in the second half of the 1890s.[23] In those years James grew tremendously concerned about the U.S. interventions in Venezuela, Cuba, and the Philippines. When the United States invaded the Philippines in the wake of the Spanish-American War, he vigorously protested. In solidarity with other eminent Mugwumps and the newly founded "Anti-Imperialist League," James passionately denounced the annexation of the Philippines as "the most incredible, unbelievable, piece of sneak-thief turpitude that any nation ever practiced."[24] With that act of "piracy," James wrote, the United States had "once for all regurgitated the Declaration of Independence" and betrayed the "old American soul."[25]

James, however, was a "singular mugwump," and his political self-definition leaves considerable room for interpretation.[26] James T. Kloppenberg sees James as a proto "social democrat," but observes that the "traces of James's political preferences are too faint to provide more than a tentative outline of his ideas."[27] Others scholars instead ascribe to James more precise political sympathies. For some, he was committed to "radical participatory democracy" and to "communitarian liberalism," while for others James supported "populism" and "petty-producerism."[28] Deborah J. Coon, whose work has done much to unearth James's political engagement, argues that chiefly in response to mounting American imperialism James became an anarchist. James confessed such feelings to William Dean Howells, revealing that, in the face of recent events, he found himself to be growing "more individualistic," even "anarchistic."[29] Coon argues that such claims must be taken at face value and documents James's intellectual affinity with a tradition of "communitarian anarchism."[30] She also stresses that James's opposition to American imperialism was often expressed in terms that suggested a parallel hostility to the institutions of capitalism. To James the rhetoric of "big national destinies" deployed by McKinley and Roosevelt to justify the annexation of the Philippines deprived the Filipinos of their just aspiration to "self-control" and self-government, in the same way as "trade-combines" and "department-stores" threatened the self-directedness of the individual.[31] James's anarchism, Coon argues, stemmed from his passionate defense of self-governance both for the individual and for ethnic groups.

While for Coon and others James strenuously resisted capitalism and the rational bureaucracy of the corporations, some scholars alternatively depict James as an ally of capitalism. Lewis Mumford started that trend in the mid-1920s when he accused James's pragmatism of "acquiescing" to modern industrialism and the world of finance. For Mumford, James's pragmatism emanated the unpleasant "smell of the Gilded Age."[32] More recently, making a virtue of what for Mumford was a sin, James Livingston praises James and other pragmatists for creating a "frame of acceptance" for "corporate capitalism," a hybrid form of capitalism that embraced at once both the older proprietary capitalism and socialism.[33] Livingston argues that James and his pragmatist friends plotted a path that enabled them to "navigate" the transition "from proprietary to corporate capitalism." Since these thinkers "recognized" that "the development of capitalism" created the "necessary condition of a passage beyond class society," Livingston concludes that James was a "socialist."[34]

These divergent conclusions suggest that, though it is an important exercise, the task of pinpointing the exact nature of James's political affiliation may ultimately elude us. However, despite disagreements with regards to that specific issue, James scholars have found common ground. Most commentators, in fact, agree that James's much-celebrated "individualism" was tempered by a complementary emphasis on solidarity and community.[35] In one of James's own favorite works, an address he presented to student audiences, he stressed that when we look at other people from the position of the "external spectator," as we ordinarily do, we are bound to remain blind to the inner significance of their lives.[36] That "ancestral blindness" was the source of many conflicts, including the mounting tensions between labor and capital. James (notoriously) ascribed these tensions, in part, to the inability of workers and capitalists to "sympathize" with the point of view of the other. Yet, James continued, sometimes a vision of the inner secrets of other people's lives comes on us suddenly, as in a mystical revelation. In these sudden experiences we step out of ourselves, away from our external point of view, and become able to commune with a larger life: the life of the universe, the life of nature, or the life of other people. From this displaced, ecstatic position we become able to intimately appreciate other people's ideals and feelings, and feel a deep sympathy for them. In such moments, James continued, the self "is riven and its narrow interests fly to pieces."[37]

In another address to students, James confessed to have experienced one such sudden sympathetic "flash of insight." One day he was traveling on a train toward Buffalo lost in his thoughts when suddenly "the sight of a workman doing something on the dizzy edge of a sky-scaling iron construction" brought him "to [his] senses": "I perceived, by a flash of insight, that I had been steeping myself in pure ancestral blindness, and looking at life with the eyes of a remote spectator." He suddenly realized that the lives of laborers struggling to build a railway bridge or a fire-proof tower, or toiling "on a freight-train, on the decks of a vessel, in cattle-yards and mines," were replete with courage and meaning. The "scales fell from my eyes," he continued, and "a wave of sympathy greater than anything I had ever before felt with the common life of common men began to fill my soul."[38]

To be sure, James's newly acquired sense of vision retained some short-sightedness. Despite the sudden revelation that hit him on the train, James continued to be remarkably blind to the inner meanings of other people's lives, especially those of the workers. As George Cotkin notes, workers "fell from [James's] pantheon of true heroes . . . because they lacked ideality" and because "they selfishly desired material comfort and security."[39] Not surprisingly, some commentators of the time, including John Dewey, depicted James as "an aristocrat" who had "no real intimation" of the labor problem.[40] These limitations, however, did not prevent James from drawing an honestly meant social lesson from his discussion of "ancestral blindness." It was a lesson of democratic tolerance, respect for individuality, and noninterference with other people's "own peculiar ways of being happy."

James made this lesson central to his "pluralistic, individualistic philosophy."[41] Yet implicit in his discussion was another social message: an invitation to go beyond "tolerance" and to practice a form of solidarity and intimacy. James was inviting his student audiences to sympathize with other people and engage with them in more intimate ways.[42] That call for empathy was central to James's anti-imperialism. As Robert Beisner observed, while other anti-imperialists emphasized the economic circumstances that backed up U.S. expansionistic politics, James psychologized imperialism. American imperialism for James stemmed from a predatory "war" instinct that was intrinsic to human nature, combined with the staggering inability of American politicians to imagine in sympathetic ways the inner lives and ideals of the Filipinos. American leaders, James complained, had framed the relationship between the United States and the Philippines as a relationship between "two corporations": "a big material corporation against a small one." Such business relationships, James emphasized, were purely "legal" relationships; they excluded a priori a consideration of the minds of the people involved and resulted in the inability to consider the Filipinos as "psychological quantities."[43] To remedy that situation, James invited American politicians and his fellow citizens to handle things "psychologically" and to try to "connect with the 'Philippine soul.'"[44]

How does James's invitation to deep mental connection, sympathy, and solidarity square with his self-description as an individualist? Today, from the vantage point of new forms of solidarity and cosmopolitanism that, as David Hollinger suggests, enable people to reconcile loyalty to larger social groups with loyalty to their own individualities and personal perspectives, it may be hard for us to see that these two goals might have appeared antagonistic at the turn of the twentieth century.[45] And yet many philosophers, psychologists, psychiatrists, and biologists of the time found that individualism did not easily square with solidarity and cooperation.[46] James himself, as we will see, was concerned that many of the available plans for solidarity and altruism required the annihilation of the individual. That has led some scholars to argue that James's individualism implied a denial of "community,"[47] and others to conclude that, because of his strong emphasis on the sociability of humans, James was no individualist. How then did James approach the tension between individualism and communitarianism?

Looking at the topology of the Jamesian self, at its inner structure and the nature of its boundaries, will help us discern some answers to those questions.

### Topologies of the Self: (1) The Divided Self

In common with other psychologists of the time, James challenged the dogma of the unity and simplicity of the self. In *The Principles of Psychology* (1890), the only text in which James ever dealt systematically with the notion of the self, he split the self into two parts: the Ego, or the principle of felt personal identity, and the Me, or "empirical self."[48] He immediately split the Me into a variety of subselves. These included the "material self" (our body, our clothes, our house, our children, our family, our "lands and horses, and yacht and bank-account"), a person's various "social selves," and a spiritual self.[49] The "social self" consisted in "the recognition which [a man] gets from his mates." A *"man has as many social selves as there are individuals who recognize him* and carry an image of him in their mind," James wrote. The existence of different social selves within one person depended on one's tendency to show "different sides" of oneself to different people (to one's "children" and "club-companions," to one's "customers," employees, or employers).[50] The spiritual self, instead, was the felt center of self-activity. James famously identified the "feeling of the central active self" with certain perceived motions in the head, and especially motions of the glottis, neck, and the eyeballs.[51] These various selves, including a person's potential social selves, could occasionally live peacefully next to each other, each practicing its own social role in a sort of "harmonious division of labor." More frequently, however, they would be at odds.[52] People were expected to negotiate the relationships between their various subselves and organize them in such a way as to avoid competition and tension.

Like the Me, the second pole of the self, the Ego or the principle of personal identity was not immune from division. While acknowledging the *feeling* that each person had of their personal unity, James insisted on combining that *perceived* unity with metaphysical disunity and pluralism. He identified the Ego with the "present thought" that an individual had. At any moment, James acknowledged, we are able to distinguish between thoughts that belong to us and thoughts that do not. The former are pervaded by a feeling of "warmth and intimacy" that does not accompany the latter. In sorting out thoughts that belong to it, James wrote, the self resembles the owner of a herd of cattle let loose for the winter on some wide Western prairie. As spring comes, the cattle herder is able to collect all the "beasts" that belong to him, picking out those "on which he finds his own particular brand."[53] This irreverent metaphor did justice to the "common sense" intuition that "there must be a real proprietor," something that actually unifies the self.[54] It simultaneously posited that plurality and division were intrinsic to the self, making the self into a "mixture of unity and diversity." Not only did the herd consist of a plurality of animals, but even the "herdsman," who came and performed the act of collecting the animals, dissolved into a plurality of things, a series of "herds*men*." Each of them inherited his "title of ownership" from his predecessor, thus standing as the "legal representative" of all past predecessors. James redefined the economic relationship of "ownership" that had been constitutive of much of the nineteenth-century American rhetoric of personal identity. Each current self (each "passing Thought"), he proposed, was born a free "owner" but died an "owned," since it ended as a property "possessed" by the subsequent self.[55] Self-ownership, self-possession, became an internalized and transient relation, ever to be reconfigured among shifting terms.

As has amply been discussed by other scholars, the Jamesian self was cut by even deeper lines of division.[56] Sometime in the late 1880s, James visited the Salpêtrière, the Parisian hospital that Jean-Martin Charcot had made into the most famous museum of living "hysterics." Among the spectacular symptoms displayed by Charcot's patients, a relatively modest one fascinated James: localized forms of anesthesia including blindness to certain objects and the related symptom of contraction of the field of perception. One of Charcot's younger associates, Pierre Janet, had developed the theory that hysterical symptoms were always correlated with forms of somnambulism involving a "dédoublement" of the personality. He suggested that hysteria was made possible by a weakness of psychological "synthesis," a defect in the subject's power to "gather . . . his psychological phenomena, and assimilate them to his personality."[57] Hysterical subjects were therefore incapable of sustaining a coherent personal identity. Janet led James through the wards of the Salpêtrière and James adopted his theory of hysteria.[58] James summarized it as follows: "the hysterical woman abandons part of her consciousness because she is too weak nervously to hold it together." Meanwhile, the "abandoned" parts may float around or solidify into secondary, "parasitic" or "subconscious," selves.[59]

In the 1880s and early 1890s, James studied both pathological and artificially induced dissociations in a series of experiments on automatic writing, hypnotic trance, and posthypnotic suggestion. For example, in the late 1880s he conducted experiments designed to test the hypothesis that in automatic writing the automatic hand could be the site of a type of local anesthesia similar to hysterical anesthesia.[60] During the experiment the right hand of the subject was placed on a planchette (an instrument normally used in spiritualist séances) in such a way that the subject could not see it. James pricked the "automatic" hand several times. While the hand complained in writing ("Don't you prick me any more!"), the subject observed that his hand "felt asleep." James concluded that the consciousness of the subject was split into two incommunicable consciousnesses: a "mouth-consciousness" and a "hand-consciousness," or "automatic consciousness."[61] In 1886, as a member of the Committee on Hypnotism created by the newly founded American Society for Psychical Research, James performed a series of experiments on "selective blindness." This condition could be artificially induced through hypnotic suggestion in certain subjects who would become temporarily "blind" to specific visual stimuli. The ASPR committee's experiments repeated, with some variations, experiments previously performed by Alfred Binet, Janet, and others. These experiments confirmed that the hypnotic blindness was "false." The images that the hypnotic subject failed to see were "felt" or "apperceived" by someone else, a secondary consciousness (self), which seemed to alternate with the waking consciousness.[62]

James was familiar with other experiments with posthypnotic suggestion performed by his friend Edmund Gurney, a leading British psychical researcher. Gurney would hypnotize his subject and ask him to perform a complex task, such as the solution of an arithmetical problem. He would then immediately wake the subject, place the subject's hands on the planchette, and keep the subject's "normal self" busy with conversation and other tasks. After a little while, the planchette would write down the correct solution of the problem that the subject had been asked to solve, or an answer closely approximating it. Gurney concluded that far from being "automatic acts" (reflex actions), the answers written by the automatic hand revealed the presence of an active consciousness, a "latent" secondary self or conscious-ness "segregated" from primary consciousness yet "simultaneous" with it.[63] From his own and Gurney's experiments James drew the conclusion that the "secondary" consciousness could not only "alternate" with the waking consciousness but also "coexist" with it.[64]

Like other psychologists of the time, James wondered whether that type of inner division was confined to the realms of the artificial and pathological.[65] Pierre Janet, for example, denied what to many seemed to follow directly from his investigations. He saved, at least temporarily, the unity of the normal self and posited a link between dissociation and hysteria.[66] Other experimental psychologists had no such hesitations. Thus Théodule Ribot, whom James met in Paris in 1882, stated unambiguously that the unity of personality stemmed not from an underlying metaphysical principle but from an empirical process. This process, largely biological, recapitulated the evolutionary processes that led to the emergence of a central consciousness in higher animals. It began with multicellular organisms consisting of physically juxtaposed identical cells, each endowed with its autonomous psychic life. Then the process went through the stage of "colonial organisms," polyps and hydrae for example, in which a new centralized consciousness made its first appearance.[67] This higher consciousness, which could temporarily harness for common goals the autonomous consciousnesses of the members of the colony, came to symbolize for Ribot the precarious unity of the human self, which required a constant effort of "coordination." Thus division was a constitutive feature of human nature.

James agreed with Ribot. The unity of personality was the result of empir-ical processes and did not preexist those processes. In the mid-1890s he used the analogy of colonial organisms ("polyzoism/polypsychism") to repre-sent the condition of dissociation found in pathological cases and challenged Janet's reluctance to extend dissociation to healthy individuals.[68] James wrote that he knew a "non-hysterical woman" who could fall into trance and dis-play telepathic powers.[69] Her case clearly proved that dissociation could be found in at least some healthy people.[70] Pierre Janet questioned that claim. In a private letter he urged James to subject the woman to a thorough ex-amination of her vital parameters: hysteria was most likely present but had gone undetected.[71] James's answer to Janet has not survived, but he did not take up the suggestion or change his mind. By the turn of the new century he had come to look at the fault line separating a normal self from a subliminal self as a feature of the human self.[72]

## Topologies of the Self: (2) Stretching the Boundaries of the Self

James liked to think of the present state of consciousness as a visual field with
its center and margins. The center represents that of which we are actively
conscious, while the margin (a "penumbra" or "halo") represents things of
which we may be dimly aware or even unaware, but which, if we refocus
our attention, could shift to a central position.[73] The topology of the field and
the distribution of light and shade are transient and can be reversed. What
is central can become peripheral and what is peripheral can become central
and luminous.[74] James used the same metaphor to describe the self. At any
moment the self has a center and a periphery. The center is occupied by
our "hot" beliefs, those that are sources of energy and direct our activities.
We identify our individual selves with the center of the field, but changes
can occur. Thus, during the course of our lives, we may suddenly find that
the inner balance has shifted. We then discover that what was central no longer has any importance and that we are now ready to identify our innermost
self with what was once marginal. The metaphor was suggestive because it
allowed one to visualize the "indetermination of the margins" of consciousness and of the self.[75]

James asked where the self ends, echoing a question that Madame Merle
asks Isabelle Archer, the heroine of his brother Henry's *Portrait of a Lady*
(1881).[76] What do the boundaries of the self look like? How far do they
stretch? To James these were the important questions raised by automatic
writing and other psychic phenomena.[77] As a member of the Society for
Psychical Research, James had come to know very well the work of F. W. H.
Myers, one of the society's founding members. Myers believed that telepathy
and hypnotism at a distance indicated that the subliminal self of an individual could have "direct relations of intercourse . . . with the consciousness of
other men."[78] Projection of the double, bilocation, phantasms appearing to
entire groups of people, "traveling clairvoyance" or "telaesthesia" in dreams,
and crystal-gazing suggested that the subliminal self was by no means confined to the region occupied by the body and its immediate surroundings.
This self could actually step out of the body and invade physical space.
Mediumistic trance and cases of what would once have been interpreted as
demonic possession seemed to indicate that the subliminal self might also
communicate with spirits of the dead and a "cosmic environment."[79] James
was skeptical of Myers's spiritualist conclusions,[80] yet that did not prevent
him from sitting outside of his hotel room in Rome where Myers was dying,
James, pencil in hand, ready to jot down otherworldly messages.

As a psychical researcher James investigated "veridical hallucinations,"
mediumistic trance, and telepathy, the only psychic phenomenon the reality

of which James took to be adequately supported by empirical evidence.[81] He also investigated an episode of clairvoyance that resulted in the solution of a case of accidental death. Moreover, he knew a few respectable people who believed they had managed to project their "doubles."[82] In February 1906 he himself had uncanny, fearful experiences with dreams. He woke in the middle of the night with the distinct impression that he had been dreaming someone else's dreams and that dreams dreamt by someone else had been "telescoping" into his own dreams. Was he telepathically *getting into other people's dreams*? Or was he experiencing "an invasion of double (or treble) personality," and "losing hold of [his] 'self'"?[83] James, furthermore, was famous among his closest friends for conducting experiments on himself with nitrous oxide, chloral hydrate, amyl nitrite, hashish, and other anesthetics that appeared to lower the threshold separating the normal waking self from the subliminal self.[84] These and similar experiments conducted by some of his acquaintances, such as the self-taught philosopher and mystical writer Benjamin Paul Blood, also seemed to indicate that the self could communicate with a larger mental region, a region of consciousness separated from "our normal waking consciousness" only by "the filmiest of screens."[85] Mystical ecstasies, which James studied extensively, supported that conclusion. The subliminal self was not enclosed but, as James surmised when discussing Myers's achievements, seemed to have "windows of outlook and doors of ingress" through which it could open up to "an indefinitely extended region of the world of truth."[86] The dogma of the isolated impermeable self was untenable: the boundaries of the self were permeable and leaky.[87]

## The Unification of the Self: How to Make Strong Citizens

Josiah Royce observed in 1901 that supporters of "extremer forms of ethical individualism" often found it convenient to resort to "realistic" theories of the self. These theories made the self into a substance logically, ontologically, and psychologically independent of the existence of other selves. Realistic theories of the self, Royce went on, were attractive to ethical and political individualists because they preserved in a direct way "the dignity," "the freedom," and "the rights of the Self."[88] Despite his self-proclaimed individualism, however, James never essentialized the self. He depicted a metaphysically weak self, threatened by inner division, surrounded by porous boundaries, and only precariously whole. This was a self that seemed hardly compatible with the claims of individualism. Yet, I suggest, James transformed the weak and divided self into a tool that individuals could use in order to achieve renewed strength and agency.

James repeatedly lectured his contemporaries on the need to fight inner enemies: inward division, loss of self-mastery, resignation, and a sense of weakness. At this juncture his account of the self intersected with the so-called "New Thought," which one of its proponents defined as "both a philosophy of life and conduct and a mode of healing." It also met up with a range of medical and religious practices for the cultivation of the self.[89]

Mind curers, mental hygienists, and followers of New Thought all advertised techniques that would enable ordinary individuals to eliminate inner division and obtain confidence, energy, inner harmony, and self-mastery. James was intensely fascinated by the culture of self-help and thoroughly familiar with many of the practices recommended by mental healers and spiritual therapists.[90] For example, he tried breathing exercises recommended by a Yogi teacher, Swami Vivekânanda, whom James had met in 1896.[91] Through such exercises individuals could learn how to control breathing and thus develop powerful forms of self-mastery. James practiced breathing exercises "somewhat perseveringly" around 1906, hoping to bring his insomnia under control. The exercises unfortunately failed to produce any "soporific effect." James found that they "got terribly against the grain with [him]," perhaps because he was "so rebellious at all formal and prescriptive methods," yet he seemed persuaded that in some cases such exercises could indeed increase "vital tone and energy" and wake up "different levels of will-power."[92] James was also familiar with the New Thought meditation technique known as "entering the silence," a practice that consisted in averting thought from the external world and "draw[ing] the diffused powers of thought" until they were focused on the soul and the divine Spirit. This technique, as the New Thought leader Horatio W. Dresser advertised it, would heal inner division and lead to the "development of spiritual poise," "spiritual self-control," and the "ability wisely to direct one's thought forces."[93] While James doesn't seem to have personally found meditation congenial, he nevertheless closely followed the activity of Dresser, who had been a student of his at Harvard and acknowledged James's influence on the development of New Thought.[94]

James was also a vocal supporter of Horace Fletcher, a mental hygienist who specialized in dietetics. Fletcher had devised a system for curing dyspepsia that consisted in chewing each morsel of food thirty-two times. He had also found a way of escaping fear and anxiety. The trick consisted in believing that those negative emotions could be overcome; once this was accepted, it became easy to see that those emotions must and would be overcome. The method was advertised as a cure for pessimism, selfishness, and violence, and as a means to attain serenity, health, and self-control.[95] As with other mental hygiene practices, James's experiments with prolonged chewing did not give the results he hoped.[96] Yet he was persuaded of the importance of

Fletcher's techniques of self-mastery and inner unification. James carefully read books and pamphlets by several other mental hygienists and Christian preachers. For over twenty years, he occasionally turned again with mixed results to mind healers in order to deal with various ailments that afflicted him, including especially fatigue and his intractable insomnia.[97]

James's account of the self was a contribution to such mental therapeutics. Perhaps the passages of *Principles of Psychology* that evoked the most enthusiastic popular response were those that advised his readers on how to cultivate good habits and eliminate bad ones. As seen in chapter 4, to James, as to most mental physiologists of the time, habits were concatenated series of automatic reflex-action processes, mechanical actions that originated from what initially had been conscious purposeful actions and had become "grooved" in the brain.[98] James believed we are responsible for the habits that we acquire and ultimately, since habits are such an important part of our nature, for the type of persons we are. We are constantly "fashioning our character" and "spinning our own fates, good or evil, and never to be undone." The physiology of habits explained why it was so difficult to get rid of an inveterate habit and to acquire a better one. At the same time, the plasticity of the nervous system ensured that if one really tried, one could fashion a new character and become a new person. One merely needed to strive "with as strong and decided an initiative as possible" and "never suffer an exception to occur" until the new habit set in. James exhorted his readers to cultivate the "faculty of effort" and keep it "alive . . . by a little gratuitous exercise every day." He stressed the importance of daily exercise of the "habits of concentrated attention, energetic volition, and self-denial in unnecessary things."[99] James's moralizing indictment of impolite habits conveyed a tremendously hopeful message. Precisely because it was not metaphysically fixed once and for all, the self could be made and remade, woven and rewoven.

### The Unification of the Self: Practices of Attention and Spiritual Exercises

Central to James's therapeutics of the self was the goal of the integration of personality. For James this was not only a medical and a hygienic necessity but also a moral ideal, in much the same way as integration of mental faculties into a harmonious whole had been a crucial goal earlier in the century for scores of moral philosophers and educators.[100] Making the self whole was a moral duty.

We all begin our lives in a chaos, James wrote. Conflicting feelings and impressions are all mixed up. Each of us needs to sort out that confusion and to organize opposite tendencies in some "stable system." That was what character formation was about. A person of character was a person who had managed to "straighten" and "unify" the self. The antithesis was represented by "heterogeneous personalities" or "hysterical temperament[s]": individuals whose inconsistent, zigzag behavior was socially disruptive as well as individually painful.[101] James toyed for a moment with the hypothesis that heterogeneity of impulses could be the result of heredity and that the divided self could be the powerless passive battlefield where "incompatible and antagonistic" ancestral tendencies struggled with each other.[102] He quickly dismissed this theory, however, and instead suggested that heterogeneous personalities resulted from a failure in the normal evolution of character. Thus inner division could be more than a pathological mark: it could also indicate a moral failure, a failure of diligence and resolve in maintaining the self.

Sometimes the unification of the self could occur spontaneously, in ways inexplicable to individual consciousness. That happened in religious conversions, especially instantaneous conversions, which James regarded as the culmination of subliminal processes through which the divided self eventually found itself rearranged around a new center of interest and gained unity and inner peace.[103]

Whether spontaneous or laboriously cultivated, the unity of the self nevertheless remained precarious. Many of the techniques of unification that James explored were fundamentally techniques of "attention," techniques that required the subject to focus attention on some object, content, or movement. But, as Jonathan Crary has shown, practices of attention such as those involved in late nineteenth-century psychological therapy and hypnosis always produced marginality and distraction. In fact, they contributed to the fragmentation of the modern self.[104]

Indeed, the unifying techniques with which James engaged ultimately generated residual materials, reproducing some of the divisiveness they attempted to heal. Examples included both the attentive, consciously synthetic, practices of breathing and meditation such as the "method of concentration" of the powers of mind preached by Vivekânanda, and the spontaneous processes underlying conversion.[105] Both involved a refocusing of attention, resulting in redefinition of the center of the field of experience and selfhood and marginalization of previously central material. The "unified self" required marginality and diversity; the topology of the field with a center and margins remained constitutive of the self. As John McDermott aptly writes, for James the unified self remained a "bundle of relations," a plurality of things.[106] Thus, when James treated with hypnotic sessions a man suffering from split personality, he did not try to eradicate one of the two personalities. He instead attempted to "introduce" the two personalities to each other so they could acknowledge each other's presence and live peacefully in the same body.[107] The therapy failed.

James's account of the principle of personal identity, the Ego, further reveals how for him the unity of the self could only be temporary. The continuity linking the "present Thought" to the previous thought and to the next, which, as we saw, provided the foundation for the feeling of personal identity, was fundamentally illusory. It resembled the optical effect of "unbrokenness" generated by the rapidly succeeding yet discrete images in a "magic lantern."[108] As in the "dissolving views" projected by a magic lantern, the perceived continuity of the self was largely performative: it was an effect of the performance, and it lasted only as long as the performance lasted.[109] Just as one's attention could not stay focused on the same object for more than a few minutes, so the unity of the self, the spectacle of continuity, required a sustained effort and had to be continually renewed.

George Cotkin has shown that much of James's public philosophy was an intervention in the budding discourse of heroism and strenuosity, as well as an invitation to his contemporaries to increase their energy "as a mode of escaping the *tedium vitae* of modern life and entering into a heroic existence."[110] The techniques for the cultivation and unification of the self that James explored were also a means of increasing the energy of the individual and of strengthening the self. In fact, they were the source of an individual's strongest sense of self. As a young woman who corresponded with James put it: "the fragmentariness and multiplicity of life are . . . the saving of the sense of selfhood."[111] The laborious, hourly maintenance of the self, which was made necessary by the inner fragmentation and metaphysical weakness of the self, endowed the individual with self-directedness and self-determination.[112] Paradoxically, the inner divisiveness and metaphysical weakness of the self appeared to be the precondition of possibility for a sense of self and a form of agency *stronger* than those promised by traditional doctrines of the unitary, simple, well-bounded self.[113]

Recent works on William James have shown that James's psychology was meant first and foremost to play an individual therapeutic function. Jeffrey Sklansky, for example, contends that James psychologized pressing social and economic problems and offered purely psychological solutions. Rather than providing his audiences with tools they could use to regain mastery over their "conduct," James internalized the meaning of "self-mastery," interpreting it as "mastery over one's thought." He thus shifted emphasis from social change to individual reform and ultimately reconciled "mental autonomy with material dependence." Sklansky concludes that "therein lay the ironic secret of [James's] . . . success . . . in a nation dominated by wage labor and finance capital."[114]

Although I agree with Sklansky, James Livingston, and others that James did not offer practical indications for social reform, nevertheless I suggest that his account of the self meant to offer more than a therapy for individual consumption. It was also a tool that James mobilized in hopes of creating or facilitating the conditions of possibility of a new type of political participation and new kinds of human relationships. James believed that the techniques of the cultivation of the self recommended by mental hygienists could play a *social* as well as an individual function. Thus, when Horace Fletcher gave a lecture at Harvard, James urged "member[s] of the Harvard union" to attend it, on the grounds that the talk was "of fundamental importance both to the individual *and to the State*": Fletcher's "observations on diet" might prove to have "revolutionary import."[115] Likewise, for James, strengthening the self and making it whole were more than an individual moral duty: they were also social duties, for James believed that, by cultivating the self and increasing its strength, people could make themselves into agents of social change.

Indeed, strong effective individuals were crucial to James's social vision. To James, famously, the world was "a real adventure, with real danger," and large-scale change could only result from each individual "agent" doing "its own" best in "a social scheme of co-operative work genuinely to be done."[116] His friend Theodore Flournoy clearly perceived the point. James's meliorism, he observed, made people "into real entities, and real agents," and awakened "them from 'the slumber of nonentity' into which the vision of a perfectly complete, eternally saved universe puts them." In that frame-work "salvation [could] come about only piecemeal and for each element individually,"[117] as the result of personal initiative. Broad-range noncoercive social change could only be spontaneously initiated by individual actors.

James made just that point in "The Gospel of Relaxation," an address given, probably in 1897, to an audience of young women, the graduating class of the Boston Normal School of Gymnastics. James tried to enlist his audience in "a cause . . . of paramount patriotic importance to us Yankees." Americans, James complained, suffered from a national habit of "anxiety," of "over-tension, jerkiness, breathlessness, intensity and agony of expression," a "bad" habit that impaired their efficiency and their spiritual life. How could they "tone down their moral tensions" and eliminate a phenomenon that, before being physiological and psychological, was social? No social measures would work: the solution could only come through individual initiative. The principle of "imitation," which James believed lay at the very root of the social fabric, ensured that once an individual or a few people had successfully practiced "the gospel of relaxation" and reformed their habits, others would follow. This way the new habit of relaxation would spread rapidly, and society would be reformed.[118] Thus James's account of the self,

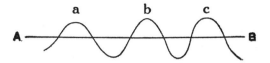

Fig. 6.1. The threshold of consciousness. From James, *Human Immortality* (1898). The image was based on Fechner's illustration of the threshold of consciousness in *Elemente der Psychophysik*. In the public domain.

while depriving the individual self of substantiality and metaphysical unity, was designed to give people agency and ultimately the ability to "revitalize" and regenerate American society.[119]

## The Open Self: Ecstasy and Community

In Western cultures conceptions of the self and practices of selfhood have often carried suggestions about how to reorganize social life and rethink the relationship between the individual and community. As Jan Goldstein notes, however, it is less obvious how any particular conception can do so.[120] James's account of the self is a case in point. James never spelled out all the social and political implications of the self that he delineated. Instead, he alluded to them obliquely with metaphors and poetical language and by using quotations that would have evoked dense webs of meanings to his contemporaries. Yet the social significance of James's account of the self emerges clearly if we place James's texts, personal experiences, and the techniques of the self that he advocated within the context of other late nineteenth-century doctrines and practices that more explicitly addressed issues that James at times only hinted at.

A good starting point is provided by an image that James adopted from Gustav Theodor Fechner and used in his 1898 booklet *Human Immortality* (fig. 6.1) to illustrate the idea of the threshold of consciousness.[121] The sinusoidal wave represents a "wave of consciousness." The horizontal line represents the threshold of consciousness, the boundary separating what we are conscious of (above the straight line) from what we are not conscious of (below the line). The horizontal line can move down or up as we become more alert (as more things enter our field of consciousness) or more drowsy. One feature of this graph has seldom been noticed: the graph illustrates not only the threshold of consciousness of one individual but also that different "organisms" could intermingle below the threshold of consciousness. James explained to his readers that the wavelets "a," "b," and "c" represented the "waves of psycho-physical activity" of three different "organisms," different individuals. James indicated that Fechner used this image to study the conditions under which a physical multiplicity (the physically many) could "contract into a psychical one," that is, could be psychically unified. If the threshold "sank low enough to uncover all the waves," the consciousness (or consciousnesses) surfacing above the threshold might also become continuous. Thus the image shows that these different individual consciousnesses, each of which took itself to be isolated from the others, might not in fact be really separated. They could be continuous below the threshold and could at times become aware of that continuity. Or perhaps they might be continuous in a larger span of consciousness, a "world-soul."

In 1909, discussing his experiences as a psychical researcher, James wrote: "Out of my experience . . . (and it is limited enough) one fixed conclusion dogmatically emerges, and that is this, that we with our lives are like islands in the sea, or like trees in the forest. The maple and the pine may whisper to each other with their leaves. . . . But the trees also commingle their roots in the darkness underground, and the islands also hang together through the ocean's bottom. Just so there is a continuum of cosmic consciousness, against which our individuality builds but accidental fences, and into which our several minds plunge as into a mother-sea, or reservoir."[122] Circumscribing and insulating the self from outside influences was part of the individual process of adaptation to the "external earthly environment." Yet the fence surrounding the self would remain "weak in spots, and fitful influences from beyond leak[ed] in, showing the otherwise unverifiable common connexion."[123]

This section and the next suggest that James's insistence on the porosity of the boundaries of the self, which he especially emphasized in the last decade of his life, was instrumental to rooting the self in community and to promoting new kinds of human relationships.

"Socializing" the self was a goal that many of James's contemporaries pursued and which they sought to accomplish precisely by opening up the boundaries of the self. James was familiar with many emerging theories of the "social self" and with the many ways in which such theories had been used to support social visions of community and cooperation.[124] In 1902, on board an ocean steamer crossing the Atlantic, he got to know closely the Midwestern journalist Henry Demarest Lloyd, whose exposés of corporate corruption were among the earliest examples of muckraking journalism. A neorepublican populist, Lloyd had made the theory of the social self central to his gospel of the "new morality." He prophesied that monopolies and private ownership of public properties such as railways would disappear and preached a social vision in which "individual self-interest" would be replaced by "*social* self-interest," egoism by altruism.[125] The theory of the social formation of personality, which distributed individuality over a network of social bonds, supported his vision of a cooperative society.

James was also well acquainted with idealistic theories of the social self. Josiah Royce, for example, made individuals "dependent on [their] fellows" not only physically but "to the very core of [their] conscious selfhood."[126] Royce's account of the self stood at the root of his social vision, one that privileged community and located the source of salvation not in individual acts of self-assertion but rather in the supreme virtue of "loyalty" to the community and to its causes.[127] James was not oblivious to his friend's efforts to maintain a role for individuals and for small, local groups in his social vision,[128] yet he suspected that Royce's vision—and his parallel idealistic metaphysics, which made the individual into a part of an absolute self—failed to strike the correct balance between individual and community. James had the same problem with other forms of monistic idealism. Francis Bradley, the most distinguished representative of British absolute idealism and "the bogey and bugbear of most of [James's] beliefs," as James amicably described him,[129] achieved the goal of completely embedding the individual in society but only at the cost of denying the reality of the self altogether and dissolving it in the absolute mind.[130] Bradley's "complete repudiation of the reality of the self," as Timothy Sprigge describes it, went hand in hand with his political opposition to individualism, his defense of a robust role for the state (against liberal laissez-faire), and "his tendency to think that one's nature is so bound up with one's community that purely personal interests are a myth or a disease."[131]

The renunciation of egoistic impulses was very explicit in numerous socialist utopias of the time, including William Dean Howells's vision of an "Altrurian" society. Howells's novel *A Traveler from Altruria* (1894) and its sequel, *Through the Eye of the Needle* (1907), brought to completion the demise of the substantial, insular, self-determined self, the diseased condition of which he had diagnosed in *A Hazard of New Fortunes*. Altruria was a communistic society that admitted neither ownership of property nor self-ownership of labor nor, indeed, of any form of individual selfhood. Life was based on renunciation of self-determination and all self-seeking impulses, and on the extreme practice of "altruism."[132] To James, Howells's utopia remained largely bourgeois; he had his reservations about the novelist's ability empathetically to portray "the inner joy and meaning of the laborer's existence."[133] Nevertheless, these novels and other broadly defined socialist literature sharpened his awareness that cooperation required—to some extent at least—a weakening of the boundaries of the individual self or an embedding of selfhood within social relations.

For some of these socialist writers, the regeneration of society rested on forms of selfhood rooted in religious or mystical experiences. James personally knew a few, his own father a good case in point. An unorthodox follower of Emanuel Swedenborg and of Charles Fourier's utopian socialism, Henry James Sr. envisioned an ideal society ("the regenerate society") premised upon "the brotherhood of each man with each man in God." In this society the individual would give up selfish tendencies, indeed would relinquish "his" very selfhood and recognize his membership in a large, cosmic, divine self.[134] To Henry James Sr. the pursuit of selfhood was the "source of all evils" and to relinquish the illusions of selfhood and substantiality was to take the path to individual and social salvation. The "regenerate man" was in essence "the social man."[135]

Ralph Waldo Trine, a self-educated mind curer and a Christian socialist whose works James read with great interest, provides yet another example. Trine denounced the corruption of big corporations and analyzed the problems plaguing the working classes in American society. He located the origin of all social evils in the petty self-interests that lead people to compete with one another. The teaching central to his political and hygienic message was that the individual self existed only as part of a larger, "infinite," all-embracing divine "spirit." The individual self thus was illusory. The realization of the insignificance of the "diminutive" personal self was to be the panacea for individual happiness and the miracle cure for all pressing social and economic problems, including the tensions between workers and capital that, following James, Trine ascribed to lack of active sympathy between workers and capitalists. Trine predicted that by banishing all self-seeking attitudes and petty concerns for their individual selves, and by looking at themselves as parts of the infinite self, workers and the common people could effectively unite and bring about a more peaceful, cooperative, and sympathetic form of social life.[136]

A third example deserves special consideration: the British mystical poet Edward Carpenter, an anarcho-socialist greatly popular with socialists of various persuasions, humanitarians, theosophists, and spiritualists. In 1900 James read Carpenter's Whitmanesque mystical poem, *Towards Democracy*, which quickly became "one of [his] favorite books."[137] In the four-hundred-page poem Carpenter had managed to strike a difficult balance between "individual fulfillment and social good."[138] In his social vision, self-seeking, self-affirmation, and love of self combined with love of others and recognition of the uniqueness and value of each person, enabling men and women to live in a "democratic society" that allowed both for complete flourishing of the individual and a rich communal life. A few years later Carpenter articulated an ambitious theory of the self. Linking together insights from evolutionary theory, Hindu religious traditions, and his own mystical experiences, Carpenter dissolved the individual self into a universal cosmic Self. He located three stages in the evolution of consciousness. The first, found in animals, children, and some primitive societies, was characterized by the lack of any distinction between self and other.[139] The second stage, that of self-consciousness and civilization, was marked by the opposition between self and other and by a morbid sense of the importance of the self. This stage, best instantiated by laissez-faire economies, was one of "competitive individualism."[140] In the third stage, all those divisions would be overcome and individual consciousness would evolve through love into "Cosmic Consciousness" (the same phrase that James used in the 1909 passage quoted above). In a flash of mystical revelation, the experience of cosmic consciousness, the individual would suddenly perceive himself to be just a part of the great Self of the universe.[141] Cosmic consciousness marked the inauguration of a stage of social harmony and offered the metaphysical and psychological framework for the socialist "brotherhood of workers" that Carpenter expressly advocated.[142] Indeed, as the British Fabian socialist Beatrice Webb promptly observed, Carpenter's booklet provided the perfect embodiment of the "metaphysics of the socialist creed."[143]

Examples of social commentators who associated the demise of individual selfhood with mystical experiences of unification and the advent of social solidarity are numerous. James's correspondent Richard Maurice Bucke, a member of Walt Whitman's circle of friends and an asylum superintendent, associated the mystical stage of "Cosmic Consciousness," a mystical unification with the infinite Spirit, with the advent of a socialist millennium.[144] A generation earlier, the social visionary Edward Bellamy posited a religious, even mystical, foundation for the social feelings of love of others and sympathy, and for the instinct of solidarity, the desire to "los[e] ourselves in others or [to] absorb them into ourselves." Through ecstatic experiences, he wrote, "we are wrapt out of ourselves" and, stepping out of our "narrow, isolated, and incommodious individuality," are able to commune with the "impersonal consciousness" of the universe.[145]

These accounts of the self were somehow disappointing to James. They suffered from the defect that James imputed to other types of socialism: they annihilated the individual in the collective. James perceived that his father's account of the self reduced the "individual man" to "nothing."[146] Likewise, in a letter to a Fabian socialist friend he complained that Carpenter had "overdone the monistic business," privileging the whole at the expense of the individual self.[147] Carpenter supported his social vision of community with the authority of mystical experiences. But mystical experiences, so James wrote, did not necessarily testify for "absolute unity," for the absolute fusion of the individual self with a universal infinite all-embracing Self. They simply testified to the possibility of "more unity," a claim perfectly compatible with the pluralistic viewpoint that there exist independent irreducible individual selves.[148] All the facts that Carpenter had described, James continued, could be accounted for by the "admission of a wider-span consciousness that envelops ours and uses ours."[149] It was Carpenter's and other mystics' "passing to the limit" that James found objectionable.

Seeking community and cooperation, James insisted on maintaining pluralism. "The ideal life must always be individualistic," James lectured his Fabian socialist friend. James famously rejected the extreme monism that underlay the dissolution of the self into the universal absolute Self. He rejected the very idea of an absolute infinite Self, both in its Western idealistic versions and in its Hindu formulations. Like his irreverent friend F. C. S. Schiller, James found that the all-enveloping absolute Self, if imaginable at all, was to be regarded as a macroscopic case of "multiple personality"—another "Sally Beauchamp"—most probably endowed with telepathic powers.[150]

While James could not accept Carpenter's or Trine's annihilation of the individual self in the cosmic whole, this literature and practices designed to help the self "step out" of itself paved the way for his own understanding of self and society. James's own experiences with nitrous oxide (and alcohol), for example, had given him an "immense emotional sense of *reconciliation*" of opposites, revealing that "the ego and its objects, the *meum* and the *tuum*, are one." Back to sobriety, he questioned the extreme Hegelian monistic conclusion that for an instant had seemed to him to be crystal clear. But he never challenged the idea that "I and Thou" could be "reconciled" and could merge together in states of ecstasy. While at times he feared the possibility that his own self might be "invaded" by foreign selves, like some of the mystical social visionaries discussed here he also realized that the ecstatic experiences of the open self could ground the possibility of a new cooperative life.[151] Thus the leaky contours of the self, a great source of anxiety, could also be optimistically imagined as a tool for the creation of a new type of society.

## The "Self-Compounding of Consciousnesses": Psychology, Metaphysics, and Society

In emphasizing how individuals could communicate below the threshold of consciousness and how they could realize their interlinking continuity through experiences of ecstasy, illumination, and transmarginal communication, James fashioned a psychological theory that promised to eliminate selfishness and isolation. His notion of the open self was tailored to allow not only for redemptive experiences with a divine Self, but also for the sharing of experiences, sympathetic understanding, and ultimately for cooperation and solidarity. This point has been missed by many scholars.[152] To see this clearly the discussion needs to take a rapid detour through a technical metaphysical problem that we encountered in chapter 2: the problem of "the self-compounding of consciousnesses" or the "problem of the compounding of selves."[153] I suggest that in James's solution to that problem the Jamesian self and the form of society that James promoted came together in a way that alleviated the tension that had run through his social vision: the tension between individuality and community.

In a series of public lectures at Oxford in May 1908, James described the problem to his audience: "I wish to discuss the assumption that states of consciousness . . . can separate and combine themselves freely" and yet "keep their own identity unchanged while forming parts of simultaneous fields of experience of wider scope."[154] The underlying question was could "many consciousnesses be at the same time one consciousness?" If so, how? Or how could individual, separate consciousnesses compound into a complex consciousness, without losing their individuality and self-perception? On a different level, how could different individual selves (you and me) "be confluent" into a higher Self (for example, the absolute Self of idealist philosophers) in such a way that would allow each to retain its individuality while being coconscious of the others?

James had spent much ink trying to solve this problem. In *Principles*, as we saw, he had declared the problem insoluble and had sharply criticized those who attempted to solve it by claiming that higher sensations are "compound sensations" (for example, the feeling of "lemonade" is a compound of the feelings of "sugar" and of "lemon"). He insisted that, far from being "compounds" of lesser sensations, higher sensations were just "new psychic facts." Their unity existed as such only for an external "bystander," namely the new mental state that unified them. In the years that followed, he deployed this argument against monistic idealists including Royce and Francis Bradley, who operated under the assumption that the absolute Self was "constituted" by the individual selves, each of which retained its own identity and self-perceptions.[155] James simply could not see how their claim could be compatible with the laws of logic.[156]

In the Oxford lectures James attacked the problem of the compounding of consciousnesses one last time. He offered a spectacular and controversial solution. This consisted in giving up one of the fundamental principles of logic, the "principle of identity," according to which, as James put it, a thing cannot be its "other" and if two things are distinct they must also be "disjoint." As James told his bewildered audience, if one gave up the axioms of logic and, following the lead of the French philosopher Henri Bergson, placed oneself *"d'amblée"* into the flux of reality, one would perceive that no absolute boundaries really separate any bit of experience from the next. Concrete pulses of experience "compenetrate" each other, "run into each other," and "coalesce" at their margins. They "interpenetrate" "where they touch" and "telescope" into each other (the very term that James had used to describe his own protoecstatic experience with dream invasion).[157] Once the axiom of identity was abandoned, there was no real problem in seeing how successive pulses of consciousness could interpenetrate and be part of higher compounds *without losing their distinctness and individual identity.*[158] By analogy it was also easy to see that different individual selves (you and I) could interpenetrate along their margins and freely compound themselves into larger wholes while retaining their individuality.[159] Neighboring individual selves (or different "pulses of consciousness") were thus not wholly separate well-bounded isolated things. They could be different without being "disjoint." They could flow continuously into each other in ways that recalled the intimate biological process of "endosmosis."

That was exactly the conclusion James had drawn through his studies and (limited) personal experiences of mystical and psychic phenomena. Indeed, as Richard Gale suggests, James's final metaphysics can be seen as an expression of a particular type of mysticism: James's unpretentious "backyard" mysticism. In this mystical-metaphysical framework, different selves could coalesce with each other by being "confluent" in a higher but not all-embracing superhuman consciousness (the "mother sea of consciousness," the "cosmic consciousness," or perhaps a finite God). Or, as Timothy Sprigge has shown, they could merge together by virtue of concatenated relations of continuity from one to the other (the type of "unity" that James probably thought was supported by mystical authority).[160]

To James, as to scores of other scientists, psychologists, and cosmologists of the time, the problem of the compounding of consciousnesses and, more generally, the problem of the one and the many, of which the former represented a version, was so important because it carried political implications.[161] It bore directly on the question of the nature of the relationships of the individual with society as well as the question of the relationships among individuals in society. The point should come as no surprise: after all, James was quite explicit in his contention that his pluralistic metaphysics "frankly interpret[ed] the universe after a *social* analogy."[162] In social/political terms, James reckoned, the metaphysical problem translated into allowance of spontaneous communities in which individuals could retain their identities and individual perspectives yet develop a sympathetic insight into other people's perspectives. These were the pluralistic yet intimate communities that

James dreamed of animating, at times by means of (mystical or, perhaps, more earthly) love with other similarly minded women and men. The latter included, possibly, Pauline Goldmark, a lovely young woman who made him so "happy"; or Henri Bergson with whom James hoped to develop a "socially and intellectually endosmotic relationship" ("endosmosis," as both of them well knew, involving the intimate exchange of bodily fluids); or a young Italian pragmatist friend, whom James met for the first time half-naked in a hotel room, in an attempt to start their friendship on an intimate tone.[163]

To James, these relationships—intimate, sympathetic, yet respectful of the individual's autonomy—promised the solution for all social conflicts. James, as we saw, identified the central source of the class tensions flaring between workers and capitalists as a lack of mutual "sympathetic" understanding, a mutual blindness that flowed from the external position that each class took vis-à-vis the other. And he believed that American imperialism stemmed from the deaf insensitivity that made the dominating imperialists unable intimately to engage with the Filipinos. The problem of facilitating a mutual understanding through creating a common consciousness thus lay at the very center of James's social thinking. His pluralistic metaphysics, allowing for the interpenetration of streams of consciousness or selves along their boundaries, showed how such understanding could arise, and how the otherwise "impenetrable" values and secrets of other people could become intimately accessible to those sympathizing with them.[164] James's solution to the problem of the "self"-compounding of consciousnesses avoided depicting society as an aggregate of disjointed individuals and made individuals into linked parts of communities that they "freely" created through spontaneous association. At the same time it allowed for a defense of individual autonomy and rights. James stressed that his metaphysics engendered "tolerance" and "democracy" and was incompatible with "slavery."[165] The best examples of such a society were small groups of sympathetic individuals (small anarchistic communities, Deborah Coon suggests) bound by friendship and working together toward the creation of a better world.[166]

At a time when philosophers and politicians resorted to monistic visions of the universe to legitimate aggressive forms of imperialism, James took great pains to emphasize that his solution was instead pluralistic.[167] The larger wholes of which an individual was a part, including, for example, larger selves or concatenated communities, resembled more a "federal republic than an . . . empire or a kingdom."[168] This federal analogy (doubtless disappointing to those who would like to press James for a more precise statement of his political position) embodied the two elements that James continually worked to accommodate into his social vision: individualist localism and community. Elsewhere, James was more daring in indicating political associations. He described his radically empiricist metaphysics as "the pluralistic, socialistic" view that was predicated on the vision of a "co-operative universe."[169]

## Utopias and Conclusions

"Utopias are the noblest work of man," James wrote to William Dean Howells in November 1907 after reading Howells's utopian novel *Through the Eye of a Needle*.[170] Yet James would certainly have agreed with his correspondent the British novelist H. G. Wells that all utopias have an "incurable effect of unreality": "that which is the blood and warmth and reality of life is largely absent; there are no individualities, but only generalized people. In almost every Utopia . . . one sees . . . a multitude of people, healthy, happy, beautifully dressed, but without any personal distinction whatever."[171] Shortly before his death, James offered his own utopia. "I will now confess my own utopia," he wrote in "The Moral Equivalent of War": "I devoutly believe in the reign of peace and in the gradual advent of some sort of a socialistic equilibrium."[172] The "socialistic future" toward which he believed the country was drifting would never be one of softness.[173] It would retain the institutions of corporate capitalism, and its economy would always involve hardiness and effort. Visions of "pacifist cosmopolitan industrialism" were too weak and were bound to be met with the contempt not only of the huge party of "militarists," but also of workers and everyone who had been trained to live a strenuous life. The foundation of the new socialistic state would still have to be provided by "manly" "virtues" such as intrepidity, contempt of softness, surrender of private interest, and obedience to command.[174]

Those "martial" virtues and the war instinct, however, would be coopted for pacifist purposes. In the new state, all the youthful population would be drafted into work for the country's industrial "army."[175] Quoting a long passage from H. G. Wells, James prophesied to his wide nonacademic audience that conscription into the industrial army would lift individuals to "a higher social plane" and place them in "an atmosphere of service and cooperation."[176] Requiring "our gilded youth" to share other people's occupations and experiences, especially hard labor and pain, would enable them to "come back to society with healthier sympathies."[177] Yet even in this socialist-industrialist utopia, one that came strikingly close to Edward Bellamy's 1888 vision of a collectivist society controlled by a coercive centralized power, James endeavored to avoid the "effect of unreality" resulting from the relinquishment of individuality. The proposed temporary enrollment in the industrial army promoted not only solidarity but also the cultivation of strenuous virtues that would strengthen individuals and endow them with agency. In the end, as James Livingston argues, James may have accepted a form of "corporate" socialism. Yet he did so only when he found a way of reconciling socialism with strong individual agency and a strong sense of personal identity.

To sum up, while James's conception of the self denied the substantiality, the simplicity, and the a priori unity of the self—those very features that nineteenth-century individualists had mobilized to defend the rights and priority of the individual—it provided the individual with a stronger form of agency and self-determination. Individuals were responsible for unifying their own selves through strenuous effort, for continuously negotiating amiable relationships among their various social selves, and for sustaining and creating afresh stable selves. This continuous effort of self-fashioning and self-sustaining is what made individuals into effective centers of initiative and social change and enabled them to cultivate, even renew minute by minute, the strength necessary to assert their autonomy in the new social-economic order.

The mystical experiences of the loosely bounded self and the region of extramarginal consciousness provided the ground for intimacy and for mutual, constructive understanding, without sacrificing individual selfhood to the goals of cooperation and solidarity as did many socialist visions of the time. They paved the way for entirely new modes of human interactions, ones that James believed would go a long way toward solving the problems of American society.

James, in the end, imagined not a new society but new forms of sociability, new ways in which people could interact with one another despite of—perhaps even within—the institutions of corporate capitalism. His pragmatic, multidisciplinary account of the self—a kind of practical inquiry that could not be contained within the bounds of any existing discipline or academic discourse—offered not only theoretical insight into the nature of the human self but also practical recipes. It afforded tools that people could use to cultivate new forms of individual and social being—new moral, mental, and social dispositions—that would enable them to transform at once themselves and their social environments, and fully flourish under the fully socialized conditions of life that James believed had come to stay.

# 7

# The Philosopher's Place
## James, Münsterberg, and Philosophical Trees

In 1904 James offered an unusual portrait of the genuine philosopher. The philosopher was a person located at the intersection of different roads: he was, in fact, "a cross-roads of truth." When James proposed this description, he had in mind as an exemplar of his ideal philosopher a bizarre nineteenth-century figure, someone whom professional philosophizers would hardly have admitted into their ranks—the German scholar Gustav Theodor Fechner, the initiator of "psychophysics" and an important school of scientific aesthetics, a novelist, and a student of physics, metaphysics, and cosmology as well as spiritual and occult matters. Describing Fechner as an "encyclopedic mind," one equally at home in a wide range of disciplines and endeavors, James presented him as a "philosopher in the great sense of the word."[1]

Here James identified the genuine philosopher in terms of the position that the philosopher occupied in a space of knowledge. The philosopher was to occupy the site at which distinct disciplinary paths could intersect; he was to *be* a point (perhaps temporary) of convergence between people who traveled along different roads. This metaphor was not simply a casual comment; instead, the question of the place in which philosophy ought to be conducted was actually central to James's understanding of what it meant to do "philosophy."

By addressing that question, I argue, James redefined the function of philosophy and reconfigured the intellectual and social geography of knowledge.

At the turn of the twentieth century, philosophers, scientists, historians, philologists, and a wide range of other humanists spent time and energy thinking about the proper arrangement of the disciplines, and how to redress the fragmentation of knowledge brought about by the visible trend toward specialization, the professionalization of inquiry, and the institutionalization of the disciplines. Debates over these questions quite frequently took the form of attempts to create or rearrange spaces of knowledge. These included physical spaces (the layout of a university campus) as well as metaphorical spaces of knowledge, such as charts of the disciplines, trees of the sciences, and other diagrams seeking to graphically order knowledge and represent its unity.

Spaces of knowledge not only embodied prescriptions about the intellectual relationships among fields but also dictated rules about the ways in which scientists and other producers of knowledge ought to interact. Indeed, as a rich literature in the humanities has shown, spaces not only function as arenas in which human action takes place but are also, in themselves, "constitutive" of modes of social relationship. They "enable and constrain" what people do in them and how they behave with one another. Spaces of knowledge (real or imaginary) are no exception.[2] Thus charts and trees of the sciences and other spatial schemes for the unification of knowledge encoded assumptions about the proper social ordering of inquiry and the ways in which power ought to be distributed in the academy. Furthermore, as participants in debates about the order and unity of knowledge quickly perceived, charts of knowledge and other configurations of the disciplines could also convey norms about the proper political order of a country or the proper nature of community. They functioned as points of convergence of multiple issues that surrounded the institutionalization of the modern disciplinary mode of knowledge production: issues over the proper configuration of the academy, and at the same time over the order of society; issues over the nature of the cosmos and the relationships among its parts; and issues over the desirable relationships among knowledge producers, among individuals in society, as well as between citizens and the state. Circulating as robust "immutable mobiles" through the pages of academic journals and pamphlets, and, in one case, as we will see, enacted through a hugely visible public performance, these arrangements of knowledge became dense shorthand for nested academic, social, and cosmological visions that participants perceived to be closely related.[3]

Examining turn-of-the-century plans for the ordering and unification of knowledge will enable us to perceive not only that James was keenly interested in questions of the proper order and unity of knowledge—which may come as a surprise—but also that, like many of his colleagues in philosophy, he addressed those urgent questions in spatial terms: that is, by thinking about the proper location of knowledge producers and the proper arrangement of spaces of knowledge. I will argue that by assigning philosophers a new place on a map of knowledge and converting them into crossroads of different disciplinary and professional trajectories, James assigned philosophers a new social role.

"Philosophy" in this chapter will largely indicate a kind of enterprise that, in the last decade or so of his life, James termed "general philosophy," a field encompassing all the special sciences; James depicted it as the "trunk of the tree of knowledge" from which all the sciences stemmed, a metaphor informing the many charts that arranged the different scientific disciplines in the scheme of a branching tree. James presented general philosophy as a "system of completely unified knowledge" (a formula he borrowed from Herbert Spencer) and contrasted it with philosophy in a narrower, and more modern, sense of the term as the technical discipline of metaphysics. James prophesied that "general philosophy," that is, "philosophy" in the older and to him "more worthy sense," would come again to prevail in the future—one of his many wrong guesses.[4]

It was the pursuers of "general" philosophy—people like Fechner—that James portrayed as the crossroads of truth. By doing so, I suggest, James reinvented the philosopher as a "go-between," an enabler of encounters among people who worked in different disciplines and professions and inhabited different social worlds. The new position that James ascribed to philosophers in the "more worthy" sense of the term made them responsible for facilitating cross-disciplinary and cross-divisional conversations, and for promoting a unity of knowledge.

The unity that James pursued, however, was very different from the grandiose philosophical syntheses of knowledge supposedly achieved by an "architectonic" super-discipline of philosophy. It was neither a methodological nor a conceptual synthesis. Instead, it was a social synthesis. Philosophy, thus, for James, would unify knowledge not on intellectual grounds but instead socially, by paving the way for the formation of open, pluralistic, yet cohesive, even intimate, communities of inquirers.

The social organization of inquiry fostered by philosophers like Fechner, and by the project of "general philosophy," reflected and incorporated the modes of social relationships that, as discussed in chapter 6, James elaborated through his discussion of social and political issues, and through his multidisciplinary and pragmatic conception of the human self. For James, as indeed for many other geographers and unifiers of knowledge, thinking about the philosopher's place and the nature of knowledge spaces was a means for reconfiguring the intellectual and social order of knowledge in a way that matched his vision for the order of society.

To best appreciate the significance of the place(s) that James ascribed to philosophers and the kinds of community that James perceived would be promoted by those knowledge spaces, part 1 of this chapter begins by presenting a vision that was a polar opposite to James's—the scheme for the classification and unification of the sciences that Hugo Münsterberg made central to the Congress of Arts and Sciences held in St. Louis in 1904, one of the pet projects of James's colleague. A quick review in the next two sections of period maps of the sciences and philosophical schemes for the unification of knowledge will make transparent the immediate and broader goals of Münsterberg's taxonomical and unifying efforts: locating the philosopher at the top of an epistemic and social hierarchy, and promoting a hierarchical, well-ordered society modeled after the German empire, which, itself, was to become the guarantor of the unity—peace—of the world. Part 2 shifts the discussion back to James, approaching him largely, although not exclusively, from the angle of his disagreements with Münsterberg on the question of the philosopher's place. These sections suggest that James and Münsterberg's divergences on that issue were central to their profound disagreements about the nature of philosophy, the geography of knowledge, and the proper organization of society.

## Part 1: Classifying And Unifying Knowledge

### The Unification of Knowledge: A Philosopher's Plans for the St. Louis World Congress of Arts and Sciences

The International Congress of Arts and Sciences accompanying the Universal Exposition of St. Louis opened, as planned, on September 19, 1904. Its goal was ambitious: to foster "the unity of human knowledge." The organizers anticipated that such a momentous project would attract prestigious scholars from across the ocean and around the world.[5] The president of the congress, Simon Newcomb, and its vice presidents, Hugo Münsterberg and the Chicago sociologist Albion Small, hoped the congress would remind scientists, busy in their specialized work, of the underlying unity of truth—a "mission" never as urgent as in a period that witnessed unprecedented levels of academic specialization. To this end, the congress organizers invited philosophers, mathematicians, historians, political scientists, physicians, astronomers, geologists, biologists, chemists, anthropologists, psychologists, sociologists, psychiatrists, physicians, engineers, architects, and students of the legal, economic, philological, pedagogical, aesthetic, and theological sciences: in short, representatives from all the human and natural sciences, pure and applied. Their task was to discuss the relations of their disciplines to neighboring and more distant fields, with a view toward fostering the unification of knowledge.

Like Minerva born fully armed out of Zeus's head, the grandiose plan was Münsterberg's baby.[6] Münsterberg wanted the congress to be a "symptom of the spirit of the time": "Our time," he wrote, "longs for a new synthesis, for a unified view of reality." Expressing that longing, the congress would offer something more than a group of disconnected specialized meetings.[7] In order to "fight against one-sidedness and to overcome the specializing narrowness of the scattered sciences," the utmost care was to be taken in organizing the event, especially in choosing the speakers.[8] "Narrow specialists" would not do; the organizers sought speakers who could combine specialized research with "the inspiration that comes from looking over vast regions."[9] Nor would amateurs and dilettantes do either. While businessmen, politicians, lawyers, farmers, miners, and other "unscholarly men" were welcome to attend the congress as auditors, their voices were not to be heard. The congress was to be a professional, scholarly event.

The planners faced the great challenge of assembling a program, for they believed that the organization of the program was itself a contribution to the common goal. In place of a "directory," or "a mere dictionary of phenomena, of events and laws," or a "university catalogue" with its "chance combinations," the program would attempt to resemble a veritable "cyclopedia," revealing that knowledge was intrinsically connected within a unified system.[10]

The search for the unity and order of knowledge was dear to Münsterberg's heart. In the 1890s the German-born psychologist had articulated a vision of philosophy as an "encyclopedia" embracing all the special sciences.[11] In his *Grundzüge der Psychologie* (1900),[12] Münsterberg devoted almost a hundred pages to a discussion of the relationship of psychology to the human and the natural sciences. In the next few years he turned to the ambitious goal of charting the whole field of knowledge in order to show its intrinsic wholeness.

Münsterberg took up that goal in his article "The Position of Psychology in the System of Knowledge," which he published in 1903 amid frenetic preparatory activity for the congress.[13] After rapidly dismissing previous classification schemes for the sciences, including the trees of knowledge devised by Francis Bacon, Herbert Spencer, Wilhelm Wundt, and Karl Pearson, Münsterberg offered his own taxonomy. His classificatory system was not based on distinguishing kinds from among the panoply of natural phenomena, for that principle, to his mind, would result only in "a pigeon-holing of scholarly work." Instead, his scheme sought to unveil the inner, "logical relations" among the sciences, and laid out the proper limits of their spheres of inquiry. The best way to accomplish this task was to use a "graphic representation," and thus the article was accompanied by a chart of knowledge (fig. 7.1).[14]

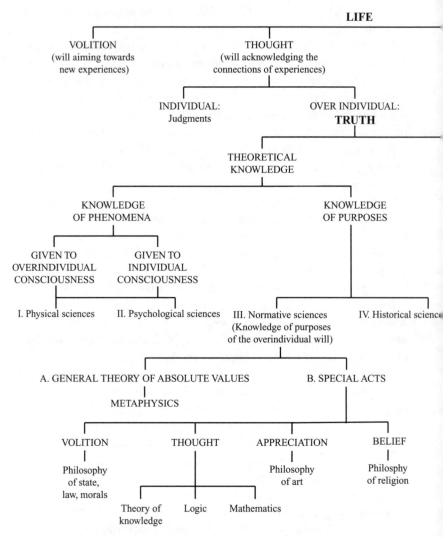

Fig. 7.1. Münsterberg's chart of knowledge. From H. Münsterberg, "The Position of Psychology in the System of Knowledge," *Harvard Psychological Studies* 1 (1903). In the public domain.

Münsterberg's map—a tree of knowledge—followed a branching scheme rooted in "Life," which he defined as "a system of purposes felt in immediate experience." One of those purposes was "Truth," and from it "stem[med] Knowledge." Knowledge, in turn, branched into two main subdivisions— theoretical knowledge and applied knowledge—each of which could be further subdivided into "knowledge of purposes" and "knowledge of phenomena."[15] The theoretical sciences of phenomena included the physical

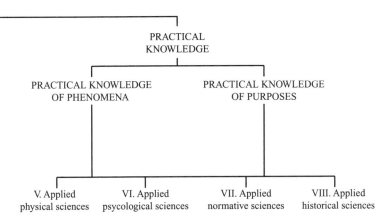

APPRECIATION
(will resting in
isolated experiences)

BELIEF
(will resting in the
suppliments of experiences)

PRACTICAL
KNOWLEDGE

PRACTICAL KNOWLEDGE
OF PHENOMENA

PRACTICAL KNOWLEDGE
OF PURPOSES

V. Applied
physical sciences

VI. Applied
psycological sciences

VII. Applied
normative sciences

VIII. Applied
historical sciences

sciences and the psychological sciences. The theoretical sciences of purpose included the normative sciences (those that dealt with "over-individual purposes") and the historical sciences (those that dealt with "individual purposes"). The normative sciences included "the general theory of absolute values," that is, metaphysics and "normative sciences of special acts." These were further subdivided into philosophy of morals, of law, and of state (under the heading of "volition"); theory of knowledge, logic, and mathematics (under the heading of "thought"); philosophy of art (under the heading of "appreciation"); and philosophy of religion (under the heading of "belief"). The historical sciences branched into the "general theory of real life" (from which derived the philosophy of history) and historical sciences of special acts, such as politics, law, and economy. The applied sciences were divided into applied knowledge of phenomena (applied physical sciences and applied psychological sciences), and applied knowledge of purposes (applied normative sciences, and applied historical sciences, for example, science of artistic production, economics, science of religious service, or practical theology). To Münsterberg this scheme presented several advantages. While the chart fell short of Münsterberg's vision of philosophy as an "encyclopedia" embracing all of the sciences, it still managed to carve out a prominent place for the philosophical sciences. These, he complained, most previous classificatory systems had either excluded or relegated into "appendix[es]" "incommensurable" with the rest of the system. By contrast, he stated, in his system of knowledge the philosophical sciences found their "natural and necessary place."[16]

Münsterberg's oversized map (it took up several folded pages) was meant to provide the "ground plan" for the congress. The final program of the congress—which was preceded by several other versions—divided all the sciences into two great groupings (knowledge of phenomena and knowledge of purposes) subdivided into seven divisions: normative sciences, historical sciences, physical sciences, mental sciences (all of which corresponded to the division of theoretical knowledge on the map), and utilitarian, regulative, and cultural sciences. These, in turn, branched into twenty-five departments, further subdivided into 130 sections, "with the possibility of an unlimited number of sub-sections."[17] Speakers for each of the seven divisions would first address the congress at 10 a.m. on September 20 in a session devoted to the themes of "the unity of the whole field" and the "relation of the special branch to other branches."[18] Münsterberg anticipated that many scientists would disagree with his program. For example, he expected mathematicians to object to being grouped with philosophers rather than, for example, with physicists. Yet one goal of the congress was precisely to displace such "popular fallacies" and educate the general public about the true relationships among the sciences.

The final plan of the conference, however, differed in some important ways from the map.[19] While Münsterberg's chart divided applied knowledge into four subdivisions, the final program divided it into only three subgroups. Most important, the two main subdivisions of the theoretical sciences (sciences of phenomena and normative sciences) were switched, so that now the "Philosophical Sciences" became "department 1" of "division A" (fig. 7.2). Thus, while the chart of knowledge placed philosophy on a par with mathematics (in fact, also with the theoretical sciences of phenomena), the congress program gave it a position of primacy. This was conveyed by the letters and numbers that indicated sessions and subsessions. The vertical arrangement of the detailed congress program distributed to all participants, by listing philosophy above all other sciences, visually presented Münsterberg's idea that it was superior to mathematics and, indeed, to all other sciences (figs. 7.3a and 7.3b).

Not only did philosophy figure at the top of the conference program, but it was obvious to many that philosophy was intended to function as the foundation on which the map of knowledge was based. Indeed, critics complained about this positioning as soon as the congress plan was made public. For example, John Dewey, who had by then abandoned idealist philosophy, asserted that the map stemmed directly out of Münsterberg's particular "school

## A. NORMATIVE SCIENCES

1. Philosophical Sciences.
2. Mathematical Sciences.

## B. HISTORICAL SCIENCES

3. Political Sciences.
4. Legal Sciences.
5. Economic Sciences.
6. Philological Sciences.
7. Pedagogical Sciences.
8. Æsthetic Sciences.
9. Theological Sciences.

## C. PHYSICAL SCIENCES

10. General Physical Sciences.
11. Astronomical Sciences.
12. Geological Sciences.
13. Biological Sciences.
14. Anthropological Sciences.

## D. MENTAL SCIENCES

15. Psychological Sciences.
16. Sociological Sciences.

## SECTIONS

1. *a* Metaphysics.
   *b* Logic.
   *c* Ethics.
   *d* Æsthetics.
2. *a* Algebra.
   *b* Geometry.
   *c* Statistical Methods.
3. *a* Classical Political History of Asia.
   *b* Classical Political History of Europe.
   *c* Medieval Political History of Europe.
   *d* Modern Political History of Europe.
   *e* Political History of America.
4. *a* History of Roman Law.
   *b* History of Common Law.
   *aa* Constitutional Law.
   *bb* Criminal Law.
   *cc* Civil Law.
   *dd* History of International Law.
5. *a* History of Economic Institutions.
   *b* History of Economic Theories.
   *c* Economic Law.
   *aa* Finance.
   *bb* Commerce and Transportation.
   *cc* Labor.

6. *a* Indo-Iranian Languages.
   *b* Semitic Languages.
   *c* Classical Languages.
   *d* Modern Languages.
7. *a* History of Education.
   *aa* Educational Institutions.
8. *a* History of Architecture.
   *b* History of Fine Arts.
   *c* History of Music.
   *d* Oriental Literature.
   *e* Classical Literature.
   *f* Modern Literature.
   *aa* Architecture.
   *bb* Fine Arts.
   *cc* Music.
9. *a* Primitive Religions.
   *b* Asiatic Religions.
   *c* Semitic Religions.
   *d* Christianity.
   *aa* Religious Institutions.
10. *a* Mechanics and Sound.
    *b* Light and Heat.
    *c* Electricity.
    *d* Inorganic Chemistry.
    *e* Organic Chemistry.
    *f* Physical Chemistry.
    *aa* Mechanical Technology.
    *bb* Optical Technology.
    *cc* Electrical Technology.

Fig. 7.2. Preliminary program, approved by the Organizing Committee of the Congress of Arts and Sciences on February 23, 1903. From *Congress of Arts and Science, Universal Exposition, St. Louis, 1904*, ed. Howard J. Rogers (Boston: Houghton, Mifflin, 1905). In the public domain.

# SPEAKERS AND CHAIRMEN

## DIVISION A—NORMATIVE SCIENCE

SPEAKER: PROFESSOR JOSIAH ROYCE, Harvard University.
(*Hall 6, September 20, 10 a. m.*)

## DEPARTMENT 1—PHILOSOPHY

(*Hall 6, September 20, 11.15 a. m.*)

CHAIRMAN: PROFESSOR BORDEN P. BOWNE, Boston University.
SPEAKERS: PROFESSOR GEORGE H. HOWISON, University of California.
PROFESSOR GEORGE T. LADD, Yale University.

**SECTION A. METAPHYSICS.** (*Hall 6, September 21, 10 a. m.*)

CHAIRMAN: PROFESSOR A. C. ARMSTRONG, Wesleyan University.
SPEAKERS: PROFESSOR A. E. TAYLOR, McGill University, Montreal.
PROFESSOR ALEXANDER T. ORMOND, Princeton University.
SECRETARY: PROFESSOR A. O. LOVEJOY, Washington University.

**SECTION B. PHILOSOPHY OF RELIGION.** (*Hall 1, September 21, 3 p. m.*)

CHAIRMAN: PROFESSOR THOMAS C. HALL, Union Theological Seminary, N. Y.
SPEAKERS: PROFESSOR OTTO PFLEIDERER, University of Berlin.
PROFESSOR ERNST TROELTSCH, University of Heidelberg.
SECRETARY: DR. W. P. MONTAGUE, Columbia University.

**SECTION C. LOGIC.** (*Hall 6, September 22, 10 a. m.*)

CHAIRMAN: PROFESSOR GEORGE M. DUNCAN, Yale University.
SPEAKERS: PROFESSOR WILLIAM A. HAMMOND, Cornell University.
PROFESSOR FREDERICK J. E. WOODBRIDGE, Columbia University.
SECRETARY: DR. W. H. SHELDON, Columbia University.

**SECTION D. METHODOLOGY OF SCIENCE.** (*Hall 6, September 22, 3 p. m.*)

CHAIRMAN: PROFESSOR JAMES E. CREIGHTON, Cornell University.
SPEAKERS: PROFESSOR WILHELM OSTWALD, University of Leipzig.
PROFESSOR BENNO ERDMANN, University of Bonn.
SECRETARY: DR. R. B. PERRY, Harvard University.

**SECTION E. ETHICS.** (*Hall 6, September 23, 10 a. m.*)

CHAIRMAN: PROFESSOR GEORGE H. PALMER, Harvard University.
SPEAKERS: PROFESSOR WILLIAM R. ° RLEY, University of Cambridge.
PROFESSOR PAUL HENSEL, University of Erlangen.
SECRETARY: PROFESSOR F. C. SHARP, University of Wisconsin.

Fig. 7.3a. Sections of the detailed final program for the Congress of Arts and Sciences.
*Congress of Arts and Science, Universal Exposition, St. Louis, 1904*, ed. Howard J. Rogers (Boston:
Houghton, Mifflin, 1905), 1:85–134. In the public domain.

SECTION F. AESTHETICS. (*Hall 4, September 23, 3 p. m.*)
CHAIRMAN: PROFESSOR JAMES H. TUFTS, University of Chicago.
SPEAKERS: DR. HENRY RUTGERS MARSHALL, New York City.
PROFESSOR MAX DESSOIR, University of Berlin.
SECRETARY: PROFESSOR MAX MEYER, University of Missouri.

## DEPARTMENT 2 — MATHEMATICS
(*Hall 7, September 20, 11.15 a. m.*)
CHAIRMAN: PROFESSOR HENRY S. WHITE, Northwestern University.
SPEAKERS: PROFESSOR MAXIME BÔCHER, Harvard University.
PROFESSOR JAMES P. PIERPONT, Yale University.

SECTION A. ALGEBRA AND ANALYSIS. (*Hall 9, September 22, 10 a. m.*)
CHAIRMAN: PROFESSOR E. H. MOORE, University of Chicago.
SPEAKERS: PROFESSOR EMILE PICARD, The Sorbonne; Member of the Institute of France.
PROFESSOR HEINRICH MASCHKE, University of Chicago.
SECRETARY: PROFESSOR G. A. BLISS, University of Chicago.

SECTION B. GEOMETRY. (*Hall 9, September 24, 10 a. m.*)
CHAIRMAN: PROFESSOR M. W. HASKELL, University of California.
SPEAKERS: M. GASTON DARBOUX, Perpetual Secretary of the Academy of Sciences, Paris.
DR. EDWARD KASNER, Columbia University.
SECRETARY: PROFESSOR THOMAS J. HOLGATE, Northwestern University.

SECTION C. APPLIED MATHEMATICS. (*Hall 7, September 24, 3 p. m.*)
CHAIRMAN: PROFESSOR ARTHUR G. WEBSTER, Clark University, Worcester, Mass.
SPEAKERS: PROFESSOR LUDWIG BOLTZMANN, University of Vienna.
PROFESSOR HENRI POINCARÉ, The Sorbonne; Member of the Institute of France.
SECRETARY: PROFESSOR HENRY T. EDDY, University of Minnesota.

## DIVISION B — HISTORICAL SCIENCE
(*Hall 3, September 20, 10 a. m.*)
SPEAKER: PRESIDENT WOODROW WILSON, Princeton University.

## DEPARTMENT 3 — POLITICAL AND ECONOMIC HISTORY
(*Hall 4, September 20, 11.15 a. m.*)
CHAIRMAN:
SPEAKERS: PROFESSOR WILLIAM M. SLOANE, Columbia University.
PROFESSOR JAMES H. ROBINSON, Columbia University.

Fig. 7.3b.  Sections of the detailed final program for the Congress of Arts and Sciences. *Congress of Arts and Science, Universal Exposition, St. Louis, 1904*, ed. Howard J. Rogers (Boston: Houghton, Mifflin, 1905), 1:85–134. In the public domain.

of metaphysics," a type of idealism that recognized the presence in the world of two main types of experiences: objects and purposes.[20] Others, including (as discussed below) William James, perceived that Münsterberg's elaborate and orderly hierarchical organization reflected his own cosmology, a monistic scheme in which all facts and laws lay in orderly compartments and were ultimately enveloped and unified under the supreme Ideal. Furthermore, Münsterberg made it abundantly clear that the unification of knowledge could ultimately be accomplished *only* by philosophers. Although he paid lip service to Simon Newcomb's belief that the goals of unifying and properly classifying knowledge required the collective cooperation of an entire "community" of scientists, he also stressed that the desired ordering and synthetic unity of knowledge could only result from "the gigantic thought of a single genius," namely a philosopher.[21] Thus the "work of the many," represented by the roughly three hundred scientists and scholars invited to participate in the congress, was to be merely preliminary and ancillary to the work of "the one" genial philosopher.[22]

At the congress Münsterberg's philosophical imperialism did not go unnoticed. In the opening remarks of the paper that he presented at the general philosophy session, the Berkeley-based philosopher George Holmes Howison focused expressly on it. "The program," Howison told his audience, "indicates, by no uncertain signs, the leading, the determining part that philosophy must have in the achievement" of the unification of knowledge. Howison perceived in the conference program "a renewed Hierarchy of the Sciences" that, "after so long a period of humiliating obscuration," placed the "figure of Philosophy" at the summit, "raised [it] anew to that supremacy, as Queen of the Sciences, which had been hers from the days of Plato to those of Copernicus." He concluded his remarks by expressing the hope that "this sign of her recovered empire" may "not fail."[23]

The International Congress of Arts and Sciences was, of course, also a political and diplomatic event. Münsterberg, skilled in diplomacy, was attentive to its political significance, especially during the organizational stage. Thus, soon after his plan was approved, he sent President Roosevelt a letter seeking his support for the momentous event, and hinting that the congress could be fruitfully mobilized for electoral purposes.[24] Speaking and writing to other leaders and audiences, he sought to use the event to improve the shaky relationship between the United States and Germany, the country where he was born and of which he remained a citizen.[25] In the summer of 1903, one year before the congress was to take place, the organizers traveled to Europe in order to personally deliver their invitations. Münsterberg traveled first to Germany, beginning in Berlin, the center of the empire, and moving outward to more peripheral places, and then to Austria and Switzerland.

He met with German philosophers, men of science, and politicians, including Kaiser Wilhelm II. During these visits, as well as in his letters to President Roosevelt, Münsterberg emphasized that the basic responsibility for the conference program was his.[26] He reiterated the point in the article "The Congress Plan," which he published in the proceedings of the congress. By emphasizing his leading position—before, during, and after the congress—Münsterberg assumed the role of the classifier and "unifier" of knowledge before an international audience comprising the leading political powers of the time. He presented himself, *qua philosopher*, as the "gigantic" figure, the genius within the international academy entrusted with charting the whole territory of knowledge, and bringing order and unity to the sciences.

To unpack the broader implications of Münsterberg's congress of arts and sciences, it is useful to take a wider look at how scholars of the era deployed charts of the disciplines and representations of the unity of knowledge. Focusing on examples from around the turn of the twentieth century with which both Münsterberg and William James were quite familiar will enable us to better understand their plans for the ordering and unifying of knowledge and will, in turn, allow us insight into the meaning of their visions for the place of the philosopher within the academy and within the wider society.

## Taxonomies of Knowledge

Münsterberg's chart of knowledge represented one very visible intervention in a much wider and intense debate over the proper ordering of the sciences and the unity of knowledge. Charts and trees of the sciences have an illustrious past that can be traced back at least to the trivium and quadrivium that prescribed courses of study in medieval universities. In the seventeenth and eighteenth centuries, debates over the order and unity of knowledge had taken the form of disputes over the proper organization of encyclopedias and dictionaries of arts and sciences, the latter almost always accompanied by charts. Historian of science Richard Yeo dates the demise of this tradition to the turn of the nineteenth century. He argues that all major nineteenth-century British encyclopedias abandoned any ambitions of ordering and unifying the sciences, stopped printing charts or trees of knowledge, eliminated most cross-references among entries for different disciplines, and adopted a format consisting—in the case of the *Encyclopedia Britannica*, for example—of separate, self-standing treatises devoted to individual disciplines and written by experts.[27] At the same time encyclopedias ceased to be understood—in contrast to seventeenth-century dictionaries of arts and sciences—as tools that would enable the gentleman to acquire well-rounded knowledge. Consequently, they also ceased to prescribe the order in which the disciplines and arts were to be learned. In view of the increasing specialization of knowledge and division of labor in the sciences, the very concept of the "encyclopedic mind," which had been so powerful in the classical age, lost its hold. The notion that one person could possibly master all the disciplines became ludicrous to most, and those who insisted on pursuing it in the nineteenth century—such as the polymath Coleridge—were ridiculed as dilettantes.

Nevertheless, throughout the nineteenth century the ideal of finding some way to order knowledge and unify the arts and sciences continued to seduce a great variety of scholars.[28] Exiled from dictionaries and encyclopedias, discussions of the order and unity of the sciences flourished in different venues. The late nineteenth and early twentieth century witnessed a plethora of classifications, tableaux, trees, charts, and other diagrams of knowledge as well as a wide range of plans for the unification of the sciences.

Anxious cartographers of knowledge obsessively assessed the merits and defects of classical taxonomies, including, especially, Auguste Comte's arrangement of the sciences in a linear sequence of increasing complexity and decreasing generality, and contrasted serial classifications with branching schemes, usually binary or ternary, modeled on evolutionary trees.[29] They produced histories of classifications and classifications of classifications, such as the one by Durand de Gros reproduced in figure 7.4.[30] They deployed a variety of classificatory principles, seeking to trace the genealogical, logical, methodological, or pedagogical ordering of the sciences, or to arrange the sciences in a system that mirrored the classification of the faculties of the human mind.

While differing in their classificatory principles and products, these taxonomies shared a common feature: turn-of-the-twentieth-century participants in debates about the proper ordering of knowledge thought of knowledge as a "space." Even those who rejected the metaphor of the map, and dismissed the idea that knowledge resembled a continent or a country that could be divided up into different regions, understood knowledge that way. Such a spatial understanding of knowledge—one with which the work of Michel Foucault and many others has made us all familiar—was, of course, not new. Think, for example, of Ephraim Chambers's metaphors of knowledge as a "terra cognita" and a common garden that needed to be freely cultivated, or of Jean Le Rond d'Alembert's elaborate *mappemonde* of the arts and sciences.[31] Yet, for many turn-of-the-twentieth-century schematizers, the metaphor of knowledge as a space, or a field, or a continent took on more concrete and local meanings. At a time that witnessed the institutionalization of the disciplinary configuration of knowledge in university

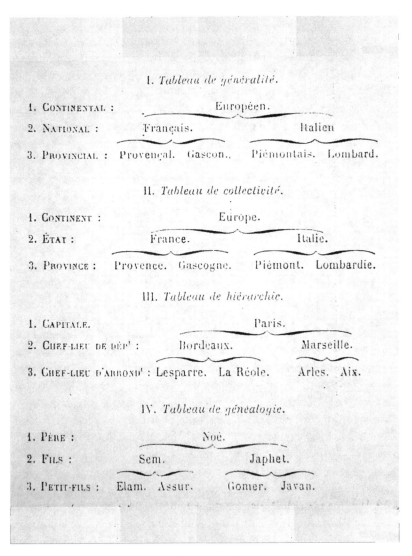

I. *Tableau de généralité.*

1. Continental : Européen.

2. National : Français. Italien

3. Provincial : Provençal. Gascon., Piémontais. Lombard.

II. *Tableau de collectivité.*

1. Continent : Europe.

2. État : France. Italie.

3. Province : Provence. Gascogne. Piémont. Lombardie.

III. *Tableau de hiérarchie.*

1. Capitale. Paris.

2. Chef-lieu de dép' : Bordeaux. Marseille.

3. Chef-lieu d'arrond' : Lesparre. La Réole. Arles. Aix.

IV. *Tableau de généalogie.*

1. Père : Noé.

2. Fils : Sem. Japhet.

3. Petit-fils : Elam. Assur. Gomer. Javan.

Fig. 7.4. Classification of classifications. From Joseph Pierre Durand de Gros, *Aperçus de Taxinomie Générale* (Paris: Alcan, 1899). In the public domain.

departments, research institutes, and professional societies, and, in North America, the creation of the research university, the question of the allotment of physical space became an issue that cartographers of knowledge often had to deal with in very concrete ways. Knowledge was a space to be divided up, occupied, cultivated, surveyed, or otherwise made use of.

It was obvious to all participants in these debates that knowledge was a nonhomogenous space; in two-dimensional schemes, center and periphery, top and bottom did not have the same value, and conveyed different implications. Geographical arrangements of this space brought with them distributions of power, and turn-of-the-twentieth-century philosophical and scientific classifiers perceived that, as Michel Serres would put it much later, "sovereignty within the city is acquired through sovereignty within the encyclopedia."[32] Thus, at the turn of the twentieth century, pigeonholing knowledge remained "an exercise in power."[33]

During this period classifiers mobilized maps of knowledge for a variety of reasons. These included, first of all, strengthening their field vis-à-vis others, attacking other fields, or promoting particular approaches at the expense of others. In such games, simple moves of inclusion and exclusion appeared to be especially advantageous, since they often allowed cartographers to rearrange power relations—something crucial at a time when the possible development or survival of an intellectual field depended on the allocation of funds and other academic resources. To give a few examples, with all of which Münsterberg and James were familiar, the sociologist Lester Ward adopted a slightly modified version of Comte's taxonomy of the sciences, which enabled him to assert that economics was "a branch of sociology" rather than, as many thought, the other way around.[34] James McKeen Cattell—a psychologist who fought for the autonomy of psychology from philosophy, and the editor of "Scientific Men of America," a turn-of-the-century Who's Who of science—divided the sciences into the physical, biological, and mental sciences. His chart included philology, sociology, and history, but excluded philosophy, thereby implying that philosophy was not a science.[35] It enabled psychology to share the prestige of the natural sciences, and presented its separation from philosophy as a matter of fact. Similarly, in a speech presented before the St. Louis congress, the Leipzig-based chemist Wilhelm Ostwald excluded philosophy from the pantheon of the sciences, counterattacking Münsterberg's transparent philosophical imperialism.[36] Ostwald presented a tripartite, neo-Comtian "scale" of the sciences that divided the sciences into three groups: mathematics, energetics (which included mechanics, physics, and chemistry), and biology (divided into physiology, psychology, and sociology). The scale was meant to reproduce the "increasing complication" and the "increasing unification" performed by each of the sciences. It also gave a prominent place to Ostwald's own "Energeticist" program, a new approach to physics intended to pass beyond attempts at mechanical reduction.[37]

Taxonomies were also useful attempts to upturn relationships of subordination and superordination, and to redefine fields. For example, many psychologists and philosophers in Wilhelmian Germany resorted to classifications of the sciences in order to redirect the field of psychology and clarify its position vis-à-vis philosophy and the natural sciences. Wilhelm Wundt, for one, divided the *Wissenschaften* into two main groups: particular sciences and philosophy. The particular sciences were subdivided into "formal sciences" (the mathematical sciences), and the "philosophical sciences." The real sciences were split into the natural sciences and the mental sciences, which further branched into the theory of the phenomena of the mind (psychology) and the sciences of the products of the mind (philology, the social sciences, and history). Philosophy was divided into "theory of knowledge" and "theory of principles" (further subdivided into metaphysics, philosophy of nature, and philosophy of spirit).[38] This scheme made psychology autonomous from both philosophy and the natural sciences. At the same time it visually represented psychology's intermediate position between the natural sciences and the human sciences, and it indicated psychology's priority with respect to other human sciences.[39]

As for Münsterberg's chart of knowledge (fig. 7.1 above), its immediate goal was to clarify the position of psychology by splitting the field into two parts and allocating them to two different divisions of knowledge. Experimental psychology fell in the same division as physics, suggesting that it was a "science of phenomena." Psychology of personality and the study of cultural products (*Volk-psychologie*) fell in the division of the historical sciences—they were "sciences of purposes." To Münsterberg the division demonstrated by his chart denounced and eliminated the ambiguity in the way the term "psychology" was often deployed to signify incongruous things.[40] Opponents of experimental psychology likewise resorted to classificatory schemes. The well-known orderings offered by the neo-Kantian philosophers Wilhelm Windelband and Heinrich Rickert entirely separated experimental psychology from the philosophical disciplines—reflecting their hopes of eliminating psychologists from chairs of philosophy.[41] Windelband's binary classification, which cut the sciences into "idiographic sciences" (that is, sciences dealing with individual events) and "nomothetic sciences" (or sciences that seek laws), allowed him to neatly separate psychology from philosophy and history, and locate it on methodological grounds in the domain of the natural sciences, at a safe distance from philosophy. Rickert divided the sciences into *Kulturwissenschaften* and *Naturwissenschaften*,[42] two groups characterized by different methods of concept formation. His classification located empirical psychology among the natural sciences, which were not concerned with values, suggesting that experimental psychology did not belong with philosophy, which instead provided a theory of values.[43]

In short, turn-of-the-century cartographers of knowledge abundantly used more or less obvious power games of including, excluding, subordinating, or repositioning fields to redistribute epistemic credibility among practitioners of different disciplines. These games were generally played *within* a taxonomy. These cartographers, however, were amply aware that the exercise itself of producing a chart of knowledge (and, a fortiori, staging public performances that enacted such charts, as happened at the St. Louis congress) could be a hugely empowering act—a point that Münsterberg fully appreciated.

Given the power conferred by producing charts of knowledge and new visual arrangements of the sciences, the question became: Who is entitled to chart knowledge? Historians, psychologists, and a wide range of social and natural scientists each claimed to be uniquely qualified for this important task. Among all participants, however, philosophers were the group who by far asserted that they alone were entitled to order the sciences. The reason behind philosophers' claims is not hard to discern. As we saw in chapter 1, in the second half of the nineteenth century, and especially at the end of the century, philosophers active in many Western countries perceived that their field was going through a deep crisis—the recent (and according to many, still ongoing) "period of humiliating obscurity" to which Howison referred with shivers at the Congress of Arts and Sciences. Philosophical diagnoses uniformly identified as a chief cause of the crisis the new prestige enjoyed by the natural and the rapidly growing social sciences, whose practitioners seemed to challenge at once the epistemic and the social authority of philosophers. Under such pressures philosophers found that it was imperative for them to redefine their field; they appreciated that charting knowledge would enable them to reconfigure philosophy in such ways that would imply a subordination of the natural and social sciences to philosophy.

The job of reordering the sciences must fall to philosophy, philosophers insisted, because philosophy could *unify* them. Philosophers alone, or so they claimed, possessed the tools and clarity of vision necessary to identify the conceptual and methodological ties linking the disciplines, and thus assemble the scattered "membra disjecta" of the specialized sciences into a well-ordered "whole."[44]

At the turn of the century, thus, plans for the philosophical unification of the sciences abounded. Each carried its distinctive epistemic, metaphysical, but also cosmological, and in some cases, political messages, offering abridged visions for a unity of knowledge, of the polity, and of the cosmos. Exploring some of these plans will disclose the many valences of Münsterberg's congress program and will enable us to perceive that his vision of the unity of knowledge was radically opposite to that developed in the same years by William James.

## Unities of Knowledge: Philosophy as "Scientia Scientiarum"

At the turn of the twentieth century the disunity of knowledge was a source of concern to a great variety of philosophers, moralists, and educators. At a time when the figures who had represented the ideal of the notion of the "scientist"—a term introduced in the 1830s to designate "a cultivator of science in general"—were being rapidly replaced by specialists, many worried that extreme specialization, which made communication among different specialists difficult, would stall the broader progress in science.[45] Philosophers worked hard in order to invert the process and make knowledge whole once again. To that end, they resurrected an older conception of philosophy as "scientia scientiarum," the science that embraced all the other sciences and ordered them. Many philosophers thereby redefined philosophy as the science of sciences, believing that that conception would enable them not only to redefine their field but also to rearrange the whole geography of knowledge in ways that would promote the primacy of philosophy.

A few examples of arrangements of the sciences produced by philosophical unifiers of knowledge will illustrate this idea. The first (figs. 7.5a and 7.5b) was produced by the scientific philosopher Herbert Spencer, an encyclopedic mind who sought to accomplish, through his own work, the task of unifying knowledge. Having spent some time investigating the meaning of philosophy, Spencer denied that philosophy had anything to do with God and what he called the Unknowable. This distinction neatly separated philosophy from theology but seemed to imply that the "domain" occupied by philosophy was the same as that occupied by science. If that was the case, Spencer surmised, it became imperative to ask in which ways philosophy differed from science, and to discuss whether there was any room for philosophy.

Spencer visualized the system of the sciences in the form of a tree of knowledge. He classified the various sciences into three groups (the abstract sciences, the abstract-concrete, and the concrete sciences), which stemmed from a common origin—the trunk of "Science" or "common knowledge."[46] He warned readers about the limitations of the metaphor of the tree of knowledge and its visual representations. Trees of knowledge, he said, were good because they conveyed the idea that the sciences had stemmed from a common trunk. Yet they were also inadequate, because they failed to illustrate that the sciences would not only further branch out (as an effect of specialization) but could also combine again in higher syntheses. To that end, he suggested that the different branches or sub-branches of the tree of knowledge could "unite again" through processes of "inosculation," giving

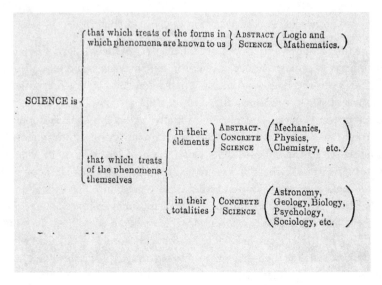

Fig. 7.5a. Herbert Spencer's tree of knowledge. From Spencer, "Classification of the Sciences" (1864), in Spencer, *Essays: Scientific, Political, and Speculative*, vol. 2 (New York: D. Appleton, 1892). In the public domain.

birth, through those unions, to higher generalizations and syntheses.[47] Spencer, however, emphasized that the processes of synthesis and generalization performed by the sciences could only be partial; and he depicted the sciences as bodies of "partially unified knowledge." "Science," he stated, meant "merely the family of the Sciences," and stood for "nothing more than the sum of knowledge formed of their contributions." It could not grasp the idea of the "fusion" of all those partial knowledges "into a whole."[48] Only philosophy could do that.

Philosophy appeared nowhere in Spencer's tree of knowledge, but its absence was quite revealing. It indicated that for Spencer philosophy was not one science among the others but rather the forum within which the missing final synthesis could be produced. Philosophy was, indeed, a "system of completely unified knowledge."[49] It alone could perform the final synthesis by showing that the most general laws of the special sciences could be derived as corollaries from one supreme, universal principle, the general law of the "concomitant redistribution of matter and motion."[50] In this scheme philosophy enjoyed a primacy that the special sciences simply could not claim.[51]

The second chart, conceived by the British philosopher Thomas Whittaker, produced a similar effect by other means. Whittaker proposed two simple, but ingenious, modifications to Comte's scheme. First, to the vertical series (in descending order) of mathematics, physics, chemistry, biology, and

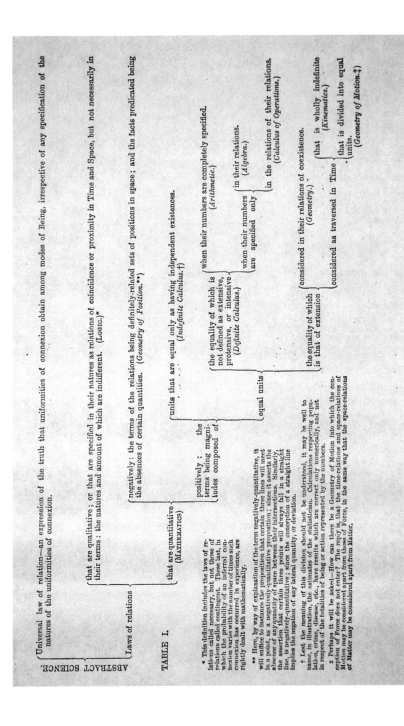

Fig. 7.5b. Classification of the abstract sciences. From Spencer, "Classification of the Sciences" (1864), in Spencer, *Essays: Scientific, Political, and Speculative*, vol. 2 (New York: D. Appleton, 1892). In the public domain.

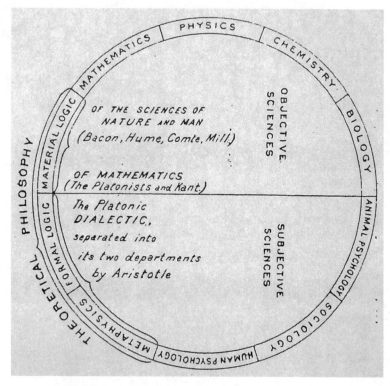

Fig. 7.6. Circle of knowledge. From Thomas Whittaker, "A Compendious Classification of the Sciences," *Mind*, n.s., 12 (1903): 21–34. In the public domain.

sociology, he added morality at the bottom.[52] He then argued that mathematics was preceded by logic, and that morality was followed by metaphysics, and, noticing that metaphysics was itself preceded by logic, bent the vertical line into a circle. This circular scheme made knowledge into a veritable "encyclopedia," he noticed, alluding to the etymology of the term (fig. 7.6). This graphical arrangement, furthermore, made philosophy into the beginning and the end of the circle of knowledge, suggesting that philosophy alone could hold the system of knowledge together.

A third scheme was conceived by the French philosopher Paul Janet, the uncle of the philosopher-psychologist Pierre Janet. One of the leading eclectic philosophers during the Second Empire and well into the Third Republic, Janet adopted a provisional "binary" taxonomy that divided the sciences into "cosmological sciences" and "noological sciences," that is, sciences of nature and sciences of mind (the latter including the historical, philological, and sociological sciences). This classification, however, was incomplete, and to those two groups Janet added a third group of sciences. These included *"epistemographie"* (or *"epistemotaxie"*), which was the philosophical science in charge of classifying and systematizing all the other sciences, and metaphysics, that is, the science in charge of explaining the unity of the universe.[53] This arrangement conveyed the idea that philosophy could unify the sciences because it belonged to a "higher order." The many representations of philosophy as an architectural science circulating at the time, including those discussed in chapter 1, similarly managed to solve the problem of the unification of knowledge and reinstate—at least on a symbolic terrain—philosophy's primacy over the sciences.

In the second half of the nineteenth century and at the turn of the twentieth, discussions about the philosophical unification of knowledge went well beyond concerns about the organization of the academy. Philosophers and other participants in these discussions soon perceived that plans for the unification of the disciplines, especially graphic or material embodiments of the unity and proper taxonomy of the arts and sciences, encoded not only intellectual programs or academic hierarchies redistributing epistemic authority but also broader cosmological and political visions. People perceived that such objects (real inscriptions in some cases, imagined spatial arrangements in others) could provide powerful blueprints for the reordering of society as well as guidelines for reconfiguring the relationships between citizens and state and between individuals and society.

As Peter Galison and others have noted, in countries striving to achieve political unification, or those recently unified, the phrase "the unity of knowledge" carried potent political overtones.[54] Thus, during the Italian *Risorgimento*, visions for the unity of the sciences, often premised on particular classificatory schemes, were mobilized to support various programs for the political unification of Italy. For example, the Piedmontese Catholic priest Antonio Rosmini proposed a tripartite project for a "Christian encyclopedia" of the sciences, which paralleled his (failed) political program for the creation of a confederacy that would unite the three kingdoms of Piedmont, Tuscany, and the Papal States under the pope.[55] His classification of the sciences became a powerful and contested object, and functioned as a polarizer for political discussion.[56]

In countries that had already attained unity, philosophers deployed schemes for a philosophical unification of knowledge in order to produce new political configurations and new forms of community. Friedrich Paulsen, the Prussian-based philosopher whom we met in chapter 1, represents an excellent example of the uses to which the idea of a philosophical unity of the sciences could be put. Paulsen portrayed philosophy as the only science that could keep alive the goal of the university, namely, to make knowledge whole.[57] He believed that the unification of knowledge required daily contacts among philosophers, natural scientists, and other researchers, and complained that the "ever-increasing division of labor" and the "resulting specialization of study" had endangered the possibility of those encounters.[58] Not only did German students find it more difficult to get the "encyclopedic" education that they needed to become teachers, physicians, lawyers, and clergymen, but too many students and faculty ended up pursuing very narrow research topics and lost a broader vision.[59] Paulsen believed that philosophy could provide the antidote to the "spirit of specialism" and the fragmentation of inquiry responsible for this state of affairs. Philosophy was uniquely able to carry out the task of the unification of knowledge since historically it had represented the trunk from which the sciences had originated. Ancient philosophers, Paulsen continued, were endowed with truly "encyclopedic minds," which had encompassed the whole of knowledge. In the twentieth century, of course, philosophers could no longer hope to master all the specialized bodies of knowledge, and perhaps they would never be able to actually unify knowledge. Nevertheless, Paulsen was confident that through their "love" of encyclopedic, all-embracing knowledge, they could keep alive the quest for unity; that was the true mission of philosophy.[60]

Paulsen's ideas about the philosophical unity of knowledge dovetailed with his cosmology, his political vision, and his vision of community. Paulsen was part of a wider German movement for "wholeness," one that, as Ann Harrington has shown, struggled at once against the "fragmentation of knowledge," modern individualism, and "the loss of community values."[61] By the time he wrote the books in which he outlined the unifying mission of philosophy, Paulsen had come to see himself as occupying a political position very close to social democracy.[62] Writing against the "Manchester school" and the "vulgar" doctrine of laissez-faire, Paulsen defended a strong state, with collective property and state control over labor and production.[63] Fascinated by the socialist Ferdinand Lassalle's picture of a state socialism, Paulsen depicted the state as something more than a legal institution. The state (as an embodiment of "Society") ought to be an all-inclusive union that worked to preserve "Society as a whole" and to promote the welfare of "all of its citizens."[64] Turning upside down "the old rationalism" that regarded "a people as an aggregation of individuals," he contended—like many other Germans of his time, ranging to opposite ends of the political spectrum—that the state was prior to individuals, the same way the organism was prior to its parts, and the individuals could exist "only through the whole."

"All in each and each in all": this slogan was central to Paulsen's cosmology and to his pantheistic metaphysics, the vision of a cosmic "social Monarchy," in which the monarch was constituted by the cosmic whole.[65]

Writing against atomism, the metaphysical doctrine that depicted the cosmos as an aggregate of "original elements absolutely independent," Paulsen represented the cosmos as an all-embracing whole within which the smallest units existed only in dynamic interaction with each other and with the whole.[66] Plurality was "an illusion," and atoms were only an "abstraction"; the unitary whole alone was "real."[67] Even the individual souls could not exist as independent elements but only as parts of a larger whole. In this cosmological framework Paulsen painted philosophy as the only activity that could uncover the underlying unity of the cosmos and of knowledge, and the interdependence of everything with everything else, and everyone with everyone else.

American philosophers, like their German counterparts, did not fail to perceive that plans for the unification (and ordering) of the sciences could foster political visions and forms of community, and that they could function as powerful tools for reconfiguring the issue of the relationship of the individual to society and to the state. For example, in his speech at the St. Louis congress, Howison proposed a metaphysical conception of the unity of knowledge and reality that encapsulated and furthered his vision for society. The proponent of a form of pluralistic idealism, Howison joined the choir of those who contended that the task of unifying knowledge belonged to philosophy. For him the unity of knowledge could only follow from a philosophical demonstration of the "genuine" metaphysical unity of the universe and from a philosophical solution of the metaphysical problem of the "One and the Many"—two tasks that, to Howison's mind, were strictly interconnected, and that he believed "could only be adjudicated before the Tribunal of Philosophy."[68] Monistic idealists described the "One" as an absolute mind that encompassed all the individual minds and from which all of reality flowed.[69] To Howison that conception was unacceptable because of its social implications: it deprived the individual minds of any autonomy from the One, and seemed to portray the Absolute Mind as an absolute monarch. Howison rejected the "oppressive" notion of an "all-inclusive Unit" that engendered the Many and that "completely determined, subjected, and controlled" them. In his pluralistic metaphysics the One was not the efficient cause of the universe but only a "final cause." This weaker form of causation still allowed for a unity of reality and for unity in the realm of knowledge.

Howison's own "purposive unity" left intact the ontological independence and individuality of the Many. In his scheme the One was a "federation" of the Many, and it embodied a "harmony among many free and independent primary realities, a harmony founded on their intelligent and reasonable mutual recognition."[70] The philosophical unification of knowledge likewise represented a goal. Howison, we may surmise, envisioned it as a sort of "federal" configuration of knowledge, one that would allow a degree of independence to the various disciplinary perspectives.

To return now to Hugo Munsterberg's plans for the St. Louis congress, we can appreciate how the Harvard philosopher used it as a powerful tool to promote what he believed was the proper social and political order. Münsterberg's approach to the congress program furthered his concept of a hierarchical, organic, orderly society, in which people were to stay in their place and perform specific functions, ones for which they were specially trained, and in which philosophers were to take a position of leadership.[71] A vocal critic of key aspects of American democracy—especially of the dangerous "dogmas" of majority vote and equality—Münsterberg publicly defended the virtues of the German monarchical government. He made the case that the Germans enjoyed a higher form of freedom precisely because they lived under "more complicated and systematized rules than the Americans," and argued that "titles and degrees and decorations representing social differentiation" were essential for civilized life and for progress.[72] The utmost attention that Münsterberg gave to the most minute details of the organization of the congress, ever revising his formidable lists of speakers, and his staunch resistance to "dilettanti" and candidates who might otherwise compromise the decorum of the event, reflected his obsession with social differentiation and etiquette.

They also mirrored his love for elaborate and carefully planned social rituals, such as the reception that he hosted at his home in Cambridge for the foreign speakers at the congress a few days after its close in St. Louis.[73] The semicomic confusions about the invitation of the Oxford philosopher F. C. S. Schiller, one of William James's closest friends, are telling. Schiller complained to James that his invitation to present a paper at St. Louis had been "revoked" on Münsterberg's order. James wrote Münsterberg about it, and Münsterberg responded that Schiller had never been invited in the first place. Münsterberg would never have invited Schiller, he explained, not because Schiller's philosophy was incompatible with his own, and not even because Schiller had publicly attacked Münsterberg's *Grundzüge der Psychologie*, but because Schiller's bad manners and questionable reputation would compromise the dignity of the event: "I should consider the level of the congress too much lowered if a personality of his type is in it."[74] The congress was to be an orderly and decorous gathering;[75] it would mimic the forms of hierarchical, formal social interactions that Münsterberg deemed most appropriate for academic communities of any size.

Münsterberg's choice of speakers also mirrored his vision of an international community in which Imperial Germany (and its army) had an important role to play. Those political overtones are indicated by Münsterberg's efforts during the preparation for the congress to reserve a large number of slots for German speakers.[76] That caused one of the several diplomatic incidents that jeopardized the entire program, as the French threatened to withdraw en masse "unless enough Frenchmen [were] invited to equalize with Germans."[77] A few years later, in his address to the Carnegie Peace Conference of 1907, Münsterberg made his political vision explicit, stating that Germany was uniquely positioned to foster harmony in the world. "Yes," he told his audience on that occasion, "if a sculptor were to create today a statue of the goddess of Peace, he might safely choose as his model fair Germania, with the Emperor's crown on her head, with a pure sword in her hand, and with mild eyes calmly looking on a serious yet happy nation of laborers who work for the eternal good of peaceful civilization."[78] Just as philosophy could unify knowledge and bring the arts and sciences together in harmonious unity, so a well-armed Germany could guarantee and promote international peace and harmony.

## Part 2: Spaces of Knowledge and the Mission of Philosophy

The social, political, and intellectual vision conveyed by Münsterberg's congress program conflicted sharply with James's vision of knowledge and society. Here I argue that whereas Münsterberg placed the philosopher at the pinnacle of an absolute hierarchy of disciplines, James located philosophers in interstitial knowledge spaces, and made philosophy—"general philosophy"—into a form of mediation between diverse modes of inquiry. Whereas for Münsterberg the philosopher was the most elite intellectual, the architect who should preside over the construction of the whole building of knowledge, for James the philosopher was a go-between, a person who should facilitate interaction between all sorts of professional investigators but also between intellectuals and ordinary people. Whereas for Münsterberg philosophy could unify the sciences by unearthing and prescribing the proper relationships among the various arts and sciences, for James philosophy could help bring about a very different type of unification of knowledge. The issue of the place in which philosophical inquiry was to be carried out lay at the nexus of these radical oppositions between James and Münsterberg.

### Ordering Knowledge and Nature: Italian Gardens and Classical Buildings

James was directly or indirectly familiar with most of the taxonomies and unification plans discussed or mentioned above, including those of Bacon,

Comte, Rosmini, Pearson, Windelband, Rickert, Cattell, and Spencer.[79] Despite his lack of sympathy for Spencer and his work, he appropriated Spencer's formula, which painted philosophy as a "system of completely unified knowledge."[80] James thoroughly enjoyed Friedrich Paulsen's *Einleitung in der Philosophie*, which represented an important source for his own conception of "general philosophy" and of the philosopher.[81] He was also on close terms with George Holmes Howison, who sent James a copy of his article "Philosophy and Science," in which he trumpeted the unifying function of philosophy.[82]

Above all, James was perfectly familiar with his Harvard colleague Münsterberg's taxonomy of the sciences and the philosophical unification of knowledge. He did not fail to note that Münsterberg's scheme for the St. Louis congress replicated the arrangement of knowledge and the position of philosophy in the territory of knowledge that Münsterberg had been championing in the same years through different projects. Readers will recall that, as noted in chapter 1, in 1901, at the same time that Münsterberg had started putting together a plan for the St. Louis congress, he had also begun to implement his vision for a new philosophy building at Harvard, Emerson Hall. In a circular addressed to a Visiting Committee, he had petitioned for a noble, "worthy monumental building," in Greek style, to be built in a "quiet central spot in the Harvard campus."[83] Not only would the building provide Harvard philosophers with a much needed "dignified, beautiful home," but, by virtue of its "central position," it would indicate that philosophy alone could unite the sciences. Philosophy, Münsterberg had written to the committee, was "more than one science among other sciences." Philosophy was indeed "the central science which alone has the power to give inner unity to the whole university work."[84] The location of Emerson Hall on the Harvard campus, a material space of knowledge, along with the position of philosophy at the summit of the hierarchy of the sciences, in the metaphorical space of knowledge embodied in the St. Louis congress program, provided twin emblems of philosophy's unifying function and of its supremacy vis-à-vis the sciences.

Examining James's reactions to Münsterberg's twin projects will disclose the significance of James's own vision for the philosopher's place. I suggest that, by assigning philosophers and philosophy, including his own pragmatist philosophy, certain places, James reconceptualized the nature and purposes of the philosophical unification of knowledge. In so doing, he redefined the meaning of philosophy.

Münsterberg invited James to give a talk at the congress. Indeed, he even went so far as to list James's name in a published roster of speakers for the psychology and philosophy sections.[85] James, however, was determined not to attend. "I am very sorry to be so persistently disobliging," he wrote to Münsterberg, "but I have nothing, absolutely nothing for which that congress seems a proper frame. . . . As for my brother [Henry James] he is less available for St. Louis than I am. A pity for your *Uberburdung*."[86] In letters to other correspondents, James emphasized that "personally [he took] no interest either in the form or the content of the whole enterprise,"[87] and confessed that, in fact, he had "taken particular pleasure in refusing to have anything to do with it."[88] He even tried to organize a miniboycott by discouraging his own correspondents from participating. He warned them that the weather in St. Louis at the end of the summer was unbearable, and predicted that "all the Europeans [would] die of it, and have to be sent home in their coffins."[89]

James's attitude toward Emerson Hall was similar. As we saw in chapter 1, when called upon he agreed to take part in the official planning and to help collect funds for the new building, even though he was lukewarm about the project.[90] Indeed, James expressed reservations about the enterprise from the beginning. After reading a copy of the circular that Münsterberg had addressed to the Visiting Committee, asking for construction of a new, "dignified" building for philosophy, James wrote to Münsterberg that he was not "eager" to see the project realized. "Philosophy, of all subjects," he argued, "can dispense with material wealth. And we seem to be getting along very well as it is,—I was not aware til I read your circular how great our popularity among the advanced students is."[91]

In a concerned letter to his colleague George Herbert Palmer, James wrote that he dreaded "becoming an accomplice in another architectural crime," and urged Palmer and the committee to reconsider both the initial choice of the architect and the site.[92] Emerson Hall was to be built between Sever Hall and Robinson Hall (a new building, which had been designed by the architect Charles McKim and was completed in 1904). The two existing buildings were very different, James noticed, and "to introduce a third heterogeneity and discord there, would . . . be an absolutely unpardonable outrage on the public eye." He proposed either to build the new philosophy hall in a different, more remote spot, or to design Emerson Hall as an "almost identical mate to Robinson Hall, opposite to it."[93] He implored Palmer "not to lend [himself] to architectural villainy," and asked him to share his letter with both Münsterberg and Harvard's president. The corporation decided to assign the job to a different architect, but neither of James's other suggestions was followed. These suggestions were polar opposites of Münsterberg's plans for expressing the centrality and supremacy of philosophy.

The new building, inaugurated on December 27, 1905, during the joint meeting of the American Philosophical Association and the American Psy-

Fig. 7.7. Emerson Hall. From *Harvard Graduates Magazine*, December 1905 (Harvard University Archives).

chological Association, which was discussed in chapter 1, fulfilled Münsterberg's plans for a "spacious, noble, monumental hall" (fig. 7.7). It was erected at a (relatively) central location, and while the construction materials—bricks and limestone—were chosen and distributed in such a way as to balance "in mass and design" Robinson Hall, it was by no means an identical copy of it. With its "columns of Ionic order," its "freize over the colonnade, on the front," its "impressive and simple Doric proportions," the "whole building" was "Greek, in feeling and detail." The author of the anonymous article describing the building for the readers of the *Harvard Graduates Magazine* praised the nobility, simplicity, and purity of form of the building, using phrases that Münsterberg would certainly have approved of.[94]

To James, both Münsterberg's Emerson Hall and the St. Louis program became the material embodiments of a style of philosophizing that James utterly disdained. Both the physical building and the metaphysical plan proclaimed, in different ways, the supremacy of philosophy within the academy and within the wider universe of knowledge. Both represented attempts to bureaucratize philosophy and to fix, in institutional terms, hierarchical power relations—something James found unacceptable. He expressed the point forcibly in a passage that is long but worth quoting in full:

Münsterberg's Congress-program seems to me, e.g., to be sheer humbug in the sense of self-infatuation with an idol of the den, a kind of religious service in honor of the professional-philosophy-shop, with its faculty, its departments and sections, its mutual etiquette, its appointments, its great mill of authorities and exclusions & suppressions which the waters of Truth are expected to feed to the great class-glory of all who are concerned. To me "Truth," if there be any truth, would seem to exist for the express confusion of all this kind of thing . . . and to be expressly incompatible with officialism. Officials are products of no deep stratum of experience. M-g's congress seems to be the perfectly inevitable expression of the system of his Grundzüge, an artificial construction for the sake of making the authority of professors inalienable, no matter what asininities they may utter, as if the bureaucratic mind were the full flower of Nature's self-revelation. It is obvious that such a difference as this, betw. me & M-g, is a splendid expression of pragmatism. I want a world of anarchy, M. one of bureaucracy, and each appeals to "Nature" to back him up. Nature partly helps & partly resists each of us.[95]

Emerson Hall, as James perceived it, expressed the same very undemocratic "officialism," providing a physical embodiment of the "Ph.D." Münsterberg had stressed just that point in his circular to the Visiting Committee, where he had invoked the symbolic way in which the title of "Philosophy Doctor" expressed the view that "all the special sciences are ultimately only branches of philosophy." By virtue of its "quiet central spot," Münsterberg had argued, Emerson Hall would provide a concrete embodiment of the "truth of that symbol," which had "faded away in the academic community."[96]

James had little respect for Münsterberg's deference to titles, hierarchy, and decorum. In a letter to Münsterberg written about six months after Münsterberg's circular, James needled him about "the greediness of the Germans for titles of distinction—very non-democratic to an American, and quite like the English knighthood, which is so paltry a thing." "But the french [sic]," he continued, "are almost as bad, and we are making a beginning with our Ph.D."[97] In 1903 James publicly mocked the growing obsession with academic titles and "decorated scholarship" in an article entitled "The Ph.D. Octopus." "Other nations suffer terribly from the Mandarin disease," he wrote: "are we doomed to suffer from it as well?" He bemoaned the institutionalization of educational differences, which he feared would "develop a tyrannical Machine with unforeseen powers of exclusion and corruption."[98] He envisioned the university as a bastion against the bureaucratic octopus: "Our universities at least should never cease to regard themselves as the jealous custodians of personal and spiritual spontaneity. They are indeed its only organized and recognized custodians in America today. They ought to guard

against contributing to the increase of officialism and snobbery and insincerity as against a pestilence; they ought to keep truth and disinterested labor
always in the foreground, treat degrees as secondary incidents, and . . . make
it plain that what they live for is to help men's souls, and not to decorate
their persons with diplomas." As a physical emblem of the meaning of the
title "Ph.D." Emerson Hall represented the invasion of bureaucratic and corporate mentality within the university, against which James indefatigably
fought.[99]

Both Münsterberg's chart of knowledge for the St. Louis congress and
the Greek architectural style of Emerson Hall were, for James, expressions
and symbols of an artificial approach to nature. Indeed, James denounced
the "whole St. Louis scheme" as "an almost insane piece of abstract schematization," a "case of the pure love of schematization running mad."[100] As
he confessed to his students in 1903–4, James found "certain philosophical
constructions," such as "Kant's whole scheme" and "Münsterberg's classification scheme," to be "subjective caprices, redolent of individual taste,"
out of touch with the "temperament of Nature itself."[101] Far from being an
expression of genuine universalism, James found that, like any other philosophical Weltanschauung, Münsterberg's chart of knowledge was just the
expression of a particular person's individual temperament and aesthetic
preferences. Concerning the latter, James had no doubts: Münsterberg's absolute taxonomies of knowledge and his "artificial," schematic vision of an
absolutely orderly nature exuded "bad taste."[102]

Münsterberg's plans conflicted sharply with James's vision of the position
of the philosopher and the philosopher's role in the academy and society,
as well as his understanding of the nature of philosophy. In the wake of
Münsterberg's (neo)classical plan for Emerson Hall, James began to deploy
architectural metaphors—of gothic and classic styles—to refer to cosmolo-
gies that he liked and disliked. His use of such metaphors became more
frequent as plans for the new building materialized. Thus James coupled classical style in architecture with "absolute schematisms" such as Münsterberg's
taxonomies of knowledge and of nature: "All classic clean, cut & dried, 'noble', fixed, 'eternal' Weltanschauungen seem to me to violate the character
with which life concretely comes & the expression which it bears, of being,
or at least of involving, a muddle and a struggle, with an 'ever not quite' to
all our formulas, and novelty and possibility forever leaking in."[103] On July 4,
1903, in a letter to George Howison, James sarcastically announced that he
would soon start his "new 'system der Philosophie.'" It would be "a genooine
[sic] empiricist pluralism and represent the world in such gothic shape that
people will wonder how any philosophy of classic form could ever have
been believed in. You, dear H., are a classicist, in spite of your pluralism."[104]

In *Pragmatism* (1907), James described rationalistic philosophical systems in words that echoed the words Münsterberg had used to describe and praise Emerson Hall, except that James used the terms "noble," and "classical," associated with classical architectural style, to indicate a kind of philosophy that he utterly disliked: rationalistic philosophies that sought to replace the real, muddy world of experience with a refined—but artificial—outline. As he famously wrote in *Pragmatism*, "The world [to] which your philosophy-professor introduces you is simple, clean and noble. The contradictions of real life are absent from it. Its architecture is classic. Principles of reason trace outlines, logical necessities cement its parts. Purity and dignity are what it most expresses. It is a kind of marble temple shining on a hill." In James's view such a philosophical vision completely missed the "confused and gothic character" of reality.[105] It would never "satisfy the empiricist temper of mind," and would appear to be "a monument of artificiality."[106] James conveyed the point directly to Münsterberg in a letter of 1906. "Were it not for my fixed belief that the world is wide enough to sustain and nourish . . . many different types of thinking, I believe that the wide difference between your whole Drang in philosophizing and mine would give one a despairing feeling. I am satisfied with a free wild Nature; you seem to me to cherish and pursue an Italian Garden, where all things are kept in separate compartments, and one must follow straight-ruled walks. Of course Nature gives material for those 4 hard distinctions which you make, but they are only centres of emphasis in a flux for me; and as you treat them, reality seems to me all stiffened."[107]

James's metaphor of the Italian garden—however dismissively meant—actually found favor with Münsterberg, who wrote back to James that if the world of immediate experience was not an Italian garden, it ought to be made into one, with straight paths and "flowerbeds." Indeed, Münsterberg ran wild with James's metaphor, demonstrating what James feared: there was no limit to Münsterberg's obsession with metaphysical order and tidiness. The task of the philosopher, Münsterberg wrote, was precisely to *order* the cosmos. "Our life's duty makes us gardeners, makes us to unwed the weeds of sin and error and ugliness and when we finally come to think over what kind of flowers were left as valuable and we bring together those which are similar—then we have finally indeed such an Italian garden as the world which we are seeking, as the world which has to be acknowledged as ultimate."[108]

To James no absolute classificatory scheme could ever capture a constantly growing universe. "Novelty" crept into the real world at every moment. Boundaries and "the edges" of experience were precisely the places where such growth took place.[109] A few years earlier, James had resorted to yet another metaphor to describe his metaphysics of radical empiricism and his thesis of ontological novelty. Radical empiricism, he stated, was a "mosaic philosophy." Yet his metaphysical mosaic differed from actual mosaics, since, in the latter, "the pieces [were] held together by their bedding." In radical empiricism, instead, "there [was] no bedding"; it was "as if the pieces clung together by their edges, the transition experiences between them forming their cement." In some ways, James continued, the metaphor was misleading, because "the more substantive and the more transitive parts" of experience "[ran] into each other continuously," and thus there was "in general no separateness needing to be overcome by an external cement." Yet it was also valuable, because it "symbolize[d] the fact that experience itself, taken at large, could grow by its edges." James's metaphysical thesis of the production of novelty implied that no absolute classification or rigid scheme could ever comprehend nature and experience. Reality escaped all rigid classifications, and so did knowledge, which to James was just a part of reality.

The clash between the two Harvard philosophers ran on multiple levels. It was about nature, power, and forms of sociability and interaction. Münsterberg's and James's topologies of nature, knowledge, and society were all at odds. In Münsterberg's social and metaphysical arrangements of knowledge the center was more important than the periphery, and the top more important than the bottom. The society he envisioned, like the final program for the St. Louis congress, was a hierarchically ordered space, where everything occupied a well-demarcated place. All social interaction in this space was to be governed by decorum and a sense of hierarchy. To James, instead, the most promising spaces were the margins—of knowledge, of society, of the self, of experience—as well as the interstitial regions separating disciplines, discourses, social groups, and parts of experience. It was in those boundary regions that he plied his trade, and those boundary regions took on the greatest importance in his geography of knowledge, the self, and society.

This can be readily seen in the classification of types of minds that James presented in his lectures on pragmatism, a binary classification that drew a distinction between "tough" and "tender" temperaments, and associated them with the prevailing types of philosophy—empiricism and rationalism. James's table depicts not only those two types of philosophy but also the blank space that separated them. It was in that space that James proposed to work. He advanced his philosophy of pragmatism as a scheme that might mediate between the two, which would appeal to the "'healthy human understanding' of the ordinary man or woman," the type of mind that would occupy that intermediate space.[110]

Table 1. From William James, *Pragmatism* (1907).

| The Tender-minded | The Tough-minded |
| --- | --- |
| Rationalistic (going by 'principles') | Empiricist (going by 'facts') |
| Intellectualistic | Sensationalistic |
| Idealistic | Materialistic |
| Optimistic | Pessimistic |
| Religious | Irreligious |
| Free-willist | Fatalistic |
| Monistic | Pluralistic |
| Dogmatical | Skeptical |

## James and the Philosophical Tree

In the last ten years or so of his life, James often took up the question of the relationship between philosophy and the special sciences.[111] Perhaps the clearest formulation of his position is found in the first chapter of *Some Problems of Philosophy*, published posthumously.[112] The book was meant to be an introductory text for college students, after the manner of Harald Höffding's *Philosophische Probleme*—a text that James himself frequently assigned—and various other texts entitled *Introduction to Philosophy* written by friends and correspondents including Henry Sidgwick and Friedrich Paulsen.

The book opened with a discussion of some objections that scientists typically raised against philosophy.[113] The first objection—one that, as noted in chapter 1, was constantly raised by men of science or scientific philosophers in the second half of the nineteenth century—was that whereas science made "steady progress," philosophy was stationary.[114] To answer it James, like many other philosophers of the time, launched into a historical study of the relationships between philosophy and the sciences. Resorting to a traditional metaphor—one that others, including James Ward, Shadworth Hodgson, Paul Janet, and Charles Renouvier, had mobilized for exactly the same purpose—he claimed that the "special sciences" were just "branches of the tree of philosophy."[115] As soon as a question, or a set of questions, found an accurate answer, James wrote, "the answers were called 'scientific'" and split off from the trunk, to be classified as a special science. Even today, he continued, "we are seeing two sciences, psychology and general biology, drop off from the parent trunk and take independent root as specialties." Echoing Ward and Janet, James contended that what we call "philosophy" today is "but the residuum of questions still unanswered."[116] According to that definition, philosophy concerned itself with yet unsolved problems; thus it should come as no surprise that philosophy made no progress.[117]

The metaphor of the philosophical tree of knowledge not only conveyed the idea that knowledge formed an organic whole but also suggested that the sciences had originated from the trunk of philosophy and continued to draw vital nourishment from it. Using terminology similar to that deployed by contemporary philosophical "unifiers" of knowledge, James wrote that in ancient times philosophy was conceived precisely as a universal knowledge. In antiquity the unity of knowledge was embodied in the persona of the ancient philosopher, an "encyclopaedic sage" who mastered all the scientific knowledge of his time. The notion of philosophy as encyclopedic knowledge, however, James wrote, did not disappear with Thales, Pythagoras, and Empedocles but had thrived in medieval scholastic philosophy, and had been left unchallenged by the new, antischolastic philosophy of Descartes that "preserved the same encyclopedic character."[118] Leibniz was a universal sage, and so were some of his followers. It was only with Locke and Kant that things began to change. With them, philosophers stopped investigating the world of nature and concentrated exclusively on "mental and moral speculations."[119] Yet encyclopedic minds were still to be occasionally found, and James, agreeing with Paulsen, praised them as incarnations of the true philosophical temper, regardless of their training and research interests. One modern example was the person of Gustav Theodor Fechner, the polymath whom, as we saw, James depicted as "a philosopher" in "the great sense of the term."[120] That portrayal was quite provocative, especially if one considers that, as James himself noted, Fechner was by no means "a professional philosopher." To make things worse, James's praise appeared in his introduction to the English translation of Fechner's *Little Book of Life after Death* (1904), a book on spiritualistic and occult matters that many of James's philosophical colleagues, Münsterberg first among them, declared to be far removed from the domain of reputable philosophy.[121]

Following a distinction he had drawn in the first years of the century between two meanings of the term "philosophy," James clarified that in the original acceptation "philosophy" meant the "completest knowledge of the universe." As such it "include[d] all the sciences—logic, mathematics, physics, psychology, ethics, politics and metaphysics"—and it aimed "at making of science what Herbert Spencer call[ed] a 'system of completely unified knowledge.'"[122] From this "general" philosophy James distinguished philosophy more narrowly defined as "something contrasted with the sciences." In this "more modern sense" philosophy meant metaphysics, a technical, specialized body of knowledge that addressed a well-defined group of questions.[123] James reiterated these claims in *Some Problems of Philosophy* and clarified that that book would concentrate on philosophy in the second, narrower sense. To James, however, the "older sense" was "the more worthy

sense," and he believed that "as the metaphysical questions [got] more settled," the term "philosophy" would revert to its original meaning.[124]

James's conception of general philosophy has been barely noted by scholars, perhaps because it appears to be hardly compatible with James's own
technical philosophical work. Yet it is worth investigating. What exactly did
James mean by philosophy in the more inclusive sense? In what sense could
it unite the sciences? Why did James, a precise writer who skillfully used
metaphors, resort to the metaphor of the tree of knowledge? As we saw
earlier, Herbert Spencer had described philosophy as a system of completely
unified knowledge, one that could fuse into a systematic whole the special
sciences by abstracting from them principles of the highest generality.[125]
Could James have entertained the same idea? Did he envision, like other
philosophers of the time, philosophy as a super-discipline, a science of a different, higher order? As discussed in chapter 2, in the second half of the 1870s,
at the beginning of his career, James had announced the advent of a "new era
of philosophical studies," and had sketched a conception of philosophy as an
inductive enterprise, one in which "solid philosophical conclusions" would
emerge "piece-meal" from the mass of details and results assembled by
the special sciences.[126] Was James subscribing to a similar conception now?
More important, how could James, a pluralist, even support any notion of a
"system of completely unified knowledge"?

Examining the places—both physical and social—that James assigned
to philosophers will illuminate his conception of general philosophy, and
will uncover how, for him, general philosophy could help bring about the
unification of knowledge.

## The Philosopher's Place

In 1909 Hugo Münsterberg launched his most concerted attack yet against
the pragmatists and other "modern relativists." Those would-be philosophers, Münsterberg insisted, had lost sight of the eternal moral and aesthetic
values that were the foundation of society, and their doctrines had devas-
tating moral and social consequences. If pragmatists' thinking was indeed
so unfortunate, it was because the pragmatists "stood outside the temple
of philosophy." They had failed to realize that in order to philosophize one
must "leave the street" and "enter the temple."[127]

James could never have agreed. For him philosophers did not need sanc-
tuaries or secluded spaces where they could practice their arts. As he wrote
in *Some Problems of Philosophy*, "philosophy, like life, must keep the doors
and windows open."[128] The place where James worked when he was in
Cambridge was not, technically, a "study." It was a large cozy "library" in

his home, a place designed at once for study and for social interaction. Émile Boutroux, who visited James in Cambridge just a few months before James's death, reported that James's library contained "not only a desk, tables and books" but also "couches, window-seats, [and] Morris-chairs." It "welcomes visitors at all hours of the day, so that it is in the midst of merry conversations, among ladies taking tea, that the profound philosopher meditates and writes."[129] When not at home, James philosophized in threshold spaces, such as hotel rooms, apartments in foreign cities, cabins in the woods, and his summer place in Chocorua, New Hampshire, "the most delightful house you ever saw," as he wrote to his sister, with "fourteen doors, all opening outwards."[130]

For James philosophers might as well conduct their business on the street. The philosopher's places were "relational places."[131] Indeed, in a metaphorical geography of knowledge, philosophers themselves constituted such places—as illustrated by the case of Fechner, whom, as we saw, James depicted as "one of those multitudinously organized cross-roads of truth, which are occupied only at rare intervals by children of men."[132] In saying that, James was making more than a claim concerning who is entitled to be considered a philosopher. He was making a claim about the roles that philosophers should play in the community of knowledge.

Münsterberg had advocated the construction of a philosophy building at Harvard chiefly on the grounds that it was inconvenient for Harvard philosophers to conduct their business "scattered under many roofs" throughout the campus.[133] Philosophers needed a single location where they could interact, teach their courses, and do their research. They needed to unify philosophy in the first place before they could unify the university.[134] For James, in contrast, the fact that philosophers were disseminated throughout the campus was not a problem nor was it an obstacle to the unification of knowledge. In fact, it was a boon, for it facilitated the true goal of philosophers, namely, to function as sites of intersection of multiple disciplines and pursuits. For James, as for Freidrich Paulsen, philosophers—who should be encyclopedic minds—could unify knowledge by inhabiting relational spaces, even by functioning as sites of exchange and encounter on a map of knowledge.

Pragmatists were particularly suited to carrying on such functions. In a famous passage in *Pragmatism*, James quoted from the Italian pragmatist Giovanni Papini, who had recently described pragmatism as a "corridor in a hotel." "Innumerable chambers" open "out" of the corridor. "In one you may find a man writing an atheistic volume; in the next someone on his knees praying for faith and strength; in a third a chemist investigating a body's properties. In a fourth a system of idealistic metaphysics is being excogitated; in a fifth the impossibility of metaphysics is being shown." All of these people, however, "own the corridor, and all must pass through it if they want a practicable way of getting into or out of their respective rooms."[135] It may happen, Papini added, that along the corridor hotel "guests" who otherwise would never have met would start talking together: in that case, no "waiter" would be so rude as to prevent such "conversations."[136] James fully agreed.

Pragmatism, as a "corridor-theory," was a threshold space (the intermediate blank space in James's chart of the two prevailing types of philosophy); its position symbolized the ways in which pragmatists could act as "go-betweens" and promote encounters between scientists, philosophers, mystics, and other people who were pursuing different callings and even fighting against each other.[137] Philosophers could not impose unity upon the sciences in the way Münsterberg had imagined. Unity could only be attained piecemeal by people who worked in all sorts of disciplines and professional callings, and inhabited different social and intellectual worlds.

James's vision for the unification of knowledge reflects his approach to the metaphysical problem of the "One and the Many," to him the single most important problem of philosophy. Is the world "One" or "Many"? If it is one, what holds it together? At the turn of the century scores of philosophers took on this problem. James had come to the conclusion that none of the available conceptions of the metaphysical unity of the world was acceptable. These included his friend Josiah Royce's idea of a "noetic unity" of the world (in which everything is unified in the thought of an Absolute, all-encompassing Knower), the idea of a "causal unity" (in which everything stems from the same cause), the idea of an "aesthetic unity" (in which the world "tells one story" fusing all the partial stories told by things and individuals), and George Holmes Howison's idea of a "unity of purpose" (in which everything in the universe ultimately aims at the same purpose). James found each of these notions to be ultimately unsustainable.[138] After discussing them, James came to the conclusion that he had reached many years earlier: people in which the "passion for distinguishing" is stronger than the "passion for unity" will tend to discover a lot of disunity in the world, indeed as much or more disunity than unity.[139]

To James nothing necessitated the idea of an absolute unity of the universe. One could easily imagine universes (or pluriverses) endowed with greater or lesser degrees of unity. For example, a universe in which individuals could communicate telepathically with each other would appear to be more unified than the one we think we inhabit. Conversely, a universe without causal interactions among things would appear to be more disconnected.[140] The world was both one and many, depending on how one looked at it. What one found in the world was local concatenations, piecemeal chains of relations, partial systems of connection that were always premised on human interests: "colonial, postal, consular, commercial systems," for example. The result of such systems, James wrote, was "innumerable little-hangings-together of the world's parts within the larger hangings-together, little worlds, not only of discourse but of operation, within the wider universe." Indeed, "each system exemplifie[d] one type or grade of union."[141]

These systems of concatenations were framed and assembled by human efforts and always served practical purposes. Using an analogy to physical systems—"electric, luminous and chemical influences"—James emphasized that such concatenated systems could be expanded, provided that the right type of "conductors" were chosen as intermediaries. However, whenever "opaque and inert bodies interrupt[ed] the continuity" of those physical lines of "influences," the absolute unity of the universe would be lost, at least in some particular respect.

Returning to the issue of the philosophical unification of knowledge, I suggest that James thought pragmatists and philosophers "in the most worthy" sense of the word to be such "conductors." By virtue of their location at the point of intersection of different roads and, as we will see in the next chapter, by virtue of their frame of mind and social-epistemic sensibilities, philosophers could be excellent intermediaries; they could make possible transmission and exchange of signals, and communication, facilitating a bottom-up social unification of knowledge.

In the end James's notion of the philosophical unification of knowledge was thus fundamentally social, rather than methodological, conceptual, or metaphysical. "General" philosophy would fulfill its unifying function not by inductively generating, à la Spencer, laws of higher and higher generality, nor by demonstrating the metaphysical (causal, noetic, purposive, or aesthetic) unity of the world, nor even by classifying the sciences and placing them in a neatly ordered system, as Münsterberg had hoped. Instead, it would help unify the "special sciences"—among them "technical" philosophy—by facilitating cross-disciplinary conversations, and by keeping alive the goal of creating open, inclusive communities of inquirers where a variety of people would be allowed to speak and interact freely.

Thus, at the turn of the twentieth century, when disciplinary specialization and division of labor threatened to cut off communication among people who worked in different fields, James made (general) philosophy—the trunk of his philosophical tree of knowledge—expressive of a social unity of knowledge. Philosophers could facilitate the creation of systems of connection, but unity—or unities—would come in the long run through interactions among many.

# 8

# The Philosopher's Mind

## Routinists, Undisciplinables, and "The Energies of Men"

In 1935, looking back at his friend's life and work, Ferdinand Canning Scott Schiller praised the "virgin freshness with which James's mind approached the problems of philosophy." He ascribed it to "James's exemption from . . . the dull mechanical routine of academic philosophy." "Strictly speaking," Schiller noted, James was "not a professional philosopher at all." James belonged to the "great succession of amateurs" who had "stirred philosophy and stimulated thought, in line with Descartes, Spinoza, Leibniz, Berkeley, Hume, Schopenhauer, Mill, Bentham, and Spencer." James's freedom from the routines that regulated the practice of academic philosophy, Schiller concluded, gave him "all the advantages which the amateur has over the professional."[1] By emphasizing James's freedom from professional and disciplinary routines, Schiller captured an essential aspect of James's philosophical persona, his conception of philosophy, and his plans for reordering knowledge.

In the late nineteenth century a wide range of scientists, philosophers, educators, and reformers shared the belief that intellectual training was, first of all, a means for cultivating the mind.[2] James agreed with them. Disciplines, for him, were also mental disciplines. They shaped the minds and bodies of those who were trained in them by inscribing habits in their nervous systems. Disciplinary and professional routines, like repeated actions of any kind, consolidated schemes for the perception of reality and modes of action that became habitual. These habits facilitated the carrying out of the tasks involved with a profession or a discipline, and were advantageous.

Habits, however, could also have negative effects. They could prevent the mind from looking at things differently and from approaching problems in fresh ways. They could also engender social division. "Habit," as James wrote in *Principles of Psychology*, "is the enormous fly-wheel of society, its most precious conservative agent." As such, habits "[kept] different social strata from mixing," and prevented social fluidity.

This short chapter explores the significance of James's calls for "freedom" from routines by looking at some of the tactics that James deployed in order to challenge the negative and social effects of "disciplinary and intellectual habits."[3] After a brief discussion of the ways in which James's writing style challenged his contemporaries' engrained expectations regarding the proper philosophical (and scientific) style, I turn to discuss James's conception of philosophy as a means for the cultivation of the mind. The last section examines James's address to the American Philosophical Association—"The Energies of Men"—and presents it as a public enactment of the kind of mental and social sensibilities that James ascribed to genuine philosophers. In "The Energies of Men," I suggest, James most visibly fought against all types of behavior that "stiffened" conversation and prevented interaction along disciplinary lines or other divisions. He needled his philosophical audience to help them get rid of the negative mental and social dispositions engendered by disciplines and professions, and illustrated by example his conception of the philosopher as an enabler of interactions among people inhabiting different intellectual and social worlds.

### James's Style As Boundary Work and Mental Irritant

James carried on highly technical work and was quite comfortable with the distinction between professional philosophers (or scientists) and the "common man." Nevertheless, from the leisure of his worldwide fame he enjoyed posing as an amateur and an "eccentric" philosopher.[4] He famously shunned the title of "professional" philosopher, which he preferred to reserve for his opponents, and in 1907, upon surrendering for good his professorship, he confessed that "as a professor," he "felt [he] was a sham."[5]

According to his contemporaries, James did not write the way a professional philosopher did or ought to. George Santayana recalled that James "didn't talk like a book, and didn't write like a book, except like one of his own," while Schiller found James's style "delightfully different from that of most philosophers." James's writings, Schiller noted, "read more easily than anything else in philosophic literature," and were "popular and intelligible, to all but some philosophy professors."[6] Most philosophers and academic readers, including some of James's friends, condemned James's style. Charles Sanders Peirce, for example, "laid awake several nights in succession in grief" over James's "carelessness" of formulation. Peirce, who had urged contemporary philosophers to sacrifice "literary elegance" to the stern requirements of efficiency, and avoid "using words and phrases of vernacular origin as technical terms of philosophy," accused James of using terms without "strict scientific exactitude" and "accuracy." To his mind James was guilty of the "greatest sin" that one could commit "against science." He wondered if James "mean[t] to mystify his readers," and informed James bluntly that philosophy was "or should be, an *exact science*, and not a kaleidoscopic dream." Indeed, it was either that "or a balderdash"—not exactly a compliment.[7]

G. S. Hall, as noted in chapter 2, critiqued *Principles of Psychology*—which other contemporaries judged to be James's *only* "technical book"—for violating the conventions of scientific-academic writing.[8] J. E. Creighton decidedly placed James among a group of "free-lance" writers and eccentric "essayists," who indulged in a picturesque style. He painted them as "would-be" philosophers, and decreed that their literary style placed them "outside" the boundaries of philosophy.[9] Hugo Münsterberg, who, as we have seen, found pragmatism, and "much of the American method of philosophizing," "antagonistic to the real character of philosophy," denounced nontechnical, literary philosophy. "Brilliant" and "picturesque" philosophy, he contended, was as unacceptable as a "picturesque and epigrammatic mathematics or chemistry."[10] He especially blamed James and the younger generation of pragmatists for making "concessions" to the tastes of "common sense," and for phrasing their thoughts in a "captivat[ing], "impressionistic style" that anyone could follow without the "slightest effort."

James's "popular tone" annoyed others as well. A. E. Taylor, whom we met in chapters 1 and 5, lectured the pragmatists exactly with regard to this point: "If we mean to be philosophical," he pontificated, "our main concern will be that our beliefs should be true; we shall care very little whether they happen to be popular or unpopular with the intellectual 'proletarians' of the moment."[11] Similarly, Albert Schinz, a man of aristocratic inclinations, scolded the pragmatists for their failure to keep "scientific truth . . . out of reach of the general public," and for having produced a new monster: to wit, "popular philosophy."[12]

If these readers accused James of blurring intellectual and social divides with his writing style, James in turn accused them of indulging in a technical style precisely in order to exclude whole ranges of possible interlocutors from philosophical (and scientific) discussion. In 1908, lecturing before an audience of British professional philosophers and well-educated men and women, he criticized "German" philosophers (and, implicitly, their American followers) for their "abuse of technicality" and for their "fear of popularity." For most German philosophers, James stated, "simplicity of statement" was "synony-mous with hollowness and shallowness."[13] They even prided themselves on being able to write in such a technical way that nonprofessional readers would get lost after a couple of sentences. James denounced such behavior as "bad form, not good form, in a discipline of such universal human inter-est" as philosophy, adding that "in subjects like philosophy [it was] really fatal to lose connexion with the open air of human nature, and to think in terms of shop-tradition only."[14] For the same reason James praised Friedrich Paulsen's unusual "untechnical," "literary" style, one that matched Paulsen's attacks against "over-professionalism," and his determination to avoid writ-ing for "nurseries of professionalizing youngsters."[15] James also enjoyed in other writers that "carelessness" of formulation that his friend Charles San-ders Peirce found abominable in his, and praised precisely that quality in the work of the "anti-philosophical," "anti-professorial" Giovanni Papini and the "Florentine Pragmatist Club," a group of young self-trained philosophers and psychologists who worked outside of, and in opposition to, academic philo-sophical and scientific circles. James contrasted their "extravagant," "imper-tinent" style with the pedantry of the "papers" milled by many "bald-headed" and "bald-hearted" young American aspirants for the PhD as they churned along—part of the machinery of professionalization and exclusion.[16]

Of course, James was aware that the style he deployed in some of his books, although by no means in all of his published work, was unorthodox. In the preface to *The Will to Believe* (subtitled "Essays in Popular Philosophy"), he predicted that his "professionally trained *confreres*" would "smile . . . at the artlessness of his essays in point of technical form."[17] Later, he remarked that *Pragmatism*, "from the point of view of ordinary philosophy-professorial manners," was "a very unconventional utterance."[18] Indeed, James made it clear that the language he deployed in that book and elsewhere was not only "un-technical"; it was also "deliberately anti-technical" and "popular," capable of reaching out to a broad public of "philosophical amateurs."[19] He predicted that the book's "non-technicality" of "statement" would "entirely disconcert" his professional colleagues: "Professional philosophers . . . are so brought up on technical ways of handling things that when a man handles them *bare*, they are non-plussed, can neither understand, agree, or reply."[20]

"Non-plussing" was, in fact, precisely the effect that James intended to produce with some of his writings. James, of course, agreed that some philosophical topics required technical treatment. However, the technical way of writing philosophy—a style that would have been unimaginable in the United States up until a decade or two earlier, when philosophical writing consisted chiefly of textbooks for college students—was becoming routine, and young academics were trained to think and write that way regardless of the specific philosophical topics they addressed. In *Principles* James had theorized that routines and habits became "branded in" the "cerebrum[s]" of people. This was true also of "professional and disciplinary habits," which gradually became automatic, a sort of second nature, in individuals at first and then in entire professions.[21] James's thinking about character and the "philosophy of habits" suggests that he believed that technical forms of writing would replicate themselves, and thereby increase social division, whether authors intended to produce these social effects or not. Thus not only did James challenge intentional attempts at exclusion; he also tried to break the habitual circuits that resulted in the automatic replication and reinforcement of social division.

James had written on the ways in which an unexpected interruption or obstacle in a rutted habitual path might jostle the ingrained course of thought and thus result in a new path being formed. By transgressing the newly emerging norms of academic writing, he meant to "break" the "inertia" that was binding the nervous systems and constraining the thought patterns of professional academics. To thwart the possibility that these writings and thought patterns might become ingrained and fixed, James posed an unexpected, and for that reason annoying, "block" that would prevent his contemporaries from mindlessly settling into a technical style from which they would find it very difficult to dishabituate themselves. Acting as an irritant, James wanted to invite his philosophical colleagues to consider new possibilities.

## Philosophy As Mental Culture

In 1903 James paid tribute to his friend Thomas Davidson, who had died a few years earlier. He praised Davidson as a champion of "individualism." This character trait did not translate into selfishness or lack of interest in others—for, as James remarked, Davidson had been an immensely sociable man. He had had a tremendous capacity to form new friendships with all sorts of people, and had started up heterogeneous communities within which people could share a common life. Davidson's individualism instead translated into a "rule" of conduct, one that James found fascinating: the rule to "form no regular habits." "When he found himself in danger of settling" into a habit, "even a good [one]," James wrote, "Davidson made a point" of getting rid of it. He believed that "habits and methods ma[de] a prisoner of a man, destroy[ed] his readiness," and kept a person "from answering the call of the fresh moment."[22] "Individualist à outrance," James stated, "Davidson felt that every hour was an unique entity, to whose claims one should lie open."

Davidson had also been strongly hostile to what he called "ignoble academicism"—and James, despite the fact that he had been the victim of one of Davidson's tirades against it, appreciated that. Davidson believed that belonging to an academic institution imposed great constraints on a person, and his command of the skills necessary to enter and survive academia was minimal. For a while, James had thought Davidson could have been appropriate for a professorship in the Greek department at Harvard. Nevertheless, Davidson managed to offend the whole department by savagely attacking its research methods so Harvard never hired him. To James, this had been a great loss since he believed that, for the growth of a university, "a few undisciplinables like Davidson [would be] infinitely more precious than a faculty-full of orderly routinists."[23]

I suggest that the undisciplinable Davidson embodied the kind of mind that, by the turn of the twentieth century, James assigned to the genuine philosopher: a mind that could rescue itself from routines, and could look at things and think in fresh ways, unconditioned by previous schemes of thought and perception.

As we saw in chapter 2, back in the 1870s James had offered a pedagogical conception of philosophy as a kind of mental training. He had argued that the philosophical cultivation of the mind engendered in students "the habit of always seeing an alternative, of not taking the usual for granted, of making conventionalities fluid again," and "of imagining foreign states of mind."[24] At the turn of the twentieth century, James resurrected that idea. The student graduating from a professional or technical school, James observed, would be a "first-rate instrument for certain specific jobs," but would lack "all the graciousness of mind suggested by the term liberal culture." Such a person was bound to "remain a cad, and not a gentleman, intellectually pinned down to his one narrow subject, literal, unable to suppose anything different from what he has seen, without imagination, atmosphere, or mental perspective."[25] The philosophical training of the mind, in contrast, would "break down our baked prejudices"; it would endow the mind with openness, and imagination, and would make it capable of appreciating novelty.[26] It engendered habits, but those habits, rather than forever wedding minds to established modes of perception, thought, and action, would help students avoid the rut of rigid patterns.

When James praised the philosophical cultivation of the mind, the kind of mental culture he had in mind was very different from the professional training that young philosophers increasingly received at American universities. Instead, it represented an antidote to that. The undisciplinable Davidson, a philosopher who had never actually managed to fully expound his philosophical views, or to secure a stable academic position, represented its best product. His resistance against habits, itself a habit, was the best fruit of a proper philosophical disciplining of the mind. For those reasons, as Douglas R. Anderson suggests, James saw Davidson's character and life as a "philosophical achievement."[27] By the first years of the new century, Davidson's unconventionality and freedom from disciplinary routines—the traits that Schiller, years later, would praise in William James–came to represent to James the virtues essential for the social function that, as we saw in chapter 7, he ascribed to the philosopher. Those were the mental attitudes that, in combination with the dispositions generated by philosophical culture of the mind, would enable philosophers to facilitate cross-disciplinary conversations and promote heterogeneous, cross-divisional communities of inquirers.

## "The Energies of Men"

James's conception of the proper philosophical mind, his understanding of the philosopher as a "cross-roads of truths," and his vision of philosophy as the activity that could enable a social unification of knowledge came together in his spectacular and controversial presidential address before the American Philosophical Association, "The Energies of Men." There, James illustrated by example the kind of work that philosophers should carry on, and showed how disciplinary routines and rutted conventions could interfere with the proper goals of philosophy and the philosopher.

The problem that James addressed in his paper was one familiar to everyone, and quite practical: How could people increase their physical, mental, and moral energy? Everyone experiences the phenomenon of "feeling more or less alive on different days," James told his audience. Sometimes we feel as if we were only half-awake, "oppressed, unfree." And yet at these times we may be using "only a small part of our possible mental and physical resources." Empirical evidence indicates that there might exist reservoirs of energy of which we are not aware, and which, sometimes, we unexpectedly manage to "release."[28]

James believed that strong emotions and unusual excitement could produce such an effect; however, the "normal opener of deeper and deeper levels of energy" was "the will." Ideas, including ideas that denied ingrained ideas, were a "third great dynamogenic agent," for, if they managed to prevail over competing ideas, they would be discharged into action. Ideas, however, might fail to be efficacious, and the will needed to be trained. James discussed several mental and bodily disciplines that could help direct the will and arouse beliefs that, in turn, could help people to "liberate" higher and higher amounts of energy. These included ascetic disciplines—Ignatius Loyola's spiritual exercises, Hatha Yoga, particularly breathing exercises, fasting, and prayer. He also discussed the use of stimulants, such as opium and alcohol, "suggestive therapeutics" under hypnosis, as well as religious or other types of conversions, especially conversions to optimistic forms of "spiritual philosophy" such as Christian Science and the "New Thought."[29]

These techniques could be tremendously helpful; however, they did not work always or for everyone. It was thus necessary and urgent to develop a more systematic approach to the problem of the maximization of physical and mental energy. To that end, James continued, it would be necessary to chart the powers of men and their limits, and to construct a methodical inventory of the paths of access to those powers. This was the twin program that James urged on his philosophical audience: it was a program at which "anyone in some measure [could] work," and one "well worthy of the attention of a body as learned and earnest as this audience."[30]

A subtext ran throughout James's address—one that, no doubt, expressly irritated many in his audience. James pointed out that "life's routine[s]" and habits, including intellectual habits, built "barriers" around people, and thus prevented people from "liberat[ing]" sources of energy "habitually not taxed at all."[31] "Social conventions" and the rules of "scientific" and "intellectual respectability" had particularly negative inhibiting effects. Some people, James told his audience, were so oppressed by notions of academic respectability that one could not "converse" with them "about certain subjects," or even "mention" such subjects "in their presence." James confessed that some of his dearest friends fell in that class, and he complained that with them he had never been able to carry on a conversation about some of his favorite writers—George Bernard Shaw, H. G. Wells, or the anarchist/socialist and gay mystical writer Edward Carpenter.[32] With those friends, James found, he just "had to be silent"; no meaningful conversation was possible.

"An intellect thus tied down by literality and decorum," James continued, "makes on one the same sort of impression that an able-bodied man would who should habituate himself to do his work with only one of his fingers, locking up the rest of his organism and leaving it unused."[33] The intellectual inhibitions that paralyzed his friends, James also subtly implied, were not dissimilar from pathological cases of "habit-neuroses"—diseased states characterized by the obsessive repetition of a habitual action or exclusive use of a part of the body. Patients affected by such morbid conditions could, for example, just "walk, walk, and walk," or "eat, eat, and eat," or constantly pull out their hair.[34] As in these functional diseases of the nervous system, disciplinary and academic routines made it impossible for scholars to carry on other activities and to play with new ideas. They constrained their minds and made them unimaginative. James suggested that people thus sealed in their habits could particularly benefit from an "eccentric activity," a "spree" even. Such unusual activities could be even "medicinal," help people to break down the habitual barriers that surrounded them, and liberate untapped sources of energy.[35]

This comment could well apply self-reflexively to James's own talk—one most unusual for a philosophical meeting. No doubt, to many of his listeners James's talk appeared eccentric, and quite remote from what they took to be the center (and the conventions) of the field. By broaching subjects almost unmentionable in North American philosophical circles, James invited philosophers to step out of their ever-narrowing academic routines and professional restrictions; break, if only for a moment, their disciplinary habits; open their frame of mind; and engage freely in the discussion of an energizing topic.[36] Indeed, the whole talk was a provocative gesture meant to expose the narrowmindedness and shortsightedness of James's professionalizing philosophical colleagues, an audience whose "field of vision"—to deploy an analogy that James used for a different purpose in his lecture—was as contracted as that of an "hysteric subject."[37]

By shocking ingrained expectations and interrupting philosophical routines, "The Energies of Men" also illustrated the kinds of alternative projects that philosophers could engage in. For example, it illustrated how philosophers could participate in and promote the creation of an integrated, crossdivisional science of human nature—one that could help human beings cultivate wholeness and live up to their utmost capabilities. The topic James chose—human energy—was well suited to that goal, for it functioned as a "boundary object." Such objects, as sociologists of knowledge have suggested, are things (or notions) over which different social groups have "partial jurisdiction." They are plastic enough to "inhabit several intersecting [social] worlds" (different disciplines, for example) and, at the same time, robust enough to be "recognizable" as shared objects by people working in different fields. Scientists deploy them to facilitate interaction among those different groups or, sometimes, to "engineer agreement."[38] Human energy, as James well knew, lay at the center of attention of clinical psychologists, but it was also important to the work of physiologists and psychologists with physiological interests. It was of great interest to psychiatrists, neurologists, physicians, psychical researchers, and social reformers of various stripes, and to all the ordinary folks who were wrestling in their lives with the problem of depleted energy.

James hoped that the topic of mental energy could also be of interest to professional philosophers, if they would only be willing to reimagine philosophy as a "theory of action." This provocative, antiacademic conception, one that James borrowed from Giovanni Papini, made philosophy into a tool that would enable people to "unstiffen" their ideas and beliefs and empower their actions. James used the concept of energy in a way that would enable him to enlist people belonging to different disciplines or working in different fields in a common project, the creation of a psychological-philosophical-medical-naturalistic-spiritual science of human energy, which he hoped would offer not only theoretical insights into human nature, but also practical guidelines on how people could improve their lives.

Finally, in his APA presidential address, James exhibited the social sensitivities that he considered essential to the proper conduct of philosophy. In the course of his address, James expressed and recounted the interests, experiences, insights, and concerns of all sorts of people, quite literally lending them his voice. For example, he read aloud a long letter from his friend Vincenti Lutoslawksi, an eccentric Polish philosopher who recounted how he had found mental peace in yoga. James took more time to read a letter from a British colonel who survived the siege of Delhi (1857) by drinking prodigious amounts of brandy. Needless to say, these were not the philosophical voices that many thought should be heard at a professional philosophers' meeting. By incorporating them into his address, James exposed his audience exactly to the kinds of cross-divisional conversations that he believed philosophers ought to promote. In fact, he transformed his solo presidential address into a multivocal performance, and enacted before his audience the role that he ascribed to philosophers: that of go-betweens, crossroads of knowledge, and enablers of conversations beyond intellectual and social divides.

# CONCLUSIONS

The "Energies of Men" sums up many of the threads discussed in this book. It illustrates how through his transgressions of disciplinary, professional, and social divides James elaborated and furthered new conceptions of what it means to do philosophy and what it means to do science. And it reveals how promoting certain moral, epistemic, and social sensibilities was central to James's efforts to define a new intellectual and social order of knowledge.

Throughout his career James seldom missed a chance to challenge the divides that separated professional from amateur discourse, technical from popular discourse, and academic disciplines from one another—especially the barriers that separated philosophy from the human and social sciences. At times James pursued these goals through direct attacks against the dominant academic modes of knowledge production and the styles of social and intellectual interaction associated with them, as when he challenged the "PhD octopus" and the newly established practice of conferring academic appointments only on PhDs, or when he publicly defended mind curers whom professionalizing regular physicians were trying to ban from medical practice.

More often, however, James resorted to local, improvised tactics of resistance: subtle types of boundary work that frequently employed the very materials, institutions, codes, and spaces posited by orthodox and professional practitioners but twisted them in a manner that would undo their regimes of knowledge from within. In his battles James resorted to a variety of techniques, which included, for example, mobilizing certain technical philosophical claims in order to challenge established disciplinary divisions, and adopting untechnical, even "anti-technical," styles of writing (chapters 5 and 8). He privileged nonofficial boundary spaces that would allow philosophers to intermingle with people working in other fields, and reveled in unconventional behaviors and provocative gestures (chapter 7). Thus working in a style that contravened approved codes, as he did in his 1906 American Philosophical Association talk, he undermined the boundaries of the disciplines that he was purportedly leading and directing. James, who had initially declined to join the American Philosophical Association, but later relented, did not attack the association directly nor did he question the legitimacy of the American Psychological Association. Instead, he used those institutions, and the temporary positions of authority they assigned him, for his own purposes. By dramatizing before captive audiences conceptions of psychology and philosophy antagonistic to the ones those professional associations were trying to establish, he promoted alternative forms of inquiry (chapters 2 and 8).

James's boundary work was not something he conducted on the side. It was central to his philosophical and scientific practices. It shaped technical aspects of his philosophy (chapters 5 and 6) as well as his views on the nature of philosophy. Because the academy was rapidly changing, many of the tactics that James deployed in attempts to reshape the meaning of science and philosophy, as well as redefine the roles of scientists and philosophers, took the form of interventions in intense debates over the mutual relationships among the disciplines, especially philosophy and the special sciences (chapter 1). His views of the goals of philosophy and its role vis-à-vis the sciences evolved from the late 1870s, when James considered philosophy a field of inquiry that should inductively abstract general principles from facts or lower-level principles discovered by men of science (chapter 2), to the mid-1890s, when he depicted philosophy as the overarching framework wherein the fundamental assumptions on which the various special sciences rested could be opened up and discussed, and where the conflicts among the different sciences could be adjudicated. By the early years of the new century, James had come to consider philosophy—"general philosophy"—as an activity that could, and should, enable conversations and exchange across disciplines both within and outside the university, and facilitate a social unification of knowledge (chapter 7).

James also redefined the boundaries of the human self and its inner topology, in a way that enabled him to rethink the relationships among different individuals and imagine heterogeneous, yet intimate social communities. In parallel, he promoted kinds of philosophical and scientific selves that would embody the mental and epistemic-social dispositions necessary to the production of crossdisciplinary and open inquiry (chapters 3, 6, and 8).

When viewed against the background framed by the shifting arrangements of knowledge of his day, and with a focus on James's boundary efforts, James's manifold activities appear to have been cut from the same cloth. His attempts to reform and promote moral and social economies of knowledge, his framing of new forms of selfhood, the scientific and philosophical personae he championed, and even his conceptions of the proper practice of science and philosophy, all conduced toward reforming not only the intellectual and academic order but also the wider social order.

Through his boundary work James promoted, both in the academy and society, the creation of spontaneous concatenations of partially connected individuals as well as deep, sympathetic relationships that would allow for meaningful conversation among the widest range of people. These social and intellectual clusters would be more than simple juxtapositions of disjointed individuals. Instead, they would instantiate new modes of social being and foster the mental, moral, and epistemic attitudes that James believed would go a long way toward solving the social tensions and fixing the problems of the academy: open-mindedness, inclusiveness, tolerance, antidogmatism, respect for different points of view, as well as the ability to imagine other people's mental states and value their proposed modes of life.

The communities of citizens that James envisioned would mount local resistance to overpowering social-economic institutions. However, as others have seen, James did not aim to specifically foment groups that would directly subvert those institutions. Instead, I suggest, James aimed to change the ways in which people socialized. He worked to frame modes of human interaction that differed profoundly from the distant, impersonal, and purely "legal" modes of sociability enforced by "big institutions" and demanded by the prevailing regime of scientific knowledge. His goal thus was to help people engage one another and interact in more human ways. In these communities people coming from different classes and social groups could interact freely, even intimately, yet they could retain their individual, social, and even class distinctions—something with which James was comfortable, as evidenced, for example, by the patronizing attitude he sometimes adopted toward his "cranks" and by the limitations of his conception of democracy and of his sympathies for "the underdogs."[1]

Likewise, the cross-divisional communities of inquirers that James promoted as a philosopher, psychical researcher, psychologist, and student of
human nature did not aim to radically subvert the new disciplinary order
of knowledge. James has sometimes been named as a precursor of post-
modern projects of "post-disciplinarity" and "anti-disciplinarity." Scholars as
diverse as Richard Rorty, Louis Menand, and Bruno Latour have deployed
his writings (and American pragmatism more generally) as means to blur
in provocative ways disciplinary boundaries, announce the "meltdown"
of disciplinary structures, and to promote new, exciting configurations of
knowledge.[2] However, James did not aim to eliminate the disciplines. Division of labor in the academy as in society was, for him, crucial. He retained
the disciplines but wished that practitioners could interact in different ways.
In his communities of inquirers people were not at all required to give up
their disciplinary identities, although they were required to suspend—if only
for a while—those disciplinary habits that would prevent them from engaging novel perspectives, unusual topics, and people who worked in different
fields or belonged to different social-cultural environments.

Thus just as in the wider social sphere James championed not radical social changes, but a new conception of the self and new styles of interaction,
so in the academy he promoted clusters of sensibilities and mental attitudes
that would enable people to engage other investigators and cooperate in
unsanctioned, spontaneous ways. James's efforts to promote a psychological-
medical-hygienic-physiological-spiritual pragmatic science of man—a project framed in a way that made it impossible to realize within the bounds of
any existing discipline—represented the epitome of his drive (chapters 5, 6,
and 8).

In his resistance against the social effects of professionalization, specialization, and disciplinarity, James was not entirely isolated. Some of his
closest allies—Schiller, Paulsen, Papini, and Giuseppe Prezzolini, for example—carried on similar, if much more circumscribed, battles in their own
countries, and, in the next generation, a few psychologists, scientists, and
philosophers followed James's lead.[3]

James and his allies lost their battles. Indeed, aside from a few well-known
episodes, the story of their efforts has become in large part invisible, or barely
visible. Yet studying the histories of these "resisters" and their tactics is a useful exercise. For one thing, it may help recast the emergence of the modern
regime of standardized knowledge, one premised upon the quest for universal objectivity, the requirement of impersonality, and bureaucratic interactions among knowledge producers. This regime of knowledge increasingly
came to regulate scientific practices across fields at the turn of the twentieth
century.[4] Extensively examined by historians of science, it has sometimes

been presented as the natural result of anonymous, impersonal historical forces. This book instead suggests that the processes resulting in the entrenchment of the new moral, social, and epistemic rules governing scientific knowledge were heavily contested, and that their outcomes could have been vastly different. By illustrating how William James and other "losers," who consciously located themselves at the margins of academic discourse and in the interstices separating academic groups, resisted the new regimentation of inquiry, it reveals that the processes that led to the consolidation of the modern regime of knowledge production involved a complex interwoven set of conflicts—over the characteristics of the ideal scientist and philosopher, and simultaneously over the constitution and core of the human self; over the proper organization of the academy, and at the same time over the order of society. By unveiling some of the backstages of scientific and philosophical knowledge at the turn of the twentieth century, the book fosters appreciation of what was at stake in these conflicts, and how much would be lost. It suggests that the historical recovery of orders of knowledge depends crucially on reconstructing the whole landscape, not just the topography as officially mapped by the winners.

As a combination of intellectual history, history of science, and what one might call "philosophy studies," a parallel to "science studies," this book discloses the complexity and openness, the sense of urgency, and the large stakes that participants around the turn of the twentieth century understood to lie in the question of how the academy should look, and the ways in which those broad issues shaped the practice of science and philosophy. Today, when the academy is engaged once again in intense discussions on disciplinarity and its alternatives, James's imagined geography of knowledge can offer a renewed sense of the manifold possibilities for knowledge within, across, and between the disciplines.

# NOTES

## INTRODUCTION

1. James, "Energies of Men," presidential address.

2. Cotkin, *William James*, 112–15. On the cultural values associated with the science of energy, see Rabinbach, *Human Motor*, and Smith and Wise, *Energy and Empire*.

3. In the Harvard edition of The Works of William James, "The Energies of Men" was not included in the volume *Essays in Philosophy*. Instead, it was published in the volume *Essays in Religion and Morality*.

4. Davies, "Proceedings," 201. For the lecture that James presented on May 18, 1906, at the Psychology Club at Harvard, see James, *Essays in Religion and Morality*, 199–200.

5. James, "Energies of Men," *Science* 25 (March 1, 1907): 321–32; James, "Powers of Men." For the details, see editorial notes in James, *Essays in Religion and Morality*, 247–49.

6. On the late nineteenth-century trends toward professionalization, see especially Bledstein, *Culture of Professionalism*, and H. Kuklick, "Professionalization and the Moral Order." See also Larson Sarfatti, "Production of Expertise," and *The Rise of Professionalism*.

7. James, "Energies of Men," presidential address, 144–45.

8. Ibid., 130.

9. Hollinger, "'Damned for God's Glory,'" 11.

10. See Cotkin, *William James*, 4.

11. My thanks to George Cotkin for suggesting this expression.

12. On James's "patrician elitism," see especially Westbrook, *Democratic Hope*, 59. See also Kittelstrom, "Against Elitism." Kittelstrom proposes to use James as a

"method" for studying a social and cultural world that the historian would otherwise find difficult to access. She also provides an excellent discussion of the historiography on James's social
sympathies—or lack thereof.

13. B. Kuklick, *Rise of American Philosophy*, esp. 266; Myers, *William James*, 55 and 300; Bird,
*William James*, 121; Sprigge, *James and Bradley*, 79 and chap. 2; Taylor, *William James on Conscious-
ness beyond the Margins*, 4; Lamberth, *William James*, esp. chap. 2; Leary, "William James and the
Art of Human Understanding."

14. D. Wilson, *Science, Community*, 38.

15. Seigfried, *William James's Radical Reconstruction*, 1, 39, 140.

16. Leary, "Influence of Literature," and "'Authentic Tidings.'" On the implications of
James's artistic interests for his work in psychology and philosophy, see Barzun, *Stroll with
William James*; Barzun, "William James"; Feinstein, *Becoming William James*; and Leary, "William
James and the Art of Human Understanding." This important theme and James's analogy of
"philosophies" to "pictures" are not taken up in *William James at the Boundaries*.

17. Hollinger, *American Province*, 3–22.

18. Croce, *Science and Religion*.

19. Lamberth, *William James*, 147; Gale, *Divided Self*, esp. chaps. 9 and 10.

20. Lamberth, *William James*, 235–36.

21. Menand, *Metaphysical Club*, 95.

22. Bjork specifically contends that to say that James was "essentially a philosopher" (or a
"psychologist" or an "artist") is a mistake, adding, however, that it would be equally erroneous
to say that James was all of these. Bjork, *William James*, xvii.

23. On boundary work, and related concepts of "boundary object" and "standardized
packages," see Gieryn, "Boundaries of Science"; Gieryn, *Cultural Boundaries of Science*; Star
and Griesemer, "Institutional Ecology"; Fujimura, "Crafting Science"; Jasanoff, "Contested
Boundaries"; Klein, *Crossing Boundaries*; Giri, "Transcending Disciplinary Boundaries"; Fuller,
"Disciplinary Boundaries"; Good, "Disciplining Social Psychology"; and Good, "Quests for
Interdisciplinarity." On the disciplines and disciplinarity more generally, see Hull, *Science as a
Process*; Lenoir, *Instituting Science*; Lenoir, "Discipline of Nature/Nature of Disciplines"; and
Becher, *Academic Tribes*.

24. Rouse, "Policing Knowledge," 355 (quoted in Good, "Disciplining Social Psychology,"
384). See also Fuller, "Disciplinary Boundaries"; Gieryn, *Cultural Boundaries of Science*.

25. Gieryn, *Cultural Boundaries of Science*, 15–17.

26. Good, "Quests for Interdisciplinarity," 243. See also Roberts and Good, "Introduction," 6.

27. Bourdieu, *Outline of a Theory of Practice*, 78–97.

28. I borrow the term "interpersonal styles" from Morawski, "Self-Regard and Other-
Regard." On forms of sociability and their centrality to attempts to reconfigure particular dis-
ciplines and/or science more generally, see also Shapin and Schaffer, *Leviathan*; Zimmerman,
*Anthropology and Antihumanism*; and Jurkowitz, *Liberal Pursuits*.

29. Gieryn, "Three Truth-Spots."

30. Among many excellent discussions of ways in which spaces convey norms regulating the
social interactions that take place within them, see, for example, Cresswell, *In Place/Out of Place*;
Shapin, "House of Experiment"; Latour, "Mixing Humans and Nonhumans"; and Livingstone,
*Putting Science in Its Place*. On spatial practices, see also de Certeau, *Practices of Everyday Life*.

31. See, e.g., Flint, *Philosophy as Scientia Scientiarum*, 21.

32. See Hollinger, "Inquiry and Uplift"; Hollinger, "Defense of Democracy"; Shapin,
*Social History of Truth*; Daston, "Moral Economy of Science." I discuss the moral and social
economies of science in Bordogna, "Scientific Personae." Jurkowitz has shown how Hermann
von Helmholtz, Ernst Mach, and other German-speaking scientists drew on political values in

order to craft liberal social-epistemic communities. See Jurkowitz, "Helmholtz and the Liberal Unification of Science," and *Liberal Pursuits*.

33. See, e.g., Coon, "Courtship with Anarchy," chap. 3.

34. James, *Some Problems of Philosophy*, 19–20, and James, "Herbert Spencer Dead," 101.

35. Following Roberts and Good, I use the term "cross-disciplinarity" to indicate a kind of border crossing that takes place when "individuals from different disciplines" cooperate on projects of common interest (Roberts and Good, "Introduction," 6). "Cross-disciplinarity" in this sense is different from "inter-disciplinarity," in which a single individual resorts to the tools offered by a variety of disciplines in order to achieve a certain aim.

36. James, introduction to the English translation of Fechner, *The Little Book of Life after Death*, 116–117.

37. Franzese, *L'uomo indeterminato*, esp. 12–20; Seigfried, *William James's Radical Reconstruction*, esp. chap. 6.

38. James, "A Word More about Truth," 78–79.

39. James, "Philosophies Paint Pictures," 5.

40. James, *Manuscript Lectures*, 370. For an insightful philosophical discussion of this problem, see Sprigge, *James and Bradley*, chap. 4.

41. Livingston, *Pragmatism, Feminism, and Democracy*, 11. See also Livingston, *Pragmatism and the Political Economy of Cultural Revolution*, 267.

42. In this part of my work I build on Seigfried, "James," and technical philosophical discussions offered by Sprigge, *James and Bradley*, esp. 180, and Gale, *Divided Self*, esp. 253.

43. James, "Energies of Men," presidential address, 132 and 146.

44. Skrupskelis, introduction to *Manuscript Lectures*.

45. James, "Teaching of Philosophy," 5.

46. Skrupskelis, "James's Conception of Psychology."

### CHAPTER ONE

1. H. Münsterberg, "Emerson Hall," 6.

2. Ibid., 6, 34, 35.

3. See Bjork, *Compromised Scientist*, chap. 3.

4. H. Münsterberg, "Emerson Hall," 13. In his letter to the committee Münsterberg emphasized Harvard's need for a new psychological laboratory, complaining that the existing laboratory "on the corner of Harvard square" was too noisy. On Münsterberg's complaints and the problem of noise in experimental psychology, see Schmidgen, "Time and Noise."

5. The date of the foundation of the laboratory is uncertain. See Perry, *Thought and Character*, 2: 7–10. G. S. Hall rejected the conjectured date of 1875 or 1876, on the ground that James's demonstration room—one room endowed only with "a metronome, a device for whirling a frog, a horopter chart and one or two other bits of apparatus"—was too ill-equipped to be called a laboratory (Hall, *Life and Confessions*, 218). James was the director of the laboratory continuously until 1892. In 1891 the laboratory was transferred to Dane Hall, a building that faced Harvard Yard, and housed the Bursar's Office and the Cooperative Stores. There it was confined to two large rooms on the upper story, packed between a large lecture room used for courses offered by other departments, and a philosophy reading room. The same academic year, James resigned the directorship of the laboratory, which was taken over by Münsterberg. Münsterberg increased the number of rooms by subdividing one of the original rooms and part of the main lecture room into smaller areas, a temporary solution to the problem of lack of space that hardly met the needs of those who worked in the laboratory. See also Harper, "Laboratory of William James."

6. H. Münsterberg, "Place of Experimental Psychology," printed in Münsterberg, "Emerson Hall," 31.

7. Ibid., 7, 32.

8. Ibid.

9. See H. Münsterberg, *Science and Idealism*, 1.

10. H. Münsterberg, "Emerson Hall," 13.

11. Ibid.

12. Ibid., 14.

13. Ibid.

14. See M. Münsterberg, *Hugo Münsterberg*, 61–62.

15. Anonymous, "Emerson Memorial Hall." Münsterberg himself described the new hall as a "Greek, brick building with brick columns and rich limestone trimmings" (H. Münsterberg, "Emerson Hall," 6). It was a "noble monumental building"; "at last" it provided Harvard philosophers with "imposing quarters," as well as "a new centre for Harvard's intellectual life" (M. Münsterberg, *Hugo Münsterberg*, 136). In his 1901 letter to the Visiting Committee, Münsterberg had envisioned a permanent exhibit of busts of Plato, Aristotle, Descartes, Spinoza, Bacon, Hobbes, Comte, Spencer, Helmholtz, and Darwin in the new building. See H. Münsterberg, "Emerson Hall," 6, 15.

16. The sculptor was Frank Duveneck.

17. Abbott, *System of Professions*. See also Abbott, "Things of Boundaries," and *Chaos of Disciplines*.

18. Clifford, *Routes*, 59.

19. See F. M. Turner, "Victorian Conflict," esp. 365. For a broader discussion of religious belief and science in the late nineteenth century, see J. Turner, *Without God*. On the inadequacy of historical accounts that draw the boundary between scientific "orthodoxies" and scientific "heterodoxies" during that period on the basis of class distinctions, see Winter, "Construction of Orthodoxies and Heterodoxies," 28.

20. Gieryn, *Cultural Boundaries*, 43. See also F. M. Turner, "Victorian Conflict," 363.

21. Gieryn, "Three Truth-Spots." See also C. Smith and J. Agar, "Introduction: Making Space for Science," in Smith and Agar, *Making Space*, 13; Shapin, "House of Experiment"; Outram, "New Spaces"; Ophir and Shapin, "Place of Knowledge"; Galison and Thompson, eds., *Architecture of Science*.

22. See, e.g., Forgan, "Architecture of Science," and "Architecture of Display."

23. See Gieryn, "Three Truth-Spots."

24. See, e.g., Gieryn, *Cultural Boundaries*, chap. 1. See also Yeo, *Defining Science*.

25. I have explored the concept of scientific personae in Bordogna, "Scientific Personae."

26. Rorty, Schneewind, and Skinner, eds., introduction to *Philosophy in History*, 1.

27. See, e.g., Hadot, *Philosophy as a Way of Life*, and *La philosophie*; Caizzi Decleva, *Immagini del Filosofo*; Shapin, "Philosopher and the Chicken."

28. On the production of scientific disciplines, see Lenoir, *Instituting Science*.

29. See H. Kuklick, "Boundary Maintenance." Kuklick stresses that the language of professionalization is time-and-place specific.

30. Nineteenth-century philosophers also interacted in many ways with theologians, and their mutual relationships were important to the ways in which they thought of themselves. This important theme, however, goes beyond the scope of *William James at the Boundaries*. For an excellent discussion of those relationships within the American context, see B. Kuklick, *Churchmen and Philosophers*.

31. See William James to Hugo Münsterberg, April 11, 1901, and James to George Dorr, May 1, 1903: "Dear George, I have just sent the Treasurer a check for $200 from Albert Hering of Cleveland Ohio. . . . My other victims don't bleed well" ("Emerson Hall," Harvard University Archives).

32. "To tell the truth, a part from the laboratory whose need is real, I don't feel very eager [about a building devoted wholly to philosophers]. Philosophy, of all subjects, can dispense with material wealth, and we seem to be getting along very well as it is,—I was not aware til I read your circular how great our popularity among the advanced students is. I am not sure that I should n't [sic] be personally a little ashamed of a philosophy Hall, but of course I shall express no such sentiments in public" (James to Hugo Münsterberg, April 11, 1901, in Correspondence, 9: 564). In letters to other correspondents, James referred to Emerson Hall, expressing his lack of interest for the whole plan. Thus, in April 1903, James wrote to his friend the British philosopher Ferdinand Canning Scott Schiller: "We are about to have a philosophy building, 'Emerson Hall' so called. I learn here by the papers that the subscription is secure, & work will probably commence speedily. I don't care a great deal for it myself, but it will please Palmer egregiously, as well as M—g, whose laboratory is now in very bad quarters" (James to F. C. S. Schiller, April 8, 1903, in Correspondence, 10: 232–34).

33. "Surely it isn't too late for both architect and site to be reconsidered? Longfellow [the architect initially chosen by the Harvard Corporation] is capable of any atrocity" (James to Palmer, April 3, 1903, in Correspondence, 10: 227).

34. "Must the building go on Quincy St.? Is n't the Holmes field region, with 'power' for the laboratory etc., accessible from outside, better? I think this question ought to be thoroughly threshed out, before any irrevocable step is taken" (James to Palmer, ibid.). For a period map locating the Holmes Field region, see Winsor, Reading the Shape of Nature, 33, fig. 6.

35. Münsterberg presided over the inauguration of the building. Before introducing Edward W. Emerson, he gave a short speech. He introduced John Dewey (the president of the American Psychological Association), and later, that evening, he and his wife hosted a formal reception for members of the two associations at their home. See H. Münsterberg, "Affiliation of Psychology."

36. James to Eliot, December 28, 1905, in Correspondence, 11: 129–30.

37. James to Münsterberg, December 27, 1095, in Correspondence, 11: 128–29. For a detailed account, see Bjork, Compromised Scientist; and Simon, Genuine Reality, 325.

38. See Simon, Genuine Reality, 85.

39. Ibid.

40. Ibid., 101.

41. On James's early artistic efforts and his indecisions between science and art, see Barzun, Stroll with William James; Bjork, Compromised Scientist; Croce, Science and Religion, vol. 1; Feinstein, Becoming William James; Simon, Genuine Reality, 85–86. See also Leary, "William James and the Art of Human Understanding."

42. James to Katharine James Prince, September 12, 1863, in Correspondence, 4: 81; quoted in Simon, Genuine Reality, 91.

43. On James's Brazilian adventure, see Simon, Genuine Reality, 92–96.

44. Simon, Genuine Reality, 93; and William James to Henry James, May 3, 1865, in Correspondence, 1: 61–64.

45. Simon, Genuine Reality, 100–101.

46. James, April 10, 1873, and February 10, 1873, in Diary 1 (1868–1873). (For a description of the problems of philosophy as "universal questions," see also James, "Quelques considérations sur la Méthode subjective.")

47. James, entry for April 10, 1873, in Diary 1 (1868–1873).

48. See Simon, Genuine Reality, 182.

49. Menand, Metaphysical Club, 215.

50. Peirce, preface to Peirce, Collected Papers, 5: 13. On the Metaphysical Club and its short life, see Croce, Science and Religion, esp. 1: 152–56; Menand, Metaphysical Club, 201–32; and Fisch, "Was There a Metaphysical Club."

51. These courses were listed within the Division of Natural Sciences.

52. The undergraduate course was listed under the Division of Natural Sciences until 1882. On James's early courses, see Stern, "William James and the New Psychology." On these courses and, more broadly, James's teaching career, see Ignas Skrupskelis's very informative introduction to James, *Manuscript Lectures*, xxxi–xxxii.

53. See chapter 2 below.

54. F. C. S. Schiller, "William James," 145.

55. See Skrupskelis, introduction to *Manuscript Lectures*, xlii, and chapter 2 below.

56. James wrote that by appointing him as a professor of psychology, "The Corporation . . . recognized the position of Psychology as an independent science." See James, "Psychology at Harvard University."

57. James to Shadworth Hodgson, March 28, 1892, in *Correspondence*, 7: 255. See also James to Hugo Münsterberg, February 21, 1892, ibid., 7: 243.

58. James to Carl Stumpf, December 18, 1895, in *Correspondence*, 8: 197, and James to Stumpf, November 24, 1896, ibid., 8: 210.

59. E. I. Taylor, *William James on Consciousness beyond the Margins*, 3.

60. Ben-David and Collins, "Social Factors," 461.

61. Rodgers, *Atlantic Crossings*, 4.

62. James to Carl Stumpf, December 18, 1895, in *Correspondence*, 8: 196. On James, cosmopolitanism, and travels, see Bordogna, "L'Hotel Pragmatista." On James's travels between 1878 and 1884, see also Simon, introduction to James, *Correspondence*, 5:xxi–xlix.

63. George Croom Robertson to William James, May 13, 1878, in *Correspondence*, 5: 7–8. On Croom Robertson and the early years of *Mind*, see Neary, "Question of 'Peculiar Importance.'"

64. Robertson, "Prefatory Words."

65. Hall, "Philosophy in the United States."

66. H. Sidgwick, "Philosophy at Cambridge." Sidgwick was elected to the Cambridge chair of Moral Philosophy in 1883. The prestige of the Moral Sciences Tripos remained low through the end of his life. See A. Sidgwick and E. M. Sidgwick, *Henry Sidgwick*, 304. On the relationship between Sidgwick and James, see Perry, *Thought and Character*, 1: 594. Andrew Warwick observes that by 1830 mathematical studies displaced all other studies (including philosophy) at Cambridge. See Warwick, *Masters of Theory*, 180.

67. Heyck, *Transformation of Intellectual Life*, 24–29.

68. On Lewes's theatrical performances, see Spencer, *Autobiography*, vol. 1 (Osnabrück edition, vol. 20), 377. On Lewes and positivist philosophical circles in Britain, see Wright, *Religion of Humanity*.

69. See, e.g., Spencer, *Autobiography*, vol. 1 (Osnabrück edition, vol. 20), 455.

70. On Spencer's initial lack of interest in philosophy, see ibid., 378. For a description of Spencer as a distinguished philosopher, see A. Sidgwick and E. M. Sidgwick, *Henry Sidgwick*, 344.

71. J. S. Mill's *Utilitarianism*, for example, was first published in *Fraser's Magazine*, while Lewes's *Biographical History of Philosophy* first appeared "in the series of shilling volumes published by Knight, . . . one of the pioneers of cheap literature," Spencer, *Autobiography*, vol. 1 (Osnabrück edition, vol. 20), 378. On British periodicals, see Collini, *Public Moralists*, 51–52, and Heyck, *Transformation of Intellectual Life*.

72. As a philosophically inclined scholar of the time put it, "If weary you grow at your books / Or dyspeptical after you've dined, / If your wife makes remarks on your looks, / If in short you feel somewhat inclined / For fresh air and a six hours' grind / And good metaphysical talk— / With a party of writers in *Mind* / You should go for a Sabbath day's walk." (A. J. Butler, "'The Ballade of the Sunday Tramps," 1881; quoted in Perry, *Thought and Character*, 1: 595. The "'Tramps" was a British philosophical club.)

73. William James to Alice Howe Gibbens James, June 24, 1880, in *Correspondence*, 5: 107; William James to Alice Howe Gibbens James, June 27, 1880, ibid., 5: 108; and William James to

Alice Howe Gibbens James, June 28, 1880, ibid., 5: 110–11. Hodgson, James also wrote, looked frail, and had something "feminine" about him that made one feel "he must be tender to him." James suspected he was "one of the doomed victims of the theoretic fever" (ibid.). On the relationship between James and Hodgson (including James's initial reverence and, later, his detachment and reservations about the British philosophers), see Perry, *Thought and Character*, 1: 611–54.

74. James to Alice Howe Gibbens James, June 28, 1880, in *Correspondence*, 5: 111.

75. James to Alice Howe Gibbens James, July 10, 1880, in *Correspondence*, 5: 118.

76. Ibid., and James to Alice Howe Gibbens James, June 24, 1880, in *Correspondence*, 5: 107.

77. James to Alice Howe Gibbens James, January 18, 1883, in *Correspondence*, 5: 396.

78. Ibid., 398. James also met the Comtian philosopher Frederic Harrison, leader of the London Positivist Society.

79. Perry, *Thought and Character*, 1: 594.

80. On Leslie Stephen, see Annan, *Leslie Stephen*, and Collini, *Public Moralists*.

81. James to Alice Howe Gibbens James, December 16, 1882, in *Correspondence*, 5: 332.

82. "Society for Psychical Research," 3. On the foundation of the Society for Psychical Research, see Gauld, *Founders of Psychical Research*, and Oppenheim, *Other World*. On the wave of occultism in Victorian and Edwardian culture, see Owen, *Place of Enchantment*.

83. James to Alice Howe Gibbens James, December 16, 1882, in *Correspondence*, 5: 332.

84. Croom Robertson to William James, January 28, 1884, in *Correspondence*, 5: 485.

85. S. Hodgson to James, November 20, 1884, in *Correspondence*, 5: 535. On the Aristotelian Society, see also Brown, *Metaphysical Society*.

86. Cultivated outsiders, Hodgson continued, could not even understand how the Aristotelian Society could be considered as a "learned society" rather than some sort of "debating club." See S. Hodgson, "Relation of Philosophy to Science," 6, 8 (William James Papers, Houghton Library, Harvard University).

87. See S. Hodgson, "Relation of Philosophy to Science," ibid.; S. Hodgson, "Method of Philosophy"; and S. Hodgson, "Philosophy in Relation to Its History." These papers were printed for private circulation. James received a copy of each of the articles, and had them bound together in a book, which is preserved at the Houghton Library.

88. Ferrier, *Institutes of Metaphysics*, 8. James's copy of this book, one that he probably inherited from his father, is preserved at the Houghton Library. James mentioned Ferrier in James, *Pluralistic Universe*, Lecture I.

89. Green, *Prolegomena to Ethics* (quoted in Kloppenberg, *Uncertain Victory*, 31).

90. Green wondered whether, for the public, the very "existence of a Professor of Moral Philosophy [could] be justified as ministering to an independent study of real value, distinct alike from any natural science and from any enquiry into literary history" (Green, "Province of Ethics").

91. See Richter, *Politics of Conscience*, 154.

92. On Huxley, see Desmond, *Huxley*.

93. White, *Thomas Huxley*.

94. Huxley, quoted in Small, "The Era of Sociology," 12.

95. Lewes, *Biographical History of Philosophy*, xv. See also Lewes, *History of Philosophy from Thales to Comte*, 2: 749–50 and 1: xxiv–xxvii.

96. A good example is Lewes's sarcastic portrait of Hegel, one of the villains of his *History*. See Lewes, *Biographical History of Philosophy*, 601.

97. Ibid., 654.

98. J. Ward, "Progress of Philosophy," esp. 215. (James quoted this text on several occasions and assigned it to his students.) Ward, however, was also ready to recognize that philosophy could benefit from the discoveries and methods of science, and he did not hesitate to approach

metaphysical questions from a psychological and a biological point of view. (See Passmore, *Hundred Years of Philosophy*, 81–83.)

99. See, e.g., H. Sidgwick, "Review of *The History of Philosophy from Thales to Comte*, by G. H. Lewes," and H. Sidgwick, *Philosophy, Its Scope and Relations* (published posthumously by Sidgwick's wife). See also Schneewind, *Sidgwick's Ethics*, 53–54; and Schultz, *Henry Sidgwick*, 66.

100. H. Sidgwick, "Is Philosophy the Germ or the Crown of Science?"

101. Spencer's architectonic vision of philosophy is discussed in chapter 7 below.

102. H. Sidgwick, "Is Philosophy the Germ or the Crown of Science?"

103. S. Hodgson, "Philosophy and Science: I, II, and III," esp. 68.

104. S. Hodgson, "Method of Philosophy." Hodgson set himself to the task of demarcating philosophy from the sciences. He contended that, while philosophy did in fact constitute a *scientia scientiarum* and could not be completely "separated" from the sciences, it also offered a completely separate constructive endeavor. Philosophy deployed a method entirely different from the methods of the sciences, and investigated a distinct subject matter (Hodgson, "Philosophy and Science: III," 359).

105. Green, "Evidence to the Royal Commission." Richter describes Green as possibly "the first Fellow . . . in his University, to conceive of himself as a professional philosopher" (Richter, *Politics of Conscience*, 140, 155). See also Collini, *Public Moralists*, 227.

106. H. Sidgwick, "Is Philosophy the Germ or the Crown of Science?" See also Schneewind, *Sidgwick's Ethics*, 53.

107. A. Sidgwick and E. M. Sidgwick, *Henry Sidgwick*, 303–4. See also H. Sidgwick, "Philosophy at Cambridge."

108. William James to Alice Howe Gibbens James, January 13, 1883, in *Correspondence*, 5: 392.

109. S. Hodgson, "Method of Philosophy." The majority of members (and of people whom the Society hoped to attract) were men and women who "in many and perhaps most cases, have already intellectual homes and intellectual affinities elsewhere," including men of science and serious amateurs interested in a conscientious study of philosophy (ibid., 6).

110. See Richter, *Politics of Conscience*, 162; Collini, *Public Moralists*, 227; and Collini, "Ordinary Experience." Sidgwick made his voice heard within the highest political and economic echelons of the country, clearly seeing no tension between the role of the public moralist and that of the professional philosopher.

111. James to Thomas Davidson, January 2, 1883, in *Correspondence*, 5: 374.

112. On the use of the term "amateur" in late nineteenth-century Britain, see Collini, *Public Moralists*, esp. 220.

113. James to Shadworth Hodgson, September 25, 1880, in *Correspondence*, 5: 139; Perry, *Thought and Character*, 1: 465. James put Renouvier in contact with Shadworth Hodgson, hoping, unsuccessfully, that the two would become allies.

114. James to Alice Howe Gibbens James, December 5, 1882, in *Correspondence*, 3: 318–19.

115. Ribot, "Philosophy in France"; quotations on 367 and 368. For an insightful analysis of Ribot's article, see Barberis, "First *Année Sociologique*." For an insightful discussion of the educational, social, and political role of Cousinianism in nineteenth-century France, see Goldstein, *Politics of the Self*.

116. Ribot, "Philosophy in France," 372.

117. In this group Ribot included, among others, the physician Claude Bernard, the chemist Berthelot, and the philosopher, historian, and psychologist Hyppolite Taine. Ribot stressed that the two groups existed outside the "official world," and that orthodox positivists included "a number of '*proletaires*'" (Ribot, "Philosophy in France," 373).

118. See also Ribot, *English Psychology*, 328. On the approach that Ribot's journal *Revue Philosophique* adopted with regard to the issue of the proper relationships between science and philosophy, and the proper ways of mixing them, see Barberis, "First *Année sociologique*," 106, 113.

119. See Barberis, ibid., chap. 4.

120. Ibid.

121. James, Diary (1905), entry for May 28th, William James Papers, Houghton Library, Harvard University.

122. See Bergson, *Introduction to Metaphysics*; Deleuze, *L'image-mouvement*; Crary, *Techniques of the Observer*. On Bergson, see also Antliff, *Inventing Bergson*.

123. Boutroux, "La philosophie en France depuis 1867," esp. 688 (quoted and discussed in Fabiani, *Les philosophes*, 122–123). See also Ringer, *Fields of Knowledge*, 245–46.

124. William James to T. Flournoy, April 1910, quoted in Perry, *Thought and Character*, 2: 566. On Boutroux's efforts to reconcile philosophy with the sciences, see Barberis, "First *Année sociologique*," 89, 113–20. James read several works by Boutroux, including *Science et religion dans la philosophie contemporaine* and *De la contingence de lois de la nature* ("Sources of William James," 13). When James died, Boutroux wrote a moving philosophical eulogy (Boutroux, *William James*).

125. Shadworth Hodgson to James, October 17, 1882, in *Correspondence*, 5: 276. For Hodgson's views on the relationship between philosophy and psychology (and for his claim that philosophy preceded psychology both historically and logically), see Hodgson, "Philosophy and Science: II," 225.

126. Fabiani, *Les philosophes*, esp. chap. 5.

127. These included the Société francaise de philosophie (1901), an association for the defense of philosophy "against the pretensions of the social sciences" (Fabiani, *Les philosophes*, 42). James owned copies of the society's bulletin for the years 1901, 1902, 1904, 1905, and 1907.

128. Boutroux was an active participant in these debates. He worked to prevent the "modern," scientific curriculum from offering access to the baccalaureate. At the international congress of higher education in 1900, he proposed the creation of a "faculté de philosophie" where philosophy could be cultivated in its "classical and universal form," and which would function as the "link between other studies." He hoped that philosophy would retain a central role in the new organization of higher education because of its universal character. Quoted and discussed in Ringer, *Fields of Knowledge*, chap. 3. See also Fabiani, *Les philosophes*, 132–33, and Barberis, "First *Année sociologique*," 89.

129. Bergson, "Les etudes gréco-latines et la démocratie"; quoted in Ringer, *Fields of Knowledge*, 142.

130. Boutroux was, again, a case in point. See Fabiani, *Les philosophes*, 133. See also Paul Janet, *Principes*, and chapter 7 below.

131. Charles Augustus Strong to William James, January 5, 1890, in *Correspondence*, 7: 1. Charles Montague Bakewell, another former Harvard graduate student, similarly wrote James that "Philosophy in the proper sense of the term seems to be dead or dying [in Germany]" (Charles Montague Bakewell to William James, Berlin, December 23, 1894, in *Correspondence*, 7: 561).

132. There was no agreement on what exactly had caused the crisis, but here, as in France and in the United Kingdom, discussions revolved around how philosophy ought to relate to the sciences intellectually, pedagogically, and in the social/political arena. For a full description of debates concerning education, see Ringer, *Decline of the German Mandarins*, and *Fields of Knowledge*.

133. See, e.g., Wundt, "Philosophy in Germany." This article was submitted to the journal *Mind* as part of the editor Croom Robertson's effort to chart the state of philosophy. James was familiar with the work of each of these naturalists. For an insightful discussion of these figures, and their attacks against philosophy, see Gregory, *Scientific Materialism*, 145–63.

134. Haeckel, *Wonders of Life*, 452–53.

135. On the social role played by physical and metaphorical "thresholds" of fields of knowledge, see Shapin, "House of Experiment." See also Jim Johnson [pseud. of Bruno Latour], "Mixing Humans and Nonhumans."

136. Strong to James, January 5, 1890, in *Correspondence*, 7: 2. Paulsen and Ebbinghaus were on close terms. See Paulsen, *Autobiography*, 299.

137. Strong to James, January 5, 1890, in *Correspondence*, 7: 1.

138. James, review of *Über das Gedächtnis* by Hermann Ebbinghaus, 389. See also Kusch, *Psychologism*, 143–44, and Ash, *Gestalt Psychology*. Ebbinghaus held philosophy chairs in Berlin (1886–94) and Wroctaw (1894–1909).

139. For the negative reviews the book received by German philosophers, see Paulsen, *Autobiography*, 352.

140. James read Paulsen's *Einleitung in der Philosophie* in the spring of 1893. In a letter to Stumpf, he wrote: "It seems to me that if there ever is a *true* philosophy it must be susceptible of an expression as popular and untechnical as this. The man [Paulsen] is a *good* man, through and through!" (James to Stumpf, April 24, 1893, in *Correspondence*, 7: 411). James praised Paulsen's "inexactitude" in another letter to Stumpf. "There is no pretence of 'strenge' about it." Paulsen's untechnical book made "one realize" the weakness of the available alternatives, "unveiled by what I must call the humbug which a would-be 'streng wissenschaftlich' treatment generally disguises them in" (James to Stumpf, May 26, 1894, in *Correspondence*, 7: 425–26). See also James's preface to the English translation of Paulsen's *Einleitung* (James, *Essays in Philosophy*, 92).

141. Paulsen, *Introduction to Philosophy*, authorized translation by Frank Thilly, with a preface by William James. James's preface is reprinted in James, *Essays in Philosophy*, 90–93. Paulsen offered help in regard to German translations of some of James's works, including *The Will to Believe* and *Varieties of Religious Experience*.

142. Paulsen, *Introduction to Philosophy*, 17. James emphasized this point in his preface to the English translation of Paulsen's book. In his words, "On [Paulsen's] view it is impossible to separate philosophy from the natural sciences" (James, preface to Paulsen's *Introduction to Philosophy*, 91).

143. In his copy of the English edition James underlined "it is simply *the sum-total of all scientific knowledge*" and "*universitas scientiarum*" (Paulsen, *Introduction to Philosophy*, 19; William James Papers, Houghton Library, Harvard University).

144. "We get philosophy by combining all the results of these sciences for the purpose of answering the question as to the nature of reality" (Paulsen, *Introduction to Philosophy*, 19).

145. Ibid., 19, 22.

146. Conversely, one could "write histories of philosophy" and yet be completely unphilosophical (ibid.).

147. Ibid. See also C. A. Strong to William James, January 5, 1890, in *Correspondence*, 7: 1.

148. James to Alice Howe Gibbens James, November 2, 1882, in *Correspondence*, 5: 286.

149. Kusch, *Psychologism*, 141. Trained by Hermann Lotze and by the philosophical psychologist Brentano, Stumpf held philosophy chairs in Wurzburg, Prague, Halle, Munich, and Berlin. On James's friendship with Stumpf, see, e.g., James to Stumpf, May 26, 1893, in *Correspondence*, 7: 425. See also Herzog, "William James."

150. See, e.g., Kusch, *Psychologism*, 142.

151. Stumpf, "Die Wiedergeburt der Philosophie." James owned a copy of this article.

152. William Ernest Hocking, review of "Die Wiedergeburt der Philosophie" by C. Stumpf, 612.

153. James to Alice Gibbens James, November 2, 1882, in *Correspondence*, 5: 285; James to Mach, June 27, 1902, Mach Nachlass, NL 174/1577–1581, Deutsches Museum, Munich.

154. James, unsigned review of *Grundzüge der Physiologischen Psychologie* by Wilhelm Wundt, and James, "Importance to Philosophy," 14. See also Diamond, "Wundt before Leipzig," esp. 58.

155. See Ash, "Academic Politics"; Ash, *Gestalt Psychology*; Kusch, *Psychologism*; and Kusch, "Sociology of Philosophical Knowledge."

156. See Diamond, "Wundt before Leipzig," 22, 61–62.

157. See van Hoorn and Verhave, "Wundt's Changing Conceptions," and Wundt, "Über die Einteilung der Wissenschaften." Wundt emphasized that philosophy alone could provide a unified *Weltanschauung* corresponding "to the total scientific consciousness of a given time" (Wundt, "Über empirische und metaphysische Psychologie," 361). Earlier he had defined philosophy as "the general science whose business it is to unite the general truths furnished by the particular sciences into a consistent system" (Wundt, *System der Philosophie*, 21; quoted in Paulsen, *Introduction to Philosophy*, 37).

158. James to Alice Howe Gibbens James, November 18, 1882, in *Correspondence*, 5: 301; James to Titchener, February 1, 1892, in *Correspondence*, 7: 239.

159. Wundt, "Philosophy in Germany," 518. James may have read this article.

160. James to Carl Stumpf, May 26 1893, in *Correspondence*, 7: 425.

161. James's initial good impression, however, was marred by the fact that Avenarius could "hardly talk for five minutes about any subject without some groaning reference" to Wundt (ibid.).

162. For a discussion of James's thoughts on Avenarius's phenomenism, see Lamberth, *William James*.

163. There were exceptions. See, e.g., F. A. Lange's hugely popular *Geschichte des Materialismus* (1873)—a book that James very much enjoyed reading.

164. See James, Photographic Album.

165. D. J. Wilson, *Science, Community*. See also Wilson, "Professionalization."

166. B. Kuklick, *Churchmen and Philosophers*, 129.

167. Ibid., 119, 144–45.

168. Geiger, *To Advance Knowledge*, 9; Parshall and Rowe, *Emergence*.

169. See, e.g., D. J. Wilson, *Science, Community*, 57, and Veysey, *Emergence of the American University*.

170. Hall, "On the History of American College Text Books and Teaching"; quotation on 160 (emphasis added). On the emergence and institutionalization of the human and social sciences in America, see Haskell, *Emergence of Professional Social Science*; Ross, *Origins of American Social Science*. See also Ross, *Modernist Impulses*.

171. Small, "Relation between Sociology and Other Sciences," 12–13. On Small, see H. Kuklick, "Boundary Maintenance"; Abbott, *Department and Discipline*.

172. Schurman, "Prefatory Note."

173. Creighton, "Purposes of a Philosophical Association"; quotation on 232.

174. Ibid.

175. For the early history of the American Philosophical Association, see also Gardiner, "First Twenty-Five Years."

176. Peirce, "Notes on Scientific Philosophy," 1: 128. Among the philosophers who held that view were James's former student Arthur Lovejoy, and James's correspondent Paul Carus, editor of *The Monist* and *The Open Court*. Carus tried to "build up a sound and tenable philosophy, one that would be as objective as any branch of the natural sciences." See Carus, *Philosophy as a Science*, 1.

177. Olesko, *Physics as a Calling*, 367.

178. See, e.g., Daston, "Moral Economy of Science."

179. See Bordogna, "Scientific Personae."

180. Ibid.

181. Ibid., 126–28.

182. D. J. Wilson, *Science, Community*, 133; D. S. Miller, "Conditions of Greatest Progress" (*Journal of Philosophy, Psychology, and Scientific Methods*), and Miller, "Conditions of Greatest Progress" (*Philosophical Review*).

183. Creighton, "Purposes of a Philosophical Association," 236.

184. Ibid., 221.

185. Ibid., 228.

186. Peirce, "Scientific Method," 7: 87.

187. Peirce, "Some Consequences of Four Incapacities," "Grounds of Validity," and "How to Make Our Ideas Clear," esp. 5: 407. On Peirce and the community of inquirers, see R. J. Wilson, *In Quest of Community*, and Haskell, "Professionalism versus Capitalism," chap. 4.

188. Menand, *Metaphysical Club*, 151.

189. Peirce, "What Pragmatism Is," 5: 413. See also Peirce, "Backward State of Metaphysics," 6: 2, and "Vitally Important Topics," 1: 620. See also D. J. Wilson, *Science, Community*, 73.

190. Peirce, "Vitally Important Topics," 1: 623. See also Peirce, "Backward State of Metaphysics," 6: 2, and "What Pragmatism Is," 5: 413. On this point, see D. J. Wilson, *Science, Community*, 73ff.

191. Peirce, "Architectonic Character of Philosophy," 1: 176. Peirce also recalled his "laboratory life" did not prevent him from "becoming interested in methods of thinking," and contended that his "attitude," even when addressing philosophical problems, "was always that of a dweller in a laboratory . . . and not that of philosopher." See Peirce, preface to *Collected Papers*, 1: 4, and Peirce, "What Pragmatism Is," 5: 412. On this point, see Croce, *Science and Religion*, 1: 184.

192. Peirce, "Vitally Important Topics," 1: 620; Peirce, "Scientific Method," 7: 87. Accordingly, if philosophy was to become akin to science, and become truly "architectonic," it could only confer "the most subordinate role" to "thought characteristic of an individual—the piquant, the nice, the clever" (Peirce, "Architectonic Character of Philosophy," 1: 176).

193. On the social organization prevalent in Wundt's laboratory, see Kusch, "Recluse, Interlocutor, Interrogator."

194. Ladd, "President's Address"; quotation on 15. (The address was given at Columbia College in December 1893.)

195. See D. J. Wilson, *Science, Community*, 135–36.

196. Dewey, "Psychology as Philosophic Method," in Dewey, *Early Works*, 1: 157–58. Discussed in D. J. Wilson, *Science, Community*, 66–67.

197. Schurman, "Prefatory Note," 6.

198. Creighton, "Idea of a Philosophical Platform," and Cohen, "Conception of Philosophy"; both discussed in D. J. Wilson, *Science, Community*, 134. On Cohen's conception of philosophy, see Hollinger, *Morris R. Cohen*, chap. 3.

199. James met Howison in 1870, when Howison was professor of Logic and Philosophy of Science at MIT. Both belonged to an informal philosophical club that revolved around the St. Louis idealist W. P. Harris and also included Thomas Davidson.

200. Howison, "Philosophy and Science." James carefully read this address and underlined several passages. The printed version of the lecture was dedicated "To my scientific colleagues with cordial respect."

201. "This is why we who are given to philosophy in its higher ranges are of the conviction that philosophy is the supreme thing, and that science is secondary; we conclude that the relation of science to philosophy is that of means to end—that of subordination and auxiliary service. Philosophy holds the standard of values, and science gets its value *derivatively*,—because it is auxiliary to the essential human aim, which philosophy has for [its] chief object of study" (Howison, "Philosophy and Science," 7).

202. Ibid., 14.

203. O'Donnell, *Origins of Behaviorism*, 1–13. On the historical genre celebrating the institution of laboratories, see Capshew, "Psychologists on Site." See also Ash, "Self-Presentation"; Leary, "Telling Likely Stories"; and Morawski, ed., *Rise of Experimentation*.

204. On the growth of psychological laboratories at American colleges and universities, and the creation of psychological journals, see, e.g., Camfield, "Professionalization"; Ross, "Development of the Social Sciences"; and Rice, "Uncertain Genesis," 490.

205. Fullerton, "Psychological Standpoint"; quotation on 115. As a philosopher of the time noticed, "to call a psychologist a philosopher is almost an insult." See Schinz, *Anti-Pragmatism*, app. C.

206. Scripture, *Thinking, Feeling, Doing*, 293.

207. Sokal, "Origins and Early Years"; O'Donnell, *Origins of Behaviorism*, 5–6; Woodward and Ash, eds., *Problematic Science*, 347. See also Ash and Woodward, eds., *Psychology in Twentieth Century*. Richard Roberts and James Good contend that the metaphor of "emancipation" (from philosophy) often deployed by late nineteenth- and early twentieth-century psychologists is deceptive (Roberts and Good, "Persuasive Discourse," 4).

208. These included, for example, the American Society of Naturalists and the AAAS. For an account of the charter meeting, which took place on July 8, 1892, see Fernberger, "American Psychological Association," esp. 2. Although according to some accounts William James was a charter member of the American Psychological Association, historians believe that he did not attend the charter meeting, and that he was only mildly interested in the project. See also Cattell, "Our Psychological Association," and Sokal, "Origins and Early Years."

209. Scripture, *Thinking, Feeling, Doing*, 293.

210. Gardiner, "First Twenty-Five Years," 146.

211. See Fernberger, "American Psychological Association," 42, and Sokal, "Origins and Early Years."

212. Gardiner, "First Twenty-Five Years," 146.

213. "Psychology," Ladd continued, "is naturally propaedeutic to philosophy" and "implicative of both epistemology and ontology." Consequently, while conceding that the American Psychological Association was formed "in the interests of a science of psychology," and, therefore, could not be expected to "occupy its time and energies largely in the discussion of philosophical problems," Ladd encouraged members of the psychological association to "take an interest in both [psychological and philosophical] problems," to add "philosophical spirit to their scientific intent," and to be "tolerant and generous toward the various possible expressions of philosophical views." Ladd, "President's Address," 16–18.

214. Schurman, "Prefatory Note," 7. Conversely, in 1927, the aging psychologist Joseph Jastrow, recalling the early history of the American Psychological Association, narrated how psychology, that former "colonial outpost of philosophy," had forever gained its independence from the parent field (Jastrow, "Reconstruction of Psychology").

215. Buchner also claimed that "philosophy had slowly, but certainly lost psychology forever" (Buchner, "Quarter Century"; quotation on 410). See also Cattell, "Address of the President," 148.

216. H. Münsterberg, "Affiliation of Psychology."

217. Thilly, "Psychology, Natural Science, and Philosophy," 130.

218. A. E. Taylor, "Place of Psychology."

219. Hall, "Affiliation of Psychology," 297, 299; Angell, summary of paper presented at the panel "The Affiliation of Psychology with Philosophy and the Natural Sciences," in "Proceedings of the American Philosophical Association," 174–75 (1905 annual meeting).

220. On the relationship between James and Hall, see Perry, *Thought and Character*, 2: 15–22; Ross, G. *Stanley Hall*, esp. 86–88, 176–78, 232–50, 269–70; and Simon, *Genuine Reality*, 219–21, 235, 238.

**CHAPTER TWO**

1. William James to Henry James Sr., Berlin, December 26, 1867, in *Correspondence*, 4: 243; emphasis added. James's plan to do laboratory work with Helmholtz and Wundt came to nothing because of his poor health.

2. Draaisma and de Rijcke, "Graphic Strategy," 1.

3. For an excellent study of the pedagogical and political uses of psychology, see Goldstein, *Post-Revolutionary Self.*

4. Bowen, *Principles of Political Economy*, 4. Quoted in Howe, *Unitarian Conscience*, 2. Another early nineteenth-century source describes moral philosophy as the study of "the mind, the means of moral and intellectual improvement, the social nature, duties, and destination of man" ("Notices of Professor Frisbie," xxiv. Frisbie was the first Alford Professor of Natural Religion, Moral Philosophy, and Civil Polity at Harvard. Quoted in Howe, *Unitarian Conscience*, 3.)

5. John Andrews, *Compendium of Moral Philosophy*. MS Notes by John H. Hobart, taken in 1790; quoted in Schmidt, "Old Time College President," 112. Daniel Howe argues that in antebellum America "moral philosophy occupied an important . . . place in the curriculum," and that it might even be described as "*the* humanistic discipline . . . for it encompassed the whole study of human nature." See Howe, *Unitarian Conscience*, 2. See also Reuben, *Making of the Modern University*, 20.

6. H. Sidgwick, *Philosophy, Its Scope and Relations.*

7. Hamilton, *Lectures on Metaphysics*, 1: 62–63. Quoted in Sidgwick, *Philosophy, Its Scope and Relations*, 35.

8. Mill, *Auguste Comte*, 53. Sidgwick commented: "This passage, which represents Mill's mature view, clearly does not distinguish Philosophy from Psychology or Sociology" (H. Sidgwick, *Philosophy, Its Scope and Relations*, 35). William James was familiar with both Hamilton's philosophy (an attempt to mediate British empiricism with German idealism) and Mill's critique of it in *Examination of the Philosophy of Sir William Hamilton*. (See Croce, *Science and Religion*, 162.)

9. Reuben, *Making of the Modern University*, 22–23. On the function of moral philosophy in college education in the first two-thirds of the nineteenth century, see also Howe, *Unitarian Conscience*, and Schmidt, "Old Time College President."

10. As Howe has argued, in places like Boston moral philosophy provided a moral justification for the economic and political practices of the dominating mercantile aristocracy. In turn, the local financial elites recognized moral philosophers' moral and educational leadership (Howe, *Unitarian Conscience*, 135).

11. "If the worst side of the American college is the philosophical, its best is the scientific department" (Hall, "Philosophy in the United States"; quotation on 65). Hall's first attack against the philosophical education offered at American colleges had appeared in a letter to the editor of *The Nation* signed "G. S. H." ("College Instruction in Philosophy").

12. See, e.g., James, "Teaching of Philosophy."

13. James, review of *Data of Ethics* by Herbert Spencer, 347.

14. James to C. W. Eliot, December 2, 1875, in *Correspondence*, 4: 526–28.

15. At Harvard in those years the study of human fossils and artifacts was carried on by Jefferies Wyman, who, from 1866 until his death in 1874, was the curator of the Peabody Museum of American Archeology and Ethnology, located in Boylston Hall. James, who studied with Wyman, may have visited the anthropological collections. On anthropology at Harvard, see Hinsley, "Museum Origins of Harvard Anthropology."

16. Youmans, "Observations." On philosophy's "subjective" method on James, "Lowell Lectures," 34.

17. Huxley, "On the Advisableness," 32.

18. Spencer, *Data of Ethics*. Quoted in James, review of Spencer, *Data of Ethics*, 348.

19. On Büchner, see Gregory, *Scientific Materialism.*

20. Green, "Can There Be a Natural Science of Man?"; quotation on 5.

21. James, review of Spencer, *Data of Ethics*, 347.

22. An early challenge to the ideal of a classical education came from the physician Benjamin Rush, one of the signatories of the Declaration of Independence. Rush did not question the goals of the harmonization of the faculties of the human mind and the necessity to develop them in a proper hierarchical order, but contended that the determination of the proper ways of cultivating the mind (and of the proper curriculum of studies) pertained to physicians like himself. See Haakonssen, *Medicine and Morals*, 187–225.

23. For nineteenth-century debates about the relative merits of classic and scientific education in Germany and France and the underlying political issues, see Ringer, *Decline of the German Mandarin*; *Education and Society in Modern Europe*; and *Fields of Knowledge*. See also Müller, Ringer, and Simon, eds., *Rise of the Modern Educational System*. For French debates on education, see also Goldstein, *Post-Revolutionary Self*, and Fabiani, *Les Philosophes*.

In Britain the controversy between advocates of the sciences and advocates of the humanities became prominent in 1859, when Herbert Spencer published his essay "What Knowledge Is of Most Worth." Spencer argued that "the teaching of the classics should give place to the teaching of science," a thesis that was regarded "by nine out of ten cultivated people as monstrous" (Spencer, *Autobiography*, vol. 2 [Osnabrück edition, vol. 21], 36).

On Huxley's multiple interventions on pedagogical debates, see Huxley, *Collected Essays*, vol. 3, and White, *Thomas Huxley*, esp. chap. 3. On Tyndall, see, e.g., Tyndall, "Address to Students." On Arnold, see Arnold, "Literature and Science," "Culture and Its Enemies," and *Culture and Anarchy*.

More broadly on the debates, see Connell, *Educational Thought and Influence*, chap. 8; F. M. Turner, *Greek Heritage*; Super, "Humanist at Bay."

24. Arnold, "Literature and Science."

25. See Reuben, *Making of the Modern University*, esp. chap. 3.

26. See Youmans, *Culture Demanded by Modern Life*.

27. Youmans, "Progress of Rational Education." See also Youmans, "Mental Discipline in Education."

28. Porter, "Professor Tyndall's Last Deliverance," 200, 218, and "Sciences of Nature versus the Science of Man," 70. See also Porter, *Human Intellect*.

29. Porter, "Sciences of Nature versus the Science of Man," 70.

30. Ibid.

31. See Rice, "Uncertain Genesis," 490. See also G. Richards, "Edward Cox," 34. Rice emphasizes that, in the late nineteenth century, some American moral philosophers started offering courses in "psychology." Richards stresses that some proponents of the "new psychology," including G. T. Ladd, J. M. Baldwin, and G. S. Hall, studied with moral philosophers.

32. See Simon, *Genuine Reality*, 141.

33. James offered the graduate course "The Relations between Physiology and Psychology" (Graduate Course 18) in 1875–76. He offered his proposed undergraduate course in 1876–77 ("Natural History 2"). See Skrupskelis, introduction to *Manuscript Lectures*, xxix, and Skrupskelis and Berkeley, eds., *Correspondence of William James*, 4: 528 n. 1.

34. "The endless controversies whether language, philosophy, mathematics, or science supplies the best mental training, whether general education should be chiefly literary or chiefly scientific, have no practical lesson for us to-day. This University recognizes no real antagonism between literature and science, and consents to no such narrow alternatives as mathematics or classics, science or metaphysics. We would have them all, and at their best" (Eliot, "Inaugural Address," 1).

35. An exception would have been a series of lectures on psychology that Chauncey Wright offered at Harvard in 1870 (Croce, *Science and Religion*, 162).

36. B. Kuklick, *Rise of American Philosophy*, 40. Kuklick offers an insightful discussion of Bowen's philosophical views, showing that his positions were much more sophisticated than usually thought (ibid., 28–45).

37. The list of textbooks that Bowen assigned to his students for his 1870 philosophy of mind class—the first Harvard course devoted exclusively to that topic—was typical of the mental science approach to psychology; it included Locke's *Essay on Human Understanding*, Bowen's own text *Principles of Metaphysics and Ethical Science Applied to the Evidences of Religion*, Noah Porter's *The Human Intellect*, and V. Cousin's *Philosophy of the 18th Century*. See I. Skrupskelis, introduction to *Manuscript Lectures*, xix–xx. Bowen's opposition is documented by a manuscript letter that one of James's former students Francis Greenleaf Allinson wrote to R. B. Perry. Allinson described James's "vivid, inspiring teaching in that course in 'Physiological Psychology,'" and wrote that the course was "officially known as 'Nat. History 2' because (as we students were told) the head of the philosophy department, Professor Bowen, I believe, refused to give the course standing within the Dept. of Philosophy!! We did begin the course with practical *laboratory* demonstrations, on nerve-reactions in frogs etc., and were shown the nervous systems, dissected from a human infant. These lectures were given in *a* laboratory to which we were taken for this part only of the course. The rest of the course was given (with Herbert Spencer's 'Principles of Psychology' as a text-book) in our regular lecture room somewhere, I suppose, in University Hall" (bMS Am 1092.10, James Papers, Houghton Library, Harvard University). For further evidence detailing Bowen's opposition to James's course being listed in the Department of Philosophy, see B. Kuklick, *Rise of American Philosophy*, 135; Skrupskelis, introduction to *Manuscript Lectures*, xx and xxxii; Palmer to James, December 25, 1900, in Perry, *Thought and Character*, 1: 435; and Hall, "Philosophy in the United States," 62.

38. Skrupskelis, introduction to *Manuscript Lectures*, xxiv, and *Museum of Comparative Zoology Bulletin* 3 (1871–76). According to Linda Simon, in 1874 James was appointed director of the Museum of Comparative Zoology (Simon, *Genuine Reality*, 141). On the Museum of Comparative Zoology, see Conn, *Museums*.

39. *Museum of Comparative Zoology Bulletin* 3 (1871–76): Report for 1876, Senate no. 5, p. 8. That lectures of physiology and anatomy should be offered inside a natural history museum was not at all unusual. In fact, as Sophie Forgan has shown, natural history museums and museums of zoology or geology were often used in the 1860s and 1870s, as earlier in the nineteenth century, as facilities for the teaching of science. That "museological" pedagogical regime, she argues, shortly afterwards was replaced for good by teaching based on laboratory practice. See Forgan, "Architecture of Display."

40. Minutes of the Harvard Natural History Society, MS, Harvard University Archives, HUD 3599-505. James was elected to the Harvard Natural History Society in December 1874. Eliot was also a member of the society, and so was Theodore Roosevelt.

41. Santayana, "William James," 104.

42. Schiller, "William James," 145.

43. Eliot, "Inaugural Address."

44. James to C. W. Eliot, December 2, 1875, in *Correspondence*, 4: 527–28. See also Skrupskelis, introduction to *Manuscript Lectures*, xxxiii.

45. Some five years later, James made a similar point in a letter to Johns Hopkins president Daniel Coit Gilman. In his attempt to persuade Gilman to hire G. Stanley Hall for a philosophy chair, and with an eye to the possibility of being hired himself, James stated: "Now I am acquainted with no one in America except Hall and myself who are prepared to teach psychology from the physiological side and to be a connecting link between your medical & your philosophical departments" (James to Gilman, July 18, 1880, in *Correspondence*, 5: 124–25).

46. James hinted that such self-appointed science spokesmen often relied on a secondhand knowledge of philosophy.

47. James, "Lowell Lectures," 17, 30.

48. "Where two lots adjoin with disputed boundary we find each owner disposed to push back the fence. If one man owns both lots, he does not care where the fence stands. I think we ought all to assume this attitude, and cast aside as ignoble this jealousy of the man on the other side of the fence. The truth is all our own whether it be truth of body or truth of mind. What truth the physiologist may have discovered belongs by equal right to any introspective philosopher who will take the trouble to understand it. What mental facts the latter notes, are for the benefit of the physiologist as well" (James, "Johns Hopkins Lectures," 3).

49. James, "Lowell Lectures," 31.

50. Ibid., 32. Also: "The phys. and psych. professionally considered . . . Each owns a lot and two lots lie side by side. Each tries to push . . . fence & reduce . . . for benefit. But we who are neither [physiologists nor psychologists] . . . ought to aspire to attitude" (ibid., 17).

51. Ibid., 32. See also 17.

52. Ibid., 30.

53. Ibid., 17 and 32.

54. "We are each alike proprietors of a body and a mind. We are as much interested in having a sound science of the one as of the other. As proprietors of a body we ought to feel the insufficiency of every theory of the mind which leaves the body out. As owners of a mind we ought to feel the worthlessness of all explanations of our feelings which leave out that which is most essential to be explained. I confess that . . . I have felt most acutely the difficulties of understanding either the brain without the mind, or the mind without the brain. I have almost concluded in my moments of depression that we know hardly anything" (James, "Lowell Lectures," 32).

55. James, *Manuscript Lectures*, 6. Elsewhere James wrote: "Not only is the intellect not built up of sensations, says the new science [the science of 'psychophysics'] but a pure sensation is a nonentity; what we call sensation is soaked through with intellect, and the result of complex logical inference" (James, review of *Revue Philosophique de la France et de l'Étranger* by T. Ribot, 320).

56. On localization theories, see Anne Harrington, *Medicine, Mind, and the Double Brain*; Hagner, *Homo Cerebralis*; and Hagner, ed., *Ecce Cortex*.

57. "In September 1876 [Gilman] invited Huxley to lecture [at Johns Hopkins]. The fact that there were no prayers before or after Huxley's lecture gave rise to a public scandal" (Skrupskelis, introduction to *Manuscript Lectures*, xxxviii).

58. James, "Lowell Lectures," 33. As James stated elsewhere: "The entire recent growth of their science may, in fact, be said to be a mere hypothetical schematization in material terms of the laws which introspection long ago laid bare. Ideas and their association in the mind, cells and their linking fibres in the brain—such are the elements. But, whereas we directly see their process of combination in the mind [by means of introspection], we only guess in the brain what it *may* be from fancied analogies with the mental phenomena" (James, review of *Functions of the Brain* by David Ferrier; *Physiology of Mind* by Henry Maudsley; and *Le Cerveau et ses fonctions* by Jules Luys, 336). See also James, review of *Physical Basis of Mind* by George Henry Lewes, 342, and review of *Relations of Mind and Brain* by H. Calderwood. There James explicitly wrote that "the nowadays much-talked-of theory of localization of functions on the surface of the brain is but an attempt to schematize mechanically the laws of association of these images," which, according to introspective psychology, are the "bricks of mental life."

59. James, "Lowell Lectures," 33.

60. Maudsley, *Physiology and Pathology of the Mind*, vi.

61. James, *Manuscript Lectures*, 34.

62. James, *Manuscript Lectures*, 41. See also James's discussion of aphasia in chapter 2 of *Principles of Psychology*.

63. James, review of Ferrier, Maudsley, and Luys, 333.

64. James, review of Ferrier, Maudsley, and Luys.

65. For an account of the experiments with dogs, see D. Ferrier, *Functions of the Brain*, 205–6.

66. Ibid., 206.

67. James, review of Ferrier, Maudsley, and Luys, 336.

68. Ibid.

69. Ibid.

70. "Harvard College. Second Lecture of the Free Course—Professor James on Recent Investigations on the Brain," *Boston Daily Advertiser*, March 2, 1877. On attendance, see Minutes of the Harvard Natural History Society, MS, HUD 3599–505, 628th meeting.

71. James, review of Ferrier, Maudsley and Leys, 336. See also James, "Lowell Lectures," 41. James argued that all of Maudsley's cerebral physiology "was to a great extent a pure *a priori* attempt to make a diagram, as it were, out of fibres and cells, of phenomena whose existence is known to him only by subjective observation [that is, introspection]. Without this latter we should hardly be trying to-day to unravel anatomically and physiologically the brain. . . . A bad psychology, too, will suggest an impossible physiology, as could easily be shown in the case of Dr. Maudsley's theories of the substratum of memory and concepts."

72. Putnam, "Contributions." As Eugene Taylor notes, in these experiments "etherized dogs had a portion of their skulls surgically removed and a partial incision was then made around the area of the cortex to be tested, creating a flap of tissue which prevented electrical stimulation from penetrating more deeply into the brain at that site. Stimulation of the surface confirmed the recognized specificity of function, thus answering the questions of the critics." (See E. I. Taylor, "New Light," esp. 55.) See also E. I. Taylor, *William James on Consciousness*, 17.

73. James, review of *Grundzüge der Physiologische Psychologien* by Wilhelm Wundt.

74. Ibid.

75. Ibid., 296–97.

76. Franzese, *L'uomo indeterminato*, 30–44. On the importance of Lotze for James, see also Kraushaar, "Lotze's Influence."

77. Skrupskelis, introduction to *Manuscript Lectures*; James to C. W. Eliot, December 2, 1875, in *Correspondence*, 4: 528.

78. James, review of Wundt, *Grundzüge*, 298.

79. James, "Importance to Philosophy," 14 (the original was untitled). See also Skrupskelis, introduction to *Manuscript Lectures*. James made the same point in his unsigned note "The Teaching of Philosophy in Our Colleges," which he published in the summer of 1876 after Wundt had been called to the philosophy chair of the University of Leipzig. James argued that the recent discoveries in the physical and physiological sciences were bound to involve a "change in the method and *personnel* of philosophy." In order to really understand these developments and criticize them, the successful teacher must go through a serious physiological training: "Accordingly," James wrote, "we find Leipzig, now the foremost university in Germany, calling the eminent physiologist Wundt to fill its principal philosophic chair" (James, *Essays in Philosophy*, 6).

80. James, "Importance to Philosophy," 14; emphasis added.

81. For example, elsewhere James expressed his belief that a "*rapprochement* [was] going on between those who approach mental science from the introspective and physiological sides respectively" (James, review of Calderwood, *Relations of Mind and Brain*, 361).

82. James, "Importance to Philosophy," 14. In recommending an inductive approach to philosophy James was responding to a message that Helmholtz had circulated in his public speeches, in his attempt to unify German science and overcome the sectarian strife that fractioned the field. See Jurkowitz, "Helmholtz and the Liberal Unification of Science."

83. James, two reviews of *Principles of Mental Physiology* by W. B. Carpenter. James met Carpenter in 1873.

84. Ibid., 273. James was praising Carpenter's *Principles of Human Physiology*.

85. James, reviews of Carpenter, *Principles of Mental Physiology*, 273.

86. Ibid., 274.

87. Ibid., 273.

88. For Carpenter's views on the advantages of "early scientific training of the mind," see, e.g., "Radiometer," 256, and *Mesmerism*, ix. See also Winter, *Mesmerized*, and chapter 3 below.

89. James, reviews of Carpenter, *Principles of Mental Physiology*, 274.

90. Ibid.

91. In doing so, James was not alone. Moral philosophers also operated on the basis of a twofold implicit definition of mental and moral philosophy, as a field and as a type of mental gymnastic most effective in disciplining pupils' minds. Scientists did the same; many viewed science both as a field of inquiry and as a method for the training of the mind.

92. James, "Teaching of Philosophy," 5. This essay was written in response to G. Stanley Hall's "College Instruction in Philosophy," where Hall had decried the "inadequate" teaching of philosophy at American colleges.

93. Ibid., 4.

94. "If the best use of our colleges is to give young men a wider openness of mind and a more flexible way of thinking than special technical training can generate, then we hold that philosophy (taken in the broad sense . . . ) is the most important of all college studies. . . . [O]ne can never deny that philosophic study means the habit of always seeing an alternative, of not taking the usual for granted, of making conventionalities fluid again, of imagining foreign states of mind. In a word, it means the possession of mental perspective" (James, "Teaching of Philosophy," 4).

95. James, reviews of Carpenter, *Principles of Mental Physiology*, 274.

96. Perry, *Thought and Character*, 2: 48, 47. Quoted and discussed in Evans, "Introduction. Historical Context," 1:xli.

97. See, e.g., Sully, "Review of William James's *Principles of Psychology*."

98. Evans, introduction, xliii–xliv.

99. James, *Principles of Psychology*, 6, 18.

100. Ibid., 18.

101. Ibid., 19.

102. James, preface to *Principles of Psychology*, 6.

103. Ibid. On this point, see also Skrupskelis, "Psychology as a Natural Science," 74.

104. Ladd, "Psychology as So-Called 'Natural Science,'" 30. See also Ladd, "Is Psychology a Science?" esp. 393.

105. Ladd, "Psychology as So-Called 'Natural Science,'" 29.

106. According to Hodgson, James pushed philosophy out of science, "much in the same sort of way as the old materialistic atheists used to push God out of this, out of that, out of Matter altogether, and leave him hovering in dim uncertainty outside the whole dominion of Laws of Nature. God forbid that philosophy should share his unenviable position!" (Hodgson to James, August 19, 1891, in *Correspondence*, 7: 191).

107. As Eugene Taylor emphasizes, James's experimental approach stemmed not from the German nineteenth-century experimental physiology tradition but rather from the French tradition of clinical medicine and experimental physiology that was important at the Harvard Medical School, and that James absorbed through the teaching of C. E. Brown-Sequard, "Harvard's first professor of neurology." James's experimental approach was more akin to the experiments on hypnosis and pathological mental conditions carried on by French "clinical" psychologists than to the laboratory work of Wundt and his students. See E. I. Taylor, "New Light," 37–40.

108. Allegedly, after reading it, Wundt stated that *Principles* was "literature, . . . beautiful, but . . . not Psychology." See G. Richards, *Putting Psychology*, 35.

109. Scripture, "Accurate Work," 427; Benschop and Draaisma, "In Pursuit of Precision," 21.

110. One can contrast the opinion of the German-trained Scripture with that of British psychologist Sully, according to whom *Principles* displayed James's "instinct for careful observation of fact and rigorous verification of theory" (Sully, "Review of William James's *Principles of Psychology*," 396). The local meanings of "carefulness," "accuracy," and "precision" differed in different national contexts and in different experimental traditions. See Olesko, "Meaning of Precision," and M. Norton Wise, "Precision," 228.

111. Scripture, "Accurate Work," 429; Benschop and Draaisma, "In Pursuit of Precision," 9.

112. Scripture, "Accurate Work," 428. See also Scripture, *Thinking, Feeling, Doing*. Although Scripture did not indicate James's name, there is no doubt that he was referring to him (as can be inferred from the sarcastic way in which, in *Thinking, Feeling, Doing* [282], he used James's passage about Wundt's experimental approach, and described many of the problems that James had abundantly discussed in *Principles* as irrelevant to psychology).

For James's description of Ladd's work as "tedious," see James, review of *Psychology: Descriptive and Explanatory* by G. T. Ladd. Ladd, in turn, had noted that "obviously the author, while writing on psychology as a natural science, is 'bored' . . . by such much minute analysis, painstaking collection of statistics, and disproportionate meagerness of scientific results" (Ladd, "Psychology as So-Called 'Natural Science,'" 32).

113. James, *Psychology, Briefer Course*, 9–10, 395; James, "Plea for Psychology as a Natural Science," 271. See also Sokal's introduction to James, *Psychology, Briefer Course*. On the exclusion of metaphysical explanations from psychology as a natural point, see especially Skrupskelis, "James's Conception of Psychology."

114. James, "Plea for Psychology as a Natural Science," 273.

115. Ibid., 272.

116. James, *Psychology, Briefer Course*, 9. Also: "What is a natural science, to begin with? It is a mere fragment of truth broken out from the whole mass of it for the sake of practical effectiveness exclusively. *Divide et impera*" (James, "Plea for Psychology as a Natural Science," 271).

117. James, "Plea for Psychology as a Natural Science," 275.

118. Ibid., 271. See Skrupskelis, "James's Conception of Psychology."

119. James, "Knowing of Things Together," 86.

120. Ibid. See also Skrupskelis, "James's Conception of Psychology," 76.

121. My discussion of "The Knowing of Things Together" is indebted to Skrupskelis's analysis of that text. I diverge from Skrupskelis, however, since Skrupskelis argues that James's suspension of the ban of metaphysics from psychology was confined only to a particular case, whereas I believe it had broader import. Skrupskelis also suggests that, while the claims that James made on the relationships between philosophy and psychology in 1890 and 1892 (and later in 1900) seem to be at odds with those he made in 1894, they were coherent, and that "James presents us with a consistent doctrine in spite of seeming inconsistencies." His main thesis is that the inconsistencies can be explained away if one considers the distinction between metaphysical "assumptions" and metaphysical "explanations." James excluded the latter from psychology but not the first (Skrupskelis, "James's Conception of Psychology").

122. Skrupskelis, "Psychology as a Natural Science," 76–77.

123. James, *Principles of Psychology*, 148.

124. Ibid., 158. The experiment was performed by Helmholtz, and was taken by some psychologists to provide evidence of the compounding of feelings, a conclusion James found erroneous. To James, experiments such as this one provided no evidence of the compounding of mental states. Instead, he took these experiments to indicate that integration took place at the physiological level, so that one should talk about the compounding of brain states rather than the compounding of psychic states. "Nativist" physiologists also went in that direction.

125. James, *Principles of Psychology*, 148. See also James, *Psychology, Briefer Course*, 397.

126. James, *Principles of Psychology*, 158, 163.

127. Ibid., 163.

128. James, *Psychology, Briefer Course*, 397, and "Plea for Psychology as a Natural Science," 275.

129. Fullerton, "Psychological Standpoint," esp. 122.

130. James, "Knowing of Things Together," 88.

131. Ibid.

132. Ibid., 87.

133. Ibid., 89.

134. Skrupskelis, "James's Conception of Psychology," 87.

135. Ibid. James discussed the problem at great length in the manuscript later known as "The Miller-Bode Objections," where he asked how two mental states could point to the same state ("terminus"), and how *"two fields"* could *"be units, if they contain[ed]"* that *"common part"* (James, "Miller-Bode Objections, 1905–1908," 66; emphasis in the original). See also Skrupskelis's enlightening discussion in Skrupskelis, introduction to *Manuscript Essays and Notes*, xxxiii.

136. Lamberth, *William James*, 9, 75. Lamberth traces the emergence of James's radical empiricism to the mid-1890s, arguing that "the bulk of James's metaphysical ideas dated, in significantly developed form, from as early as 1895," several years before the traditionally accepted date of birth of James's radical empiricism (the radically empiricist articles of 1904 and 1905).

137. James, "Knowing of Things Together," 88. See Skrupskelis, "Psychology as a Natural Science," 76–77.

138. Lamberth, *William James*, 82–83.

139. On analogies and differences between the notions of "fringe" and "margin," see ibid., 91–92. On James's metaphysical use of his psychology, see also E. I. Taylor, *William James on Consciousness*.

140. James, "Knowing of Things Together," 78.

141. Sabine, "Discussion," 85, 88.

142. James, *Psychology, Briefer Course*, 9.

143. Ibid., 395–96.

144. James, "Plea for Psychology as a Natural Science," 273.

145. James, *Psychology, Briefer Course*, 9.

146. Ibid.

147. Ibid.

148. James, *Psychology, Briefer Course*, 395.

## CHAPTER THREE

1. Josiah Royce to William James, September 19, 1880, in *Letters of Josiah Royce*, ed. Clendenning, 89.

2. For a late nineteenth-century example, see, e.g., Pearson, *Grammar of Science*. On Pearson, see Porter, *Karl Pearson*. A famous twentieth-century example is provided by Robert K. Merton. See Hollinger, "Defense of Democracy"; see also Hollinger, "Inquiry and Uplift."

3. Shapin, *Social History of Truth*. On the ways in which social and political values, and forms of sociability, may inform scientists' methodological and epistemological commitments, see also Jurkowitz, "Helmholtz and the Liberal Unification of Science," and *Liberal Pursuits*. On the ways in which moral economies of science may also function as social economies of science, see also Bordogna, "Scientific Personae." On the link between modes of sociability and scientific practices, see, e.g., Zimmerman, *Anthropology and Antihumanism*.

4. On the intersection of epistemology and morals in the notion of evidence, see Lipton, "Epistemology of Testimony"; Shapin and Schaffer, *Leviathan and the Air Pump*; and Shapin, *Social History of Truth*. Other "classical" studies of evidence, trust, or both, include Ginzburg, "Clues," and "Checking the Evidence"; Coady, *Testimony*; Chandler, Davidson, and Harootunian, eds., *Questions of Evidence*. See also Serjeantson, "Testimony and Proof"; Morawski, "Self-Regard and Other-Regard."

5. The term "telepathy" was coined in 1882 by the British psychical researcher F. W. H. Myers. For a history of spiritualism in Great Britain, see, e.g., Oppenheim, *Other World*. On spiritualism and occultism in Victorian and Edwardian culture, see Owen, *Place of Enchantment*. On the German context, see Treitel, *Science for the Soul*. Both Owen and Treitel contend that occultism was constitutive of modernity.

6. Gurney, Myers, and Podmore, *Phantasms of the Living*, 1: xlvi. The term was introduced by Myers. Myers speculated that "supernormal" phenomena obeyed still unknown natural laws that belonged "to a more advanced stage of evolution." See Myers, "Automatic Writing, II," 30. There was no consensus on how to use the terms "supernatural" and "supernormal."

7. I use the terms "conceptions" and "regimes" of evidence to indicate a "matrix" that included epistemological, moral, and social (but also class and gender) assumptions about the nature of the trustworthy evidence and trustworthy witnesses, accompanied by a theory of the human mind and by practical prescriptions regarding the setting and attitudes necessary for the production of reliable evidence.

8. For a statement of the scientific methods of the Society for Psychical Research, see, e.g., "Society for Psychical Research: Objects of the Society," 3.

9. See, e.g., Jastrow, "Problems of 'Psychic Research,'" 77; Schiller, "Psychology and Psychical Research," 351.

10. James, "Confidences of a 'Psychical Researcher,'" 362.

11. Daston and Galison, "Image of Objectivity"; Daston, "Moral Economy of Science."

12. Self-recording instruments were automatic inscription devices that, as an international authority on such instruments put it, promised to provide the "natural," "authentic expression of the phenomena [of nature] themselves." See E.-J. Marey, "Mesure à prendre pour l'uniformisation," esp. 376; quoted in de Chadarevian, "Graphical Method," 287. On self-recording instruments, see also Brain, "Graphic Method."

13. Daston and Galison, "Image of Objectivity"; Daston, "Moral Economy of Science."

14. In Leipzig, Bowditch perfected one such instrument, "a little apparatus . . . attached to a metronome for the purpose of taking time on a revolving cylinder covered with smoked paper" (Simon, *Genuine Reality*, 131).

15. H. Bowditch, "Letter to the Editor." See also Fye, *Development of American Physiology*, 103. For late nineteenth-century characterizations of the German "ethos of pure research" (including James's), see Frank, "American Physiologists."

16. The experiments Hall conducted at this time at the Berlin Physiological Institute, including experiments designed to mechanically record muscle reactions to electrical stimuli, amply relied on graphic instruments. See, e.g., Kronecker and Hall, "Die willkürliche Muskelaction." On Hall's activity in the Berlin physiological institute, see Ross, *G. Stanley Hall*, 81–82 and 95, and Frank, "American Physiologists," esp. 34.

17. See Hall, *Aspects of German Culture*, 99, 207. Hall's ethos of science is also discussed in Ross, *G. Stanley Hall*, 93–94, and Bordogna, "Scientific Personae," 106–8.

18. "Meeting of the New York Society for Neurology and Electrology." Discussed in E. I. Taylor, *William James on Consciousness*, 17.

19. Harvard Natural History Society, December 21, 1875, Harvard University Archives.

20. James, "Reaction-Time in the Hypnotic Trance."

21. E.-J. Marey, *La méthode graphique*. See James, *Principles of Psychology*, chap. 3.

22. In 1882, while in Leipzig, James had attended a lecture by Ludwig, the inventor of the first self-recording instrument, and was very much struck by Ludwig's "absolute unexcitability of manner" and "inexhaustible patience of expression," which made him into a prototype of "the traditional german [sic] professor in its highest sense." William James to Alice Howe Gibbens James, November 11, 1882, in *Correspondence*, 5: 298.

23. James, "Will to Believe," 17.

24. James, "Quelques considérations," 23–24; translation in the same volume, 331–32.

25. Benschop and Draaisma, "In Pursuit of Precision."

26. Coon, "Standardizing the Subject"; see esp. 776.

27. James, "Reflex Action and Theism," 102, and *Principles of Psychology*, 192. The chronograph became the symbol of the Wundtian school of experimental psychology and the sensibility of precision that it promoted. It was the piece of apparatus ultimately in charge of calibrating and reducing errors generated by other machinery deployed in reaction-time experiments. See Benschop and Draaisma, "In Pursuit of Precision," 8–9.

28. See Seigfried, *William James's Radical Reconstruction*, 36.

29. For James's evolutionary theory of consciousness, see R. J. Richards, *Darwin*.

30. James, *Principles of Psychology*, chap. 5, esp. 141–45.

31. Ibid., 274.

32. "C'est donc une sorte d'objectivité toute négative. Je ne vois pas, ne sens pas, ne fais pas ce que je veux; cette limitation qui m'enserre, me contraint, m'impose une conduite, une direction indépendent des mes caprices, de mas volonté, serait l'unique manifestation de l'objectif, de la réalité" (Hébert, *Le Pragmatisme*, 31). ["It is then an entirely negative kind of objectivity. I do not see, do not feel, do not do what I want; this limitation that encloses me, constrains me, imposes on me a conduct, a direction independent of my whims, of my will, will be the only manifestation of the objective, of reality."] I made this argument in Bordogna, "Pragmatismo e oggettività in William James," and "Scientific Contexts," chap. 6. On James and objectivity, especially in the context of radical empiricism, see also Seigfried, *William James's Radical Reconstruction*, chaps. 3, 14, and 15. On pragmatism, objectivity, and the importance for pragmatists of external elements of "stubborn resistance" to human interpretations, see Rorty, "Texts and Lumps," 81–83.

33. "As if the mind could, consistently with its definition, be a reactionless sheet at all! . . . As if 'science' itself were anything else than such an end of desire, and a most peculiar one at that! And as if the 'truths' of bare physics in particular, which these sticklers for intellectual purity contend to be the only uncontaminated form, were not as great an alteration and falsification of the simply 'given' order of the world, into an order conceived solely for the mind's convenience and delight" (James, "Reflex Action and Theism," 103).

34. Morawski, "Self-Regard and Other-Regard," 288–90.

35. Hall, review of *Principles of Psychology* by William James, esp. 585. James summarized Hall's comments for a friend, reporting that "my friend G. Stanley Hall, leader of American Psychology, has written that the book is the most complete piece of self-evisceration since Marie Bashkirseff's diary" (James to F. W. H. Myers, January 30, 1891, in *Correspondence*, 7: 140). G. Trumbull Ladd also remarked on the "intensely personal" tone of James's discussion in *Principles* (Ladd, "Psychology as So-Called Natural Science," 25). On the personal tone of many of James's discussions in *Principles*, see also Leary, "Profound and Radical Change," 33, and Evans, "William James and His *Principles*," 11.

36. On this point, see Hollinger, *American Province*, 6–7.

37. James's activity as a psychical researcher is carefully discussed in E. I. Taylor, *William James on Consciousness*.

38. On James's relationship with Mrs. Piper, see James, *Essays in Psychical Research*, editorial note, 394–99.

39. See, e.g., James, "Hidden Self," 248–49. See also James, "Address of the President before the Society for Psychical Research," 131.

40. Henry Sidgwick, quoted in Gurney, Myers, and Podmore, *Phantasms of the Living*, 1: 19–20, n. 1.

41. See, e.g., H. Sidgwick, "President's Address," 9.

42. Crookes, "Experimental Investigation," 339. I offer a more detailed discussion of Crookes's experiments in Bordogna, "Geography of Evidence." On Crookes, see also Oppenheim, *Other World*, 338ff. The term "psychic force" was introduced by the British barrister Sergeant Edward W. Cox, a psychical researcher who witnessed some of Crookes's experiments. See G. Richards, "Edward Cox."

43. See, e.g., [W. B. Carpenter], "Spiritualism and Its Recent Converts," esp. 340. This article described Crookes as "incautious" and complained about his "*malus animus*" toward (presumably) other scientists (ibid., 342–43). Since both the anonymous writer and W. B. Carpenter claimed paternity to the article "Electro-Biology and Mesmerism," one can conclude that the anonymous author was Carpenter himself.

44. Crookes, "Some Further Experiments." Crookes implicitly relied upon the solution to the problem of trust in the experimental sciences that had been worked out by the Royal Society in the seventeenth century, when preexisting codes for the management of dissent were imported from genteel society into experimental science. See Shapin and Shaffer, *Leviathan*. For discussions on how the Royal Society handled disagreement during the early Victorian period, see, e.g., Rudwick, *Great Devonian Controversy*, 24–27.

45. On these accusations, see Owen, *Darkened Room*, 228–30. See also Oppenheim, *Other World*, 341–42. Despite the rumors, Crookes was knighted in 1897, received the Order of Merit in 1910, and was elected president of the Royal Society in 1913.

46. See, e.g., Carpenter, *Principles of Mental Physiology*, 279–315, and "Letter from Dr. Carpenter." On Carpenter and unconscious cerebration, see also Danziger, "Mid-Nineteenth-Century British Psycho-Physiology"; Clarke and Jacyna, *Nineteenth-Century Origins*; and Winter, *Mesmerized*.

47. Carpenter, "Radiometer," 256, and *Mesmerism*.

48. [W. B. Carpenter], "Spiritualism and Its Recent Converts," 342–43.

49. Stewart, "Mr. Crookes on the 'Psychic' Force." On electro-biology, see Winter, *Mesmerized*, 281–84.

50. For a description of some such experiments, especially experiments conducted by Alfred Binet, see James, *Principles of Psychology*, 1205–8.

51. See Beard, "Scientific Study," and "Experiments."

52. Beard, "Psychology of Spiritism."

53. On Beard and American neurologists, see Brown, "Neurology and Spiritualism." For a discussion of a similar disciplinary struggle between physicists interested in studying psychical phenomena and researchers with physiological and psychological credentials, see Noakes, "'Bridge.'"

54. Jastrow, "Problems of 'Psychic Research,'" esp. 79–81. On Jastrow's attacks against psychical research, see Coon, "Testing the Limits," esp. 148.

55. H. Münsterberg, "Psychology and Mysticism," esp. 78.

56. Gurney, Myers, and Podmore, *Phantasms of the Living*, 1: 161, 169–70.

57. Foá, Herlitzka, Foá, and Aggazzotti, "Ce que le prof. P. Foá," 268.

58. On Marey's activity in psychical research, see Braun, *Picturing Time*.

59. On Eusapia Palladino, see Blondel, "Eusapia Palladino." The late nineteenth- and early twentieth-century bibliography on Palladino is immense. For a list of sources up to 1908, see Morselli, *Psicologia e spiritismo*, 117–70.

60. William James to Henry Sidgwick, November 8, 1895, in James, *Correspondence*, 8: 97.

61. For an insightful discussion of the different effects and implications of domestic and "scientific" spaces on the investigation of occult phenomena, see Wolffram, "On the Borders of Science."

62. See Courtier, "Rapport sur les séances."

63. Ibid., 441–42.

64. Ibid., 516.

65. Ibid., 519–20.

66. Ibid., 520.

67. A. Wise Wood, letter to *New York Times*, February 1910 (Archives of the Society for Psychical Research, Cambridge University Library).

68. In other investigations of psychic phenomena, however, various members of this group did not feel bound in any way to the ethos of self-restraint and emotional detachment, and did not rely on self-recording instruments.

69. See William James to Alice Howe Gibbens James, January 13, 1883, in *Correspondence*, 5: 390; William James to Alice Howe Gibbens James, June 24, 1879, ibid., 5: 54; and William James to Granville Stanley Hall, September 3, 1879, ibid., 5: 62.

70. James never met Clifford, who died prematurely in 1879. In 1883 he paid a visit to Clifford's widow. See William James to Alice Howe Gibbens James, February 31 [*sic*], 1883, in *Correspondence*, 5: 412 (the actual date of this letter was probably January 31).

71. Huxley, "Contribution to 'A Modern Symposium'"; quoted in James, "Will to Believe," 17. Years later James had ironical words for Huxley's scientific ethos and ridiculed the obligation to "sing" truth in "huxleyan heroics" and "feel as if truth, to be real truth, ought to bring eventual messages of death to all our satisfactions." See James, "A Word More about Truth," 86–87.

72. Clifford, "Ethics of Belief"; quoted in James, "Will to Believe," 17–18.

73. James, *Will to Believe*, 77.

74. Hollinger has shown that James's interpretation of Clifford was neither charitable nor accurate. See Hollinger, "James, Clifford, and the Scientific Conscience."

75. James, "Will to Believe," 25.

76. See also James, "Quelques considérations."

77. James, "Will to Believe," 21.

78. Ibid., 22.

79. Ibid., 21.

80. William James to Leslie Stephen, February 2, 1898, in *Correspondence*, 8: 343.

81. See, e.g., Ignoramus (James), "Mood of Science."

82. Coon, "Courtship with Anarchy," chap. 3. See also Cotkin, *William James*, 123–51; Coon, "One Moment"; and Simon, *Genuine Reality*, 301–5.

83. Coon, "Courtship with Anarchy," chap. 3.

84. Ibid., chap. 4.

85. See esp. James, preface to *Introduction to Philosophy* by F. Paulsen, 93. In *Principles*, James had described "Absolutism" as "the great disease of philosophic Thought" (James, *Principles of Psychology*, 334).

86. James, *Will to Believe*, 5. For the relationships between this use of the term "radical empiricism" and James's metaphysics of radical empiricism, see Lamberth, "Squaring Logic with Life," 16.

87. James, "Will to Believe," 23.

88. Ibid., 33. Emphasis added.

89. James to Leslie Stephen, February 2, 1898, in *Correspondence*, 8: 343.

90. Gurney, Myers, and Podmore, *Phantasms of the Living*, 1: 459.

91. Ibid., vol. 2, chap. 13.

92. Peirce, "Criticism on 'Phantasms of the Living,'" esp. 154–55, and "Mr. Peirce's Rejoinder," in the same volume, esp. 181, 190–91.

93. See James, "International Congress of Experimental Psychology."

94. James, "Address of the President," 129. See also James, review of the "Report of the Census of Hallucinations."

95. James, "Address of the President," 129, 131.

96. James, review of the "Report of the Census of Hallucinations," 73.

97. Cattell, "Psychical Research." On Cattell's work in Wundt's laboratory, see Sokal, ed., *Education in Psychology.*

98. Cattell, "Psychical Research," 582.

99. Ibid.

100. James, "Psychical Research," 138.

101. Ilbert, "Evidence," 15. Ilbert's discussion of presumptions and the "burden of proof" was based on Sir James Fitzjames Stephen's authoritative *Digest of the Law of Evidence,* one of the principal textbooks on evidence in use at the turn of the twentieth century in Britain and in the United States. Fitzjames Stephen was the brother of James's friend Leslie Stephen, and James read his biography (L. Stephen, *Life of Sir James Fitzjames Stephen*).

102. Myers defined telepathy as "the ability of one mind to impress or to be impressed by another mind otherwise than through the recognized channels of sense." See, e.g., Gurney, Myers, and Podmore, *Phantasms of the Living,* 1: 6–7. James accepted that telepathy was a genuine phenomenon but did not commit himself to any explanation of it. (See, e.g., James, "Address of the President," 131.)

103. James, "Psychical Research," 138.

104. Ibid.

105. See, e.g., James, "Address on the Medical Registration Bill." For an insightful discussion of James's arguments against the bill, see Coon, "Courtship with Anarchy," chap. 4. In America, state legislatures enacting licensing laws had been established in several states as early as the late eighteenth or early nineteenth century, but by the middle of the century had been repealed almost everywhere. In the late nineteenth century, self-described "regulars"—who contrasted themselves to homeopaths, eclectics, and other medical sects—endeavored, in Massachusetts and elsewhere, to have new licensing laws approved. For a history of their efforts, see Starr, *Social Transformation,* esp. chap. 3.

106. James, "Medical Registration Bill," and "Medical Registration Act," 145–46.

107. See, e.g., James, "What Psychical Research Has Accomplished" (1892), in *Essays in Psychical Research,* 89. See also James to Münsterberg, October 22, 1909, in *Correspondence,* 12: 351.

108. Schiller, "Psychology and Psychical Research," 364.

109. Sidgwick, Sidgwick, and Smith, "Experiments in Thought-Transference."

110. Hansen and Lehmann, "Ueber unwillkürliches Flüsten." Hansen and Lehmann also suggested that the high number of guesses could be explained away in terms of the physical location of "transmitter" and "receivers" during the experiment, assuming that the transmitter "whispered" the numbers, and that the receivers were endowed with exceptional acoustic powers.

111. Titchener, "Feeling of 'Being Stared At.'"

112. Titchener to J. M. Cattell, November 20, 1898, Cattell Papers, Library of Congress, Washington, D.C. Cited in Bjork, *Compromised Scientist,* 88. On the relationships between James and Titchener, see ibid., 88–93, and Coon, "Courtship with Anarchy," 210–12.

113. James, "Messrs. Lehman and Hansen on Telepathy." See also James to Titchener, May 21, 1898, in *Essays in Psychical Research,* 481.

114. Titchener, reply to James in *Science*, May 12, 1899, reprinted in James, *Essays in Psychical Research*, 176.

115. Titchener to Münsterberg, February 1, 1904, and February 29, 1908, Münsterberg Papers, Boston Public Library, MSS acc 2191. On Titchener's exclusive attitudes, see Rossiter, *Women Scientists*; Scarborough and Furumoto, *Untold Lives*, chap. 5; and Furumoto, "Shared Knowledge." For a contrasting view, see Evans, "Scientific and Psychological Positions of E. B. Titchener," 35.

116. Titchener, *Systematic Psychology*, 30, 66–67, and "Psychology: Science or Technology?" 44. Titchener's image of the scientist was heavily gendered. Ruth Leys emphasizes Titchener's commitment to "purity" and his aristocratic conception of science. See Leys, "Correspondence between Meyer and Titchener."

117. Titchener, *Systematic Psychology*, 28, 30, 31.

118. Titchener, "Discussion and Correspondence," and "On 'Psychology,'" esp. 14.

119. Titchener to Adolf Meyer, September 19, 1909, in Evans and Leys, *Correspondence between Adolf Meyer and Edward Bradford Titchener*, 126; Titchener to Meyer, May 13, 1918, in ibid., 247.

120. See, e.g., Washburn, "Margaret Floy Washburn," 2: 340 and 344, and Boring, "Edward Bradford Titchener," esp. 495, 500, and 501. On the scientific persona that Titchener cultivated and his social attitudes, see Bordogna, "Scientific Personae," esp. 125–26.

121. Seigfried, *William James's Radical Reconstruction*, 15–16, 78.

122. James, "What Psychical Research Has Accomplished" (1892), in *Essays in Psychical Research*, 91.

123. James, "What Psychical Research Has Accomplished" (1896), in *Will to Believe*, 224, and James, "Address of the President," 133.

124. James, "Myers's Service to Psychology," 194.

125. James, "What Psychical Research Has Accomplished" (1896), 224. All of us, "scientists and non-scientists, live on some inclined plane of credulity," James also observed. "The plane tips one way in one man, another way in another: and may he whose plane tips in *no* way be the first to cast a stone!" Ibid., 236.

126. James's wife and their son Willy attended other sittings.

127. "Circumstantial evidence" of a fact was "evidence of [other] facts from which the fact in question may be inferred" (Ilbert, "Evidence," 16).

128. James, "Report on Mrs. Piper's Hodgson-Control," 281.

129. Ibid., 282.

130. Ibid. On the literary origins of the expression "sense of dramatic probabilities," see Leary, "Influence of Literature."

131. On the "personal equation" in astronomy and psychology, see Schaffer, "Astronomers Mark Time," and Schmidgen, "Time and Noise."

132. James, "Report on Mrs. Piper's Hodgson-Control," 284.

133. Ibid., 282.

134. When addressing his fellow psychical researchers, James repeatedly invited them to be cautious and dispassionate in their assessment of the evidence, reminding them that "science means first of all a certain dispassionate method." (See, e.g., James, "Myers's Service to Psychology," 195, and "Address of the President," 132.) With other audiences, however, James emphasized the limitations of the attitudes of self-restraint, disinterestedness, and emotional detachment.

135. See James, "Report on Mrs. Piper's Hodgson-Control," 284–85.

136. Ibid., 285.

137. See James to Mary Chapman Goodwin Wadsworth, October 19, 1909, in *Correspondence*, 12: 346.

138. H. Münsterberg, "Psychology and Mysticism," 79–83. See also B. Kuklick, *Rise of American Philosophy*, 201.

139. Münsterberg, "Psychology and Mysticism," 80, 81, 85.

140. Ibid., 85.

141. William James to James Mark Baldwin, October 16, 1899, in *Correspondence*, 9: 60.

142. Coon, "Courtship with Anarchy," and E. I. Taylor, *William James on Consciousness*, 148.

143. James used the term "orthodox," in this context, to refer to the scientists who would refuse to consider that the laws of physics might potentially be false. He insisted that "orthodoxy is almost as much a matter of authority in science as it is in the Church" (James, "What Psychical Research Has Accomplished" [1892], 99).

144. James to G. S. Hall, November 5, 1887; cited in Ross, G. *Stanley Hall*, 176. For Hall's criticism of British psychical researchers, see Hall, "Review of *Proceedings*."

145. James, "Myers's Service to Psychology," 192, 197; H. Münsterberg, "Psychology and Mysticism," 68.

146. See James, "What Psychical Research Has Accomplished" (1892), 102.

147. William James to Alice Howe Gibbens James, August 18, 1892, in *Correspondence*, 7: 314, and James, "Myers's Service to Psychology," 194–95.

148. James, "Myers's Service to Psychology," 194.

149. Ibid.

150. Ibid.

151. See Simon, *Genuine Reality*, 196–200. James, however, on some occasions found Ms. Piper's behavior quite annoying.

152. James, "Hidden Self," 248.

153. James, "Address of the President," 136.

154. James, "Myers's Service to Psychology," 198. Spiritualists, according to James, were "indifferent to Science, because Science is so callously indifferent to their experiences" (Schiller, "Psychology and Psychical Research," 364–65).

155. James, "Report on Mrs. Piper's Hodgson-Control," 271.

156. Ibid., 280.

157. James, "Confidences of a 'Psychical Researcher,'" 371.

158. H. Münsterberg, "Science and Education," 115.

159. James, "Address of the President," 131.

160. Peirce, "Some Consequences of Four Incapacities," 5: 317. Quoted in Haskell, "Two Observations," 29. On Peirce's community of inquirers, see also Haskell, "Professionalism versus Capitalism." Peirce's community of inquirers may have also been modeled after an international group of geodecists who worked to determine the absolute value of gravity at different points on Earth by swinging pendula. See P. Galison and L. Daston, "Wissenschaftliche Koordinations."

161. James, *Pragmatism*, 31.

162. Ibid., 32.

163. James, *Pragmatism*, 44. The mediating function of pragmatism between science and religion is not discussed in this book. On that important theme, see Hollinger, *American Province*; Hollinger, "'Damned for God's Glory'"; and Croce, *Science and Religion*.

## CHAPTER FOUR

1. See James et al. (Creighton, Bakewell, Hibben, and Strong), "Discussion: The Meaning and Criterion of Truth," 180–86. See also Bakewell, "On the Meaning of Truth"; Hibben, "Test of Pragmatism"; Creighton, "Nature and Criterion of Truth."

2. Hibben, "Test of Pragmatism," 365–66.

3. Creighton, "Nature and Criterion of Truth," 603.

4. Gardiner, "Problem of Truth." See also Gardiner to James, January 8, 1908, and James to Gardiner, January 9, 1908, both in *Correspondence*, 11: 511–15.

5. James to Schiller, January 4, 1908 (quoted in Perry, *Thought and Character*, 2: 509).

6. William James to William James Jr., April 24, 1907, in *Correspondence*, 11: 350.

7. James, *Meaning of Truth*, 10.

8. A. E. Taylor, a former student of Bradley, described Schiller as an "Aristophanic sausage-seller" (Taylor, "Truth and Consequences," 90). See also Taylor, "Truth and Practice," and H. Münsterberg, *Eternal Values*, 5.

9. James's pragmatist theory of truth is one of the most studied aspects of James's work, and a number of excellent historical accounts and philosophical analyses are available. Among intellectual historians Bruce Kuklick highlights some of the ambiguities of James's account of truth and the "unfinished" character of James's epistemology. He places James's conception of truth in the context of James's discussions with Royce at Harvard, and suggests that James developed it principally in order to bridge the gap between two worlds that Kant had separated: the word of reality, or nature, and that of human nature, with its active propensities, moral needs, and desires. James's epistemology and his conception of truth, Kuklick suggests, "brought the moral and intellectual together" (B. Kuklick, *Rise of American Philosophy*, 274). Kuklick emphasizes that James's epistemology blurred the distinctions between philosophy and psychology, and his interpretation provides an important point of departure for my own discussion. Hollinger, who forcefully suggests that *Pragmatism* be read as a "text" rather than as a container of philosophical theories, depicts James's theory of truth as part and parcel of the broader goal of James's pragmatism—that of providing an alternative that would appeal to "persons too tender to give up religion but too tough to give up science." From this perspective James's main goal in the pragmatist controversy was to "release, for fuller involvement in the enterprise of creating and maintaining beliefs . . . the intellectual resources intimidated by the 'tough'" and promoted by the "tender" (Hollinger, *American Province*, 7, 12).

Deborah Coon and George Cotkin instead situate James's pluralistic account of truth in the context of his politicization in the late 1890s, and of his mounting reaction against the growing American imperialism, while James Livingston explores the meaning of James's "persistent use of financial metaphors" in his discussion of the "credit system" of truth—an aspect of James's work that gave rise to much criticism in the twentieth century. Livingston embeds James's account of truth in a "new political economy of the sign," a "radically contingent 'credit system' constituted by corporate capitalism," and suggests that the epistemological regime promoted by pragmatism and period economic theories were far more strictly intertwined than acknowledged by more internalistic readings of James's theory of truth (Livingston, *Pragmatism and the Political Economy of Cultural Revolution,*, 208–9).

Recent philosophical analyses of James's conception of truth include those offered by Thayer, Myers, Bird, Seigfried, Putnam, Lamberth, Rorty, Sprigge, and Cooper. Thayer proposed a distinction between "cognitive truth" and "pragmatic truth" in an attempt to make James's statements about truth look more coherent (Thayer, *Meaning and Action*, 539). Myers argues that "by today's standards, pragmatism is not a genuinely technical theory of meaning or truth; it is rather a method of choosing what to believe from among philosophical and religious propositions, and in some cases of finding the motivation to believe anything at all." He contends that most of James's controversial statements about truth, which were commonly construed as his definitions or criteria of it, were intended instead to describe and dramatize its motivational role in making us believe (G. E. Myers, *William James*). Bird draws a distinction between what he calls a "strict theory of truth" and a theory of an "extended truth"—that is, truth as it relates to belief, meaning, and action. He discusses the pragmatist controversy and examines

James's notion of extended truth in light of current philosophical theories (Bird, *William James*). Seigfried challenges Thayer's distinction, and argues that James's explanation of truth can only be properly understood within the framework of James's attempt to study the "structures" that define the human "active being in the world" (Seigfried, *William James's Radical Reconstruction*, 299). Putnam shows that James's theory of truth developed together with his metaphysics of radical empiricism, and was linked to it in complex ways (H. Putnam, "James's Theory of Truth," 166). He offers a realistic interpretation of James's theory of truth, and so does Lamberth. Rorty proposes an antirealist interpretation of James's account of truth, contending that it dissolved "the traditional problematic about truth" (Rorty, "Faith, Responsibility and Romance"). For Sprigge James's desire to dispense with Josiah Royce's Absolute was the main reason lurking behind an important aspect of his conception of truth, namely the thesis that truth is an experiential relation (Sprigge, *James and Bradley*, chap. 1). Cooper splits James's account of truth into two "axioms": one postulating "absolute truth," and the second dealing with "half truth" (Cooper, *Unity of William James's Thought*).

10. H. Putnam, "James's Theory of Truth," 167.

11. James, "Remarks on Spencer's 'Definition of Mind as Correspondence,'" and H. Putnam, "James's Theory of Truth," 167. See also James, "Notes for Natural History 2," 128. James defined absolute truth as "an ideal set of formulations towards which all opinions may in the long run of experience be expected to converge." He described it as an "absolute standard" or a "regulative ideal" that would retrospectively imprint the mark of truth or falsehood on all opinions. See James, *Meaning of Truth*, 143, and James, *Pragmatism*, 106–7. Peirce famously defined truth as "the opinion which is fated to be ultimately agreed to by all who investigate" (Peirce, "How to Make Our Ideas Clear," 407). Lamberth offers a different interpretation of James's conception of absolute truth, one that makes absolute truth into a habitual impulse to continue inquiry.

12. H. Putnam, "James's Theory of Truth," 170.

13. For "concrete truths," see James, "Humanism and Truth," 54. For "half-truths," see James, *Pragmatism*, 106–7. James believed that, even though we may possess some absolute truths, we cannot know for sure that they are absolutely true.

14. H. Putnam, "James's Theory of Truth," 181.

15. Ibid., 166, 168, 170. Putnam notes that James introduced this idea as early as 1878 in an innocuous form: "human actions partially determine what will happen," and, in that sense, "what will be true of the world." In his pragmatist works, however, James reformulated the interest component in more extreme and paradoxical ways (ibid., 167, and James, "Remarks on Spencer's 'Definition of Mind as Correspondence,'" 21).

16. H. Putnam, "James's Theory of Truth," 169 (emphasis in the original). For a different interpretation, one that emphasizes the role played by interest in James's conception of truth, and does not equate Jamesian truth with "absolute truth," see Seigfried, *William James's Radical Reconstruction*, esp. chap. 3.

17. H. Putnam, "James's Theory of Truth," 170.

18. Ibid., 175.

19. James, *Pragmatism*, 97, and H. Putnam, "James's Theory of Truth," 178.

20. James, *Pragmatism*, 98. The term "ambulatory" was first used by Charles Augustus Strong, who distinguished between ambulatory relations (existential relations that served to connect experiences) and "saltatory" relations (logical relations). See Strong, "Naturalistic Theory," and James, "Word More about Truth," 80.

21. James, *Pragmatism*, 97–98. As Seigfried notes, for James the truth of an idea consisted very much in its workings, and these related to our "being-and-acting-in-the-world." For James, true ideas, she continues, "are true because they lead us to act in ways that are practically important."

Truths "are disclosed as being 'invaluable instruments of action,'" (even though "not in any narrow sense"). Seigfried, *William James's Radical Reconstruction*, 299, 309.

22. James, "Word More about Truth," 86, 89. James also noted: "You may put either word [truth or satisfaction] first in your ways of talking; but leave out that whole notion of *satisfactory working or leading* (which is the essence of my pragmatist account) and call truth a static logical relation, independent even of *possible* leadings or satisfactions, and it seems to me you cut all ground from under you" (ibid., 89).

23. James, "Report of a Discussion in Philosophy 20e," 438. James attributed the doctrine that associated truth with satisfaction to John Dewey. Dewey acknowledged James's courtesy only partially: "Since Mr. James has referred to me as saying 'truth is what gives satisfaction,' I may remark (apart from the fact that I do not think I ever said that truth is what *gives* satisfaction) that I have never identified *any* satisfaction with the truth of an idea, save *that* satisfaction which arises when the idea as working hypothesis or tentative method is applied to prior existences in such a way as to fulfill what it intends" (quoted in James, "Humanism and Truth," 39).

24. James, "Humanism and Truth," 60.

25. James, "Humanism and Truth Once More," 191. Also: "There is an infinite number of satisfactions. And all these things must be studied in the psychological part of logic" ("Report of a Discussion in Philosophy 20e"; quotation on 437).

26. James, "Abstractionism and 'Relativismus,'" 134.

27. "Of course, if you take the satisfactoriness concretely, as something felt by you now, and if by truth, you mean truth taken abstractly and verified in the long run, you cannot make them equate, for it is notorious that the temporarily satisfactory is often false. Yet at each and every concrete moment, truth for each man is what that man 'troweth' at that moment with the maximum of satisfaction to himself; and similarly, abstract truth, truth verified by the long run, and abstract satisfactoriness, long-run satisfactoriness, coincide. If, in short, we compare concrete with concrete and abstract with abstract, the true and the satisfactory do mean the same thing" (James, "Humanism and Truth," 54). In this article James's task was to explain the "humanist" conception of truth, that is, the conception of truth adopted by Schiller and Dewey rather than his own. However, both Schiller and Dewey recognized that much of the view James generously ascribed to them—especially the distinction between abstract and concrete truth (later relabeled respectively as "absolute" and "half" truths)—was James's view.

28. James, "Professor Hébert on Pragmatism," 127.

29. See, e.g., James, "Word More about Truth," 89. Also: "Every idea that a man has works some consequences in him. . . . Through these consequences the man's relations to surrounding realities are modified . . . and he gets now the feeling that the idea has worked satisfactorily, now that it has not." Truth occurs in the former cases. James, "Professor Hébert on Pragmatism," 129.

30. James, "Word More about Truth," 88.

31. Among the exceptions are B. Kuklick, *Rise of American Philosophy*, 269, and Seigfried, *William James's Radical Reconstruction*, 194–96. Seigfried, in particular, discusses the relationship between absolute truth and satisfactoriness. In her interpretation, "temporality is a fundamental fact of experience and the long range satisfactoriness that characterizes abstract truth is the sum of day by day satisfactory confirmations. It is not something additional to them. Each such retroactive judgment from a later point of view compares the earlier with the later and confirms or disconfirms the belief held" (Seigfried, *William James' Radical Reconstruction*, 310). My analysis is indebted to both Kuklick and Seigfried, but goes beyond in an attempt to clarify (below) what James meant by "satisfaction."

32. Lovejoy, *Thirteen Pragmatisms*, 19.

33. Pratt, "Truth and Its Verification," 323.

34. Boutroux to William James, December 18, 1908 (quoted in Perry, *Thought and Character*, 2: 564).

35. Perry, "Review of Pragmatism as a Theory of Knowledge," 371. James indeed stated that "satisfactions are the marks of truth's presence," and that "the truth of an idea is determined by its satisfactoriness." James, "Word More about Truth," 88.

36. Lovejoy, *Thirteen Pragmatisms*, 20. Lovejoy argued that, according to James's doctrine, a proposition that lacked "conceptual consistency" or "completeness of theoretical verification" could be true if it was congruent with "our habitual ways of belief" or if it charmed "the imagination" or if it was able to "beget a cheerful frame of mind in those who accept[ed] it" (ibid., 19; see also Lovejoy, "Thirteen Pragmatisms," pt. 2, p. 33). For James's response to this accusation, see, e.g., Perry, *Thought and Character*, 2: 468. In a manuscript composed between 1907 and 1909, James distinguished "3 ranges of satisfactoriness" in "our hypotheses": "[1] Intrinsic, as when they form a pretty picture, schemes, utopias, subjective reveries generally, language of flowers, their soul, etc. [2] Relational, as when they 'agree' with all sorts of other hypotheses and beliefs. [3] Subsequential, as when the consequences to ourselves of *believing* them are satisfactory. These subsequential satisfactions can be divided into two kinds. The first kind follows from our merely *having* the belief (religion, fool's paradises etc.) the second from its 'object' being a reality" (James, "Truth and Reality, etc. 1907–1909," 235).

37. The British philosopher H. W. B. Joseph suspected that James understood "theoretical satisfactions" as a form of "pleasure" (Joseph, "Professor James on 'Humanism and Truth,'" 40). Francis Bradley predicted that any attempt to find in "the emotional expression [of an idea] . . . the specific essence of truth and falsehood in the end must break down" (Bradley, "On Truth and Practice," 314 n. 1). J. A. Leighton rebuked Schiller for attributing to truth emotional value (Leighton, "Pragmatism," 151).

38. James, *Pragmatism*, 106; and Hébert, *Le Pragmatisme*, quoted in James, "Professor Hébert on Pragmatism," 126. For criticisms of these aspects of James's account of truth, see G. E. Moore, "Professor James's 'Pragmatism,'" and B. Russell, "Transatlantic Truth." See also, e.g., Joseph, "Professor James on 'Humanism and Truth.'" For a helpful discussion of Russell's objections to James's notion of "satisfactory workings," see Bird, *William James*, 43–44.

39. Quoted in James, "Pragmatist Account of Truth," 112. A sympathetic correspondent implored James to eliminate every reference to "the cash-value" of true ideas, since "everyone thinks you mean only pecuniary profit and loss." In response to these objections, James emphasized that the interests involved included not only practical but also "theoretical interests" (ibid.).

40. Creighton, "Purpose as a Logical Category," 287, 295; Royce, "Eternal and the Practical," 118. In his article Creighton discussed Dewey's and other Chicago pragmatists' accounts of truth. James argued that the notion of "will" "ought to play no part" in a discussion about truth (James, "Word More about Truth," 86–87).

41. James, for example, wrote: "the retardation, the bad localization of sensations of p. & p. [pleasure and pain] do point to *diffused* processes of some sort . . . to irradiation" (James, "Notes for Philosophy 20a," 208, 209). See also James, *Principles of Psychology*, 1086–87.

42. James, What Is an Emotion?" 175.

43. James, *Principles of Psychology*, 1084–85. "In all cases of intellectual or moral rapture we find that, unless there be coupled a bodily reverberation of some kind with the mere thought of the object and cognition of its quality; unless we actually laugh at the neatness of the demonstration or witticism; unless we thrill at the case of justice . . . our state of mind can hardly be called emotional at all. It is in fact a mere intellectual perception of how certain things are to be called. Such a judicial state of mind as this . . . is a *cognitive* act. As a matter of fact, however, the moral and intellectual cognitions hardly ever do exist thus unaccompanied. The bodily sounding-board is at work, as careful introspection will show, far more than we usually suppose"

(ibid.). See also James, "What Is an Emotion," 174, and Bannan, "Emotions and Biology." Carl Georg Lange (1834–1900) was a Danish psysiologist. He offered an account of emotion similar to James's.

44. "Our need for such consistency [as is afforded by theoretical truth] and our pleasure in it are just a natural outcome of the fact that we are beings that develop habits—habit is itself adaptively beneficial" (James, "Humanism and Truth Once More," 197). James also referred to the "pleasure of consistency" (James, "Aesthetics," 124).

45. James, *Principles of Psychology*, 111–12. "The most complex habits," James wrote, "are . . . nothing but *concatenated* discharges in the nerve-centres, due to the presence there of systems of reflex paths, so organized as to wake each other up successively—the impression produced by one muscular contraction serving as a stimulus to provoke the next, until a final impression inhibits the process and closes the chain. . . . [A]ny sequence of mental action which has been frequently repeated tends to perpetuate itself" (ibid., 116). The habitual paths were grooved in the brain by the repetition of certain nervous events.

46. "When an object claims from us a reaction habitually accorded only to the opposite class of objects, our mental machinery refuses to run smoothly. The situation is intellectually unsatisfactory" (James, "Humanism and Truth," 58).

47. James, "Aesthetics," 125.

48. The Chicago pragmatist Addison W. Moore noted more generally that, for the pragmatists, " the relation between thought, on the one hand, and feeling and impulse, on the other, is something more than that of mere parallelism in somebody's head." For the pragmatist, "thought and feeling and habit are mutually conditioning processes . . . in professor James's phrase, they 'lean on' each other" (Moore, "Pragmatism and Its Critics," 341).

49. Montague, "May a Realist Be a Pragmatist," pt. 2, "Implications of Psychological Pragmatism." This was part of a series of articles that Montague wrote in his attempt to show that no brand of pragmatism was compatible with a realist epistemology.

50. W. B. Carpenter, *Principles of Human Physiology* (quoted in R. J. Richards, *Darwin*, 284). Carpenter himself recognized that the theory had been set forth, in a different form, by the British physiologist Thomas Laycock, who first introduced the notion of reflex action of the cerebrum. Carpenter's "ideo-motor action" was just such a reflex action. The French philosopher Charles Renouvier adopted a similar view in 1859, followed in 1868 by Alexander Bain. See Richards, *Darwin*, 284ff.

51. James, *Principles of Psychology*, 1134. Briefly, according to James's theory of will, a voluntary act follows simply from the fact that one idea has been able to capture the mind's attention, either because the idea succeeded in predominating over other antagonistic or inhibitory ideas, or because it was actively selected in view of certain interests or purposes of the knower. In either case, action follows simply from the motor power of the idea: "where there is no blocking, there is naturally no hiatus between the thought-process and the motor discharge. *Movement is the natural immediate effect of feeling* [a word that here is synonymous with "thought"] *irrespective of what the quality of the feeling may be*. It is so in reflex action, it is so in emotional expression, it is so in voluntary life" (ibid., 1135; emphasis in the original). James's theory of the will bears some resemblance to Herbert Spencer's, who likewise incorporated the notion of motor changes initiated by ideas into his account of will: "When the automatic actions become so involved, so varied in kind, . . . when, after the reception of one of the more complex impressions, the appropriate motor changes become nascent, but are prevented from passing into immediate action by the antagonism of certain other nascent motor changes . . . there is constituted a state of consciousness which, when it finally issues into action, displays what we term volition. Each set of nascent motor changes arising in the course of this conflict, is a weak revival of the state of consciousness which accompanies such motor changes when actually performed . . . —is an idea of such motor changes. We have, therefore, a conflict between two sets of ideal motor changes

which severally tend to become real, and one of which eventually does become real; and this passing of an ideal motor change into a real one, we distinguish as Will" (Spencer, *Principles of Psychology*, 4: 496–97).

52. "The mysterious tie between the thought and the motor centres next comes into play, and, in a way that we cannot even guess at, the obedience of the bodily organs follows as a matter of course" (James, *Principles of Psychology*, 1168).

53. James, "Professor Pratt on Truth," 96 (emphasis added).

54. James, "Tigers in India," 34 (emphasis added).

55. The point has been developed in recent feminist-pragmatist epistemologies that emphasize the situatedness and bodily nature of the knowers. See, e.g., Haraway, "Situated Knowledges," and Seigfried, *Pragmatism and Feminism*.

56. Strong, "Naturalistic Theory." Strong ascribed that theory of cognition to both William James and James's former student (and, by then, critic) Dickinson S. Miller. James acknowledged his indebtedness to Miller (James, "Word More about Truth," 78–79).

57. A. E. Taylor, "Truth and Practice," 272.

58. On pragmatism and evolutionary theory, see Philip P. Wiener's classical study *Evolution and the Founders of Pragmatism*, and Seigfried, *William James's Radical Reconstruction*, chap. 3.

59. A. W. Moore, "Pragmatism and Its Critics," 341.

60. Höffding, *Outlines of Psychology*, 89. According to Ernest R. Hilgard, the threefold classification was first clearly stated by Moses Mendelssohn (*Letters on Sensation, 1755*), and subsequently adopted by Kant; it remained popular even in the passage from soul-psychology to associationism (Hilgard, "Trilogy of Mind").

61. J. Ward, "Psychology," 551; Ward, "Psychological Principles," 465, quotation on 471.

62. Sully, *Outlines of Psychology*, 22. For a similar metaphor, see W. E. Johnson, "On Feeling as Indifference," esp. 80.

63. See James, review of *Psychology: Descriptive and Explanatory* by G. T. Ladd, 479.

64. Dewey, *Psychology*; quotation on 18.

65. Spencer, *Principles of Psychology*, 4: 473.

66. Mantegazza, *Fisiologia del piacere*, 455–56. In this work Mantegazza—who would later be the founder of the Società Italiana per l'Antropologia e l'Etnologia—offered an experimental and taxonomical science of pleasure, which was to provide a foundation for a "physiological study of the moral man" and for a moral system. On Mantegazza and his "anthropology of passions," see Pireddu, *Antropologi alla Corte della Bellezza*, 132–84, esp. 137.

67. Bain, *Mental Science*; quotations on 226.

68. Stanley, "Relation of Feeling to Pleasure and Pain," 539.

69. Ward agreed with former psychologists who classified this feeling along with the "formal feelings" of Herbartian descent. J. Ward, "Psychology," 583. Marshall, *Pain, Pleasure, and Aesthetics*.

70. Dewey, *Psychology*, 300. Also: "So far as we have mastered the relations which constitute the material of knowledge, and have organized these into our mental structure, there are feelings of . . . satisfaction and . . . self-possession" (ibid., 301).

71. Höffding, *Outlines of Psychology*, 263–64.

72. See *Manuscript Lectures*, 508, note 330.33. James's annotated copy of Höffding's *Outlines of Psychology* is preserved at the Widener Library. James wrote a preface for the English translation of Höffding's *Problems of Philosophy* (see James, *Essays in Philosophy*, 140–43).

73. James read Mantegazza's *Physiologie de la douleur*. By the turn of the twentieth century, Mantegazza was well known in European and American anthropological circles, and James may well have been acquainted with other works.

74. James, "Notes for Philosophy 20a"; James, review of *Pain, Pleasure, and Aesthetics* by Henry Rutgers Marshall, 492.

75. G. E. Myers, *William James*, 278. Belief, James wrote, "is the function of cognizing reality." It was strictly allied to the emotions (since only objects that "have emotional interest" excite our belief) and was, in fact, a kind of emotion—"a sort of feeling more allied to the emotions than to anything else" (James, *Principles of Psychology*, 948). James also argued that as a mental attitude belief was indistinguishable from the will: will and belief were "a manner of attending to certain objects, or consenting to their stable presence before the mind," an attitude by which the mind turns to its objects "in the interested active emotional way." Also: "Will and Belief in short, meaning a certain relation between objects and the Self, are two names for one and the same *psychological* phenomenon" (ibid.).

76. James noted, for example, that out of the many retinal sensations yielded by a tabletop, we select the square image as objective, and disregard as mere appearance the rhombus image. He contended that the first image was no more "objective" than the second one; it was simply more convenient for practical purposes. "Objective sensations are mere sensations [whose] aesthetic characteristics appeal to our convenience or delight." On this point, see James, *Principles of Psychology*, 928–34, quotation at 934, and chapter 3 above.

77. Ibid., 941. See also Seigfried, *William James's Radical Reconstruction*, 37.

78. For the generalized reflex-arc theory, see Laycock, "On the Reflex Function of the Brain," and W. B. Carpenter, *Principles of Human Physiology*. For the British debates concerning the ramifications of the generalized reflex-arc theory for the problem of will and intelligence, and the relationships between voluntary and involuntary behavior, see Danziger, "Mid-Nineteenth-Century British Psycho-Physiology," and Danziger, "Unknown Wundt," esp. 97–101. For a detailed discussion of reflex-arc theories in Britain and Germany, see Clarke and Jacyna, eds., *Nineteenth-Century Origins*. For presentations of the reflex-arc theory recommended by James to his readers, see James, *Will to Believe*, 266, editorial note 92.38.

79. See Huxley, "On the Hypothesis That Animals Are Automata."

80. On James's application of the reflex arc to cognition and conception, see Seigfried, *William James's Radical Reconstruction*, 36. She notes that "physiological evidence supports two claims fundamental to Jamesian philosophy, namely, that reflection is literally grounded in the sensed world and that its function is to guide action" (ibid., 37). On James's teleological conception of mind and its relationship to his account of truth, see also Thayer, *Meaning and Action*, 141–45.

81. James contended that the teleological character of the mind followed more generally from the "generalized" conception of reflex action: "I am not sure that all physiologists see that [the generalized reflex-action theory] commits them to regarding the mind as an essentially teleological mechanism. . . . I mean by this that the conceiving or theorizing faculty—the mind's middle department—functions *exclusively for the sake of ends*" (James, "Reflex Action and Theism," 94–95).

82. "The willing department of our nature, in short, dominates both the conceiving department, and the feeling department; or, in plainer English, perception and thinking are only there for behavior's sake" (James, "Reflex Action and Theism," 92). Also: "The sensory impression [that is, the first phase of the reflex arc] exists only for the sake of awakening the central process of reflection, and the central process of reflection exists only for the sake of calling forth the final act" (ibid.).

83. James, *Principles of Psychology*, 941.

84. John Dewey, who offered a similar reinterpretation of the reflex-arc process, emphasized precisely that point, and observed that the reflex-arc conception was meant to eliminate the traditional distinction among faculties (feeling, thinking, willing), even though the way in which it was usually interpreted reinforced that distinction. In Dewey's interpretation, the reflex-arc process was not simply a sequence of three steps, but rather a global process in which the final step (with its purposes, interests, and needs) acted as a final cause, governing retrospectively

the first two steps and giving them unity and meaning. (See Dewey, "Reflex-Arc Concept in Psychology.") Dewey emphasized the teleological nature of the process. See also Creighton, "Purpose as a Logical Category," esp. 284. Creighton described the instrumentalist epistemology as a "new teleology."

85. James, "Aesthetics," 123–24; James, "Sentiment of Rationality" (1879), in *Essays in Philosophy*, 32–33. In his manuscript notes James jotted down: "According to a certain school the rational is nothing but the habitual. And it may therefore be that the rational owes its pleasant quality to its rest, & that therefore this may be considered a mere result of habit." On the same page James referred to "the general nervous law that an unimpeded tendency to a particular nervous discharge is agreeable whereas its check, either within or without, is unpleasant. It is true that the discharge in many actions is neutral and only the arrest painful" (James, "Aesthetics," 123–24). The phrase "sentiment of rationality" had been previously used by other scholars, including Herbert Spencer and J. S. Mill (*On Liberty*, chap. 2).

86. James, "Sentiment of Rationality" (1879), in *Essays in Philosophy*, 32.

87. See Kant, *Anthropology from a Pragmatic Point of View*, par. 74.

88. James, "Sentiment of Rationality" (1879), in *Essays in Philosophy*, 35.

89. James, "Sentiment of Rationality" (1897), in *Will to Believe*, 66.

90. See also James, *Principles of Psychology*, 940–41.

91. James, "Sentiment of Rationality" (1897), in *Will to Believe*, 72. See also James, *Principles of Psychology*, 940–41.

92. James, "Sentiment of Rationality" (1897), in *Will to Believe*, 89.

## CHAPTER FIVE

1. Kant, *Critique of Pure Reason*, 17–18.

2. Zammito, *Kant, Herder*, 221.

3. Ibid., 255. Zammito's main example is Herder.

4. Ibid., 256–57.

5. Ibid., 299.

6. On the meaning of "discipline" in Kant's times, and its foreignness to our "parameters" of disciplinarity, see ibid., 3. On the meaning of "discipline" in the nineteenth century, see Lenoir, *Instituting Science*.

7. The question of the legitimacy of "metaphysical" approaches to logic was also much debated but will not be discussed here. For a discussion of this issue in the German context, see Carl, *Frege's Theory*, 11.

8. As Zammito has shown, Kant's position was much more complex. Kant aimed to subordinate both empirical psychology and anthropology to metaphysics, rather than, as most of the Kant scholarship would have it, expel those disciplines from metaphysics.

9. For the vagueness of the meaning of "psychologism" at the turn of the twentieth century, see Kusch, *Psychologism*.

10. Lipps, "Die Aufgabe der Erkenntnistheorie," 529 (quoted in Carl, *Frege's Theory*, 14). See also Kusch, *Psychologism*, 157.

11. See Carl, *Frege's Theory*, 12, and Kusch, *Psychologism*, 131.

12. See Kusch, *Psychologism*, 3; Sigwart, "Logische Fragen"; and Carl, *Frege's Theory*, 22.

13. Carl, *Frege's Theory*, 6–47; Sigwart, *Logic*, vol. 1, par. 1.3–4; 1.6; 1.9; par. 3.1 (also discussed in Carl).

14. Stumpf, "Psychologie und Erkenntnistheorie"; quotation on 468 (quoted and discussed in Kusch, *Psychologism*, 103). James's copy of this article is preserved at Houghton Library (WJ 783.89.4).

15. Ibid. (discussed in Ash, *Gestalt Psychology*, 30).

16. Stumpf, "Psychologie und Erkenntnistheorie," 483 (discussed in Kusch, *Psychologism*, 104–5). See also Kusch, ibid., 89. During the pragmatist controversy Stumpf accused James of psychologism and distanced himself from pragmatism.

17. Jerusalem, *Introduction to Philosophy*, vi, and Jerusalem, *Der kritische Idealismus*, 78. See also Kusch, *Psychologism*, 115.

18. C. A. Strong to William James, January 19, 1905, in *Correspondence*, 10: 528–29.

19. Windelband, *Präluden*, 266. Quoted in Kusch, *Psychologism*, 102–3. Among contemporary attacks against the genetic method those of the logician G. Frege are probably best known. Frege condemned the replacement of mathematical definitions by genetic descriptions of the processes through which a mathematical concept is grasped. He also ridiculed attempts to reduce the concept of number to empirical generalizations or psychological facts (Mill's "gingerbread or pebble arithmetic"), and especially those physiological approaches that would reduce numbers to "motor phenomena." (See Frege, *Foundations of Arithmetic*, esp. vii.) Frege is relevant to a discussion of the pragmatist account of truth because some of pragmatism's opponents resorted to his ideas.

20. Rickert, preface to *Der Gegenstand der Erkenntnis*. On Rickert's theory of values and notion of absolute truth, see, e.g., Willey, *Back to Kant*, chap. 6, and Oakes, *Weber and Rickert*.

21. "μετάβασις εις άλλο γένος"; Husserl, *Logical Investigations*, 1: 55.

22. Husserl, ibid., 145 and 94.

23. Ibid., 145.

24. For Husserl's thoughts on Sigwart, see ibid., 146–54.

25. According to Sigwart, Husserl wrote, "it is . . . 'a fiction that a judgment could be true if we abstract from the fact that some intelligence thinks such a judgment'" (ibid., 148). See also Kusch, *Psychologism*, 78.

26. Truth, Husserl continued, was "eternal," or "beyond time." It made "no sense to give truth a date in time, nor a duration which extends throughout time." Husserl, *Logical Investigations*, 1: 148. On this point, see also Kusch, "Criticism of Husserl's Arguments."

27. Sigwart argued, in particular, that "our" rules of thought (including the principles of noncontradiction and of identity) and the rules for the "formation of perfect judgments" are grounded in the nature and functions of human thought. From that it followed that truth "would arise and perish" with such thinking intelligences, or "with the species at least, if not with the individual" (Husserl, *Logical Investigations*, 1: 151).

28. Ibid., 139–40, 148.

29. Ibid., 161–65.

30. Ibid., 159.

31. Ibid., 183.

32. See Kusch, *Psychologism*, 79. Frege made a similar distinction. See Baker and Hacker, "Frege's Anti-Psychologism," 75.

33. Husserl argued that psychology inquired into the causal connections between mental processes, whereas logic was exclusively concerned with the "truth-content" of propositions, and had no interest in tracing the causal origins or the consequences of intellectual activities. Husserl, *Logical Investigations*, 1: 94.

34. Kusch's reconstruction of the German psychologistic debate, an exercise in the "sociology of philosophical knowledge," rests on the assumption that philosophical concepts and positions may be largely shaped by "social" factors. Mitchell Ash, however, seems to question that assumption, when he insists that the funds allocated to psychology laboratories in the period under consideration remained modest. See Ash, *Gestalt Psychology*, 17–19.

35. Wundt, as mentioned in chapter 1, was trained in medicine and physiology, and was appointed professor of philosophy in Zurich and then in Leipzig. Lipps held philosophy positions in Bonn, Wroclaw, and in Munich, where he founded the Munich Psychological Institute, while

Sigwart was a professor of philosophy at Tübingen. Stumpf was professor of philosophy at Würzburg, Prague, Munich, and Berlin.

36. On Wundt's changing definitions of psychology, see Hoorn and Verhave, "Wundt's Changing Conceptions."

37. Stumpf, "Antrittsrede"; translated and discussed in Ash, *Gestalt Psychology*, 34. See also 31, 35.

38. Pressed by the government to hire a psychologist, Berlin philosophers recommended Stumpf because, in contrast to other candidates for the position, he did not seem to challenge their interests. See Ash, *Gestalt Psychology*, 31–32. When the pragmatism controversy raged, Stumpf accused James of psychologism.

39. Kusch, *Psychologism*, chap. 7.

40. Windelband, *Die Philosophie*, 92; quoted in Ash, *Gestalt Psychology*, 43.

41. Husserl, "Philosophie als strenge Wissenschaft," quoted in Ash, *Gestalt Psychology*, 45.

42. The petition was addressed to the faculties of philosophy and ministries of culture and education. See Kusch, *Psychologism*, 190–92, and Ash, *Gestalt Psychology*, 47–49.

43. Howison, "Psychology and Logic," esp. 653.

44. Stratton, "Relation between Psychology and Logic," esp. 315 and 317. Stratton emphasized that a descriptive psychology of judgment and reasoning should "inevitably get interested in *all* phases of these facts," including especially "the tone of feeling" accompanying them. Logic, instead, was "indifferent" to "all these things" (315). "So far as the logical worth of a proposition or of a train of reasoning is concerned, it makes no difference . . . if the process is accompanied by this feeling or by that, or by no feeling at all" (317).

45. Ibid., 319. Stratton lamented the "frequent evidence of hazy or ill-observed boundaries" in the work of many psychologists and logicians.

46. "There are, to be sure, tinkers & metaphysicians in one, but the combination is rarer than is either separate element. Münsterberg and Wundt are such, and so may be Stratton for aught I know" (James to Howison, April 2, 1894, in *Correspondence*, 8: 497).

47. Wundt divided psychology into two main parts: experimental psychology and *Völkerpsychologie*. The former studied "the 'lower' psychological processes of sensation and perception," whereas the latter "investigated the 'higher' processes of thinking, as well as language, myth, and custom." According to Wundt, thinking was too "complicated" to be "open to investigation under controlled experimental conditions." See Kusch, "Recluse, Interlocutor, Interrogator," 424. See also Danziger, "History of Introspection Reconsidered," 247.

48. Howison, "Psychology and Logic," 657. In Howison's view, only the rational psychology of the absolute mind (an absolutely autonomous consciousness) could account for "the source" of logical canons and the "credentials of their authority" (ibid., 656).

49. See, e.g., Hammond, "Relations of Logic to Allied Disciplines." Summoning up Kant's attacks against "psychological," "metaphysical," and "anthropological" approaches to logic, Hammond, who taught philosophy at Cornell, argued that logic and cognitive psychology dealt with the same objects (concepts, judgment, etc.) but approached them with different aims. Psychology ought to describe actual thought processes and provide empirical explanations—including genetic explanations—of cognition. Logic, on the contrary, ought to determine ideal norms of thinking (ibid., 12, 19, 20). Columbia philosopher Frederick J. E. Woodbridge similarly attacked the "perplexing theories" that approached logic from a psychological and a biological point of view. See Woodbridge, "Field of Logic," and Woodbridge, *Nature and Mind*, 57–78, esp. 60. For similar separatist arguments, see also, e.g., Ritchie, "Relation of Logic to Psychology, I and II," esp. 585.

50. A. E. Taylor, "Truth and Practice," 266. The first question belonged to the "theory of formal logic," while the second was "coextensive with the whole field of science."

51. Ibid., 267. Husserl had made the identical point.

52. Ibid., 279.

53. Ibid., 286.

54. Ibid., 268.

55. Ibid., 277.

56. James to F. C. S. Schiller, July 15, 1907, in *Correspondence*, 11: 387.

57. Joseph, "Professor James on 'Humanism and Truth,'" 31, 37. This article was a response to James's 1904 article "Humanism and Truth," in which James defended Schiller's and Dewey's accounts of truth. See also James to Schiller, May 26, 1905, in *Correspondence*, 11: 48.

58. J. E. Russell, "Some Difficulties"; quotation on 406. James read the article, annotated it, and sent it back to Russell. On James's and Russell's disagreements, see their published correspondence in James, *Essays in Radical Empiricism*, 145–53.

59. "When I make this attempt," Russell wrote, "I seem to lose sight of what is essential to a clear understanding of both these sciences. *The features which should be distinctive of each run together into a blurred conception, which is to my mind neither intelligible as logic nor as psychology.*" Russell "could not dispossess [his] mind of the conviction that logic and psychology must deal with thinking and knowing in ways too diverse to admit of the sort of connection which as [he] underst[oo]d it, [the pragmatist] epistemology tries to maintain between these sciences" (J. E. Russell, "Some Difficulties," 406; emphasis added). Russell ended up converting to pragmatism. Examples of philosophers expressing similar criticisms could be easily multiplied. For example, Cornell philosopher George Sabine extended the accusation to James's metaphysics of radical empiricism—a kind of philosophy that, he charged, was obviously trying to "make psychology do the work of logic and metaphysics." That was the "essential weakness" of James's philosophy. See Sabine, "Discussion," 85, 88. See also Sabine, review of J. M. Baldwin, "The Limits of Pragmatism."

60. See, e.g., Hoernlé, "Pragmatism v. Absolutism," esp. 301, and Pratt, "Truth and Its Verification," 321. Hornlé was a protégé of Schiller's. See also Leighton, "Pragmatism."

61. Strong, "Pragmatism and Its Definition of Truth."

62. Bradley, "On Truth and Practice."

63. B. Russell, "Pragmatism."

64. H. Münsterberg, *Eternal Values*, vii, 34.

65. See, e.g., B. Russell, "Pragmatism," 111. Russell directed his attacks mostly against Schiller. Concerning an argument proposed by Schiller, Russell claimed that "the reader who will, throughout [Schiller's] essay on the ambiguity of truth, substitute 'butter' for 'truth' and 'margarine' for 'falsehood,' will find that the point involved is one which has no special relevance to the nature of truth" (ibid.).

66. See also Thayer, introduction to James, *The Meaning of Truth*, xxvi. For a recent formulation of that accusation, see G. E. Myers, *William James*, 298.

67. Dewey, *Studies in Logical Theory*, x. The book, written with the "co-operation of members and fellows of the Department of Philosophy," was dedicated to William James.

68. Titchener himself contended that structural psychology served to "enforce . . . the thesis that psychology is a science, and not a province of metaphysics." He also stressed that structural psychology must precede functional psychology, and that, if the order of study was inverted, psychology might end up putting "herself for the second time . . . under the domain of philosophy." See Titchener, "Postulates of a Structural Psychology," 453–54.

69. Angell, *Relations of Structural and Functional Psychology*, 17, 21. See also Angell, "Province of Functional Psychology."

70. Angell, *Relations of Structural and Functional Psychology*, 11.

71. Pragmatism and functionalism "sprang from similar logical motivation and relied for their vitality and propagation upon forces loosely germane to one another." Angell, "Province of Functional Psychology," 68. See also ibid., 81–86.

72. Schiller to James, August 12, 1902, in *Correspondence*, 10: 165.

73. Schiller to James, September 25, 1907, in *Correspondence*, 11: 450; James to J. E. Russell, January 30, 1908, in *Essays in Radical Empiricism*, 231.

74. Schiller to James, June 20, 1904, in *Correspondence*, 10: 417 and 418.

75. Schiller to James, April 2, 1903, in *Correspondence*, 10: 194. James too had no more sympathy for Bertrand Russell's "logical gymnastics" than he had for Taylor's "diseased abstractionism."

76. Schiller, *Studies in Humanism*, 72.

77. Ibid., 77–79. Schiller stated that logical values (for example, truth) had an empirical origin: they had been sorted out from experience because they proved to be valuable in practical circumstances. He also stated that logic was normative, and psychology descriptive.

78. Schiller, "Can Logic Abstract"; quotation on 224–25.

79. Schiller, *Studies in Humanism*, 73. Schiller painted intellectualists and the formalist logicians as busy "cultivat[ing] an hortus siccus" of nonexisting objects. The metaphor implied that formalistic and intellectualistic approaches to logic were not legitimate, because they dealt with a nonexistent subject matter. See also Schiller, "Can Logic Abstract," 226. Years later Schiller went so far as to claim that formal approaches to logic were suicidal. See Schiller, *Formal Logic*.

80. Schiller, "Can Logic Abstract," 235, 237.

81. Ibid., 225.

82. "Our thinking depends for its very existence on the presence in it of (a) interests, (b) purpose, (c) emotion, (d) satisfaction, and . . . the word 'thought' would cease to convey any meaning if these were really and rigidly abstracted from" (Schiller, *Studies in Humanism*, 81). Schiller also observed that the so-called "'strictly logic' sense [of truth, necessity, and other logical conceptions] is *continuous* with their psychological senses" (Schiller, "Can Logic Abstract," 230).

83. See "Sources of William James," 10. James also likely read Sigwart's clearest formulation of his thoughts about the relationship between logic and psychology, in Sigwart, "Logische Fragen."

84. James owned copies of several of Lipps's works. The two met in Rome in 1905 at the international Congress of Psychology. On that occasion Lipps accused James of "psychologism." Later he apologized, describing himself as a man in arms. See James to Pauline Goldmark, May 24 [or 25], 1905, in *Correspondence*, 11: 46. See also James, *Essays in Radical Empiricism*, 259.

85. Jerusalem sent James copies of his *Einleitung in die Philosophie* and *Urtheilsfunction*. For James's comments (as reported by Jerusalem), see Jerusalem to James, June 30, 1903, in *Correspondence*, 10: 273.

86. Jerusalem to James, June 22, 1907, in *Correspondence*, 11: 380. Jerusalem's translation of James's *Pragmatism* appeared in 1908.

87. See Kusch, *Psychologism*, 102. See also James to Münsterberg, March 16, 1907, in *Correspondence*, 11: 328.

88. James to James Jackson Putnam, December 3, 1906, in *Correspondence*, 11: 290. James read Rickert's *Die Grenzen der naturwissenschaftlichen Begriffsbildung* (1902). (His copy was much marked. See "William James's Sources," 20.) He also owned copies of both the first (1892) and the second (1904) edition of Rickert's *Der Gegenstand der Erkentniss*, which he read sometime before April 1906 (James's copy is preserved at the Houghton Library, WJ 776.13). James also owned a copy of Rickert's *Zwei Wegen der Erkenntnistheorie* (Houghton Library, WJ 700.5).

89. James, "Abstractionism and 'Relativismus.'" See also James to Münsterberg, March 16, 1907, in *Correspondence*, 11: 328–29; James, *Pragmatism*, 113 n. 1. For Rickert's attacks against

relativism, see Rickert's *Der Gegenstand der Erkentniss* (1904), 132–41. In the third revised edition of *Der Gegenstand der Erkenntniss* (1915), Rickert included a polemical reference against James.

90. James sent Husserl an offprint of his article "The Knowing of Things Together." He carefully read and marked Husserl's pamphlet *Psychologische Studien zu Elementaren Logik*. On the relationship between James and Husserl, see Herzog, "William James."

91. Husserl credited James with helping him "release" himself from the "psychologistic standpoint" (Husserl, *Logical Investigations*, 1: 420 n. 1). The source of the gossip was Walter B. Pitkin, who translated into English Husserl's *Logical Investigations*. (See Herzog, "William James," 32.)

92. On the problems that statements about the past posed for James's theory of truth, see H. Putnam, "William James's Theory of Truth," 182.

93. Against Husserl's charge that Sigwart made the law of universal gravitation false before Newton's formulation of it, Sigwart replied that truth and reality were different things. "Of course," he wrote in 1904, "the planets did move, already long before Newton, in a way that conforms to the law of gravity." Nevertheless, the law of gravitation could not be true before Newton's time, since nobody had ever thought of it. In 1905 Jerusalem similarly turned the tables on Husserl, and accused him of failing to see that truth and reality were different things. See Kusch, *Psychologism*, 51.

94. See J. E. Russell, "Some Difficulties," and James, *Essays in Radical Empiricism*, 274. James was discussing the satisfaction involved in the process of verification.

95. Pratt, "Truth and Its Verification," 321.

96. "What that something is in the case of truth psychology tells us: the idea has associates peculiar to itself, motor as well as ideational. . . . According to what they are, does the trueness or falseness which the idea harbored come to light. These tendencies have still earlier conditions which, in a general way, biology, psychology, or biography can trace. . . . This whole chain of natural causal conditions produces a resultant state of things in which new relations . . . can now be found . . . the relations namely which we epistemologists study, relations of adaptation, of substitutability, of instrumentality, of reference and of truth" (James, "Professor Pratt on Truth"; quotation on 96).

97. See, e.g., Hébert, *Le Pragmatisme*.

98. James, "Pragmatist Account of Truth," 108.

99. James, "Professor Hébert on Pragmatism," esp. 124.

100. James, "Word More about Truth," 79–80.

101. James, ibid., 86.

102. B. Kuklick, *Rise of American Philosophy*, 267.

103. Although in *Principles of Psychology* he did not include "truth" among the objects of logic, he seemed to do just that on other occasions, thus aligning himself with many other philosophers who loosely located questions of knowledge and truth in the domain of "logic." It was in that sense of the word that James characterized his own account of truth as "logical." James also used the term to refer to formal logic and the theory of science (a theory of inferences that scientists ought to follow). For this type of logic James, notoriously, had lukewarm feelings; he did not "get" it, and did not enjoy teaching it.

104. James, *Principles of Psychology*, chap. 28, esp. 1246.

105. Seigfried, *William James's Radical Reconstruction*, 304. Seigfried is quoting from James, *Pragmatism*, 109. See also ibid., 111.

106. On this point, see Seigfried, *William James's Radical Reconstruction*, 310.

107. James, "Sentiment of Rationality" (1879), 33. On this point see also B. Kuklick, *Rise of American Philosophy*, 266.

108. B. Kuklick, ibid.

109. A. W. Moore, *Pragmatism*, 19. The passage continues: "It is interesting to note, in passing, what unlimited license the psychologist holds. Under the aegis of psychology one apparently may preach any sort of doctrine on any subject without being taken very seriously."

110. Schiller, "William James," 147; emphasis added.

111. Creighton, "Purpose as a Logical Category."

112. Woodbridge, "Field of Logic" (discussed in Shook, ed., *Pragmatism*, 64).

113. Woodbridge, "Problem of Metaphysics," esp. 370 and 371 (discussed in D. J. Wilson, *Science, Community*, 131).

114. Montague, "May a Realist Be a Pragmatist," 543.

115. For example, attacking James's article "Humanism and Truth," Joseph argued that James had "been betrayed into the attempt to write a natural history of the mind." In short, he had slipped out of philosophy and into natural history. The position that James defended, Joseph continued, was "familiar to readers of . . . *Principles of Psychology*. . . . But, in an article headed 'Humanism and Truth' we should expect the standpoint of Logic," not that of psychology (Joseph, "Professor James on 'Humanism and Truth,'" 31, 33).

116. Shadworth Hodgson to James, June 18, 1907, in *Correspondence*, 11: 378–79.

117. H. Münsterberg, *Eternal Values*, 4, 35.

118. A. E. Taylor, "Truth and Consequences," and Schinz, *Anti-Pragmatism*. Schinz was a French philosopher, and taught at Bryn Mawr College.

119. Hibben, "Test of Pragmatism," 365.

120. Creighton, "Nature and Criterion of Truth," 596.

121. Gardiner, "First Twenty-Five Years," 145 (Gardiner was president of the American Philosophical Association in 1907). See also Sokal, "Origins and Early Years," 119–20, and Wilson, *Science, Community*, 108.

122. Wilson, ibid., 109.

123. See, e.g., Creighton, "Purposes of a Philosophical Association," 232.

124. See "Proceedings of the American Philosophical Association: The Fifth Annual Meeting, Emerson Hall, Harvard University, Cambridge Mass., Dec. 27–29 1905," 173.

125. A. E. Taylor, "Place of Psychology," 385. See also Taylor, "Truth and Practice," 276.

126. W. H. Davies, "Proceedings," 203–4. See also Sokal, "Origins and Early Years."

127. Cohen, "Conception of Philosophy." As a symbolic date for the beginning of this conception of philosophy as a special science, narrowly defined, Cohen chose the foundation of Woodbridge's *Journal of Philosophy, Psychology, and Scientific Methods* (1904).

128. The point was clear to Cohen. As Hollinger writes, Cohen warned American philosophers that, "by carving out a rigidly specialized area within the larger sphere of knowledge, philosophers . . . were on the verge of abandoning the claims to comprehensiveness that gave philosophy its enduring appeal" (Hollinger, *Morris R. Cohen*, 45).

129. Howison, "Philosophy and Science," 14. James's copy of this address is heavily marked.

130. Münsterberg, "Psychology and Mysticism," 80, 85. Discussing the relationships between psychology and the normative philosophical discipline of ethics, Münsterberg added: "It is bad enough when the psychological categories are wrongly pushed into ethics by the overextension of psychology, but it is still more absurd when ethics leaves its home in the real world and creeps over to the field of psychology" (ibid., 81). See also Münsterberg, *Psychology and Life*.

131. Schiller, "Psychology and Knowledge," 248. Schiller's article was written in response to Prichard, "Criticism of the Psychologists' Treatment of Knowledge." See also Schiller to William James, March 31, 1907: "that is the sort of rubbish that our sages . . . urge against the Humanist epistemology!" (*Correspondence*, 11: 337).

132. Schiller, "Logic or Psychology?" 400.

133. James, "Knowing of Things Together," and "Experience of Activity." In the Harvard edition of the Works of William James neither of these essays has been included in the volume *Essays in Psychology*.

134. Scripture, *Thinking, Feeling, Doing*, 283. Scripture did not mention James explicitly, but there is no doubt that the passage referred to James, implying that James was a "man of the old school" (ibid.).

135. James, "Energies of Men." See the introduction to this book, and chapter 8.

136. The topic of James's APA presidential address—as members of the Association could not have failed to notice—was the same as the topic on which Wilhelm Ostwald lectured before the American *Psychological* Association the previous year.

137. It was "purely logical" because it aimed exclusively to provide a "definition" of truth, and because the question it addressed—"what [would] truth . . . be like if it did exist?"—was "not a psychological, but rather a logical question." See James, "Pragmatist Account of Truth," 100, and James, "Existence of Julius Caesar," quotation on 120.

138. Seigfried, *William James's Radical Reconstruction*, 22, 30, 43.

139. Franzese links James's work to subsequent investigators in philosophical anthropology, such as Max Scheler and A. Gehlen, both of whom referred to William James in their efforts to frame a philosophical science of man. James read Kant's *Anthropology from a Pragmatic Point of View* in 1868. Franzese, *L'uomo indeterminato*, 12–15.

140. Ibid., 20.

141. Ibid., 32. James's heavily annotated copy of the English translation of this work (Lotze, *Microcosmus*) is preserved at the Houghton Library. James also read Lotze, *Medicinische Psychologie*.

142. James, "Teaching of Philosophy," 5.

143. James, "What the Will Effects," 216–17.

144. James, "Plea for Psychology as a Natural Science," 272.

145. Ibid., 272–73.

146. Even moral philosophers such as James McCosh, president of Princeton, and Noah Porter, president of Yale, realized that the attempt to study human nature and the human mind required "metaphysicians" to "enter the physiological field." (See James, "Teaching of Philosophy," 6.)

147. James, "Plea for Psychology As a Natural Science," 272.

148. See Seigfried, *William James's Radical Reconstruction*, 303.

149. Ibid., 307.

150. Conant, "James/Royce Dispute"; quotation on 199.

151. H. Putnam and R. A. Putnam, "William James's Ideas," 227.

152. James, *Pragmatism*, 11.

153. Conant, "James/Royce Dispute," 201.

154. See, e.g., *Dictionary of Philosophy and Psychology*, ed. J. M. Baldwin, s.v. "Temperament," by J. M. Baldwin, and Höffding, *Outlines of Psychology*, 349. The association that James posited between temperament and the physiology of the nervous system appears clearly from his discussion of the "explosive temperament" in *Principles of Psychology*. James associated that kind of temperament—the temperament exhibited, for example, by drunkards, persons addicted to opium or chloral hydrate, and, most clearly, hysterics—to a failure in the nervous system. The constitution of the "neural tissues" resulted in an inability of the nervous system of such individuals to properly check, inhibit, or even time the "impulsive" motor discharge leading to action. In such people, James stated, the "paths of natural impulse" were so pervious that the "slightest rise in the level of innervation produce[d] an overflow"; the phase of "latency" or "nascency" of action was very short, resulting in explosive, uncontrolled behavior. James, *Principles of Psychology*, 1145, 1148.

155. On this point, see Seigfried, *William James's Radical Reconstruction*, 36–38, and Bordogna, "Physiology and Psychology of Temperament."

156. See James, *Varieties of Religious Experience*, 37.

157. Ibid., 67.

158. See, e.g., James, *Principles of Psychology*, 1265. The locus classicus of James's claim that aesthetic passions (such as the passion for unity and simplicity, and the passion for clarity and distinction) contribute to shaping a person's philosophical positions is James, "Sentiment of Rationality" (1879). (See also James, "Sentiment of Rationality," in *Will to Believe*, 75–76.) For an analysis of these essays and of the role of aesthetic passions in shaping peoples' metaphysical choices, see Seigfried, *William James's Radical Reconstruction*, 29–32.

159. For the analogy between the creation of a work of art and the creation of a philosophical theory, and the claim that both stemmed from physiologically engrained tendencies rooted in the emotional constitution of the creator, see, e.g., Ribot, *Creative Imagination*, and Paulhan, "L'invention," 232. James followed with interest developments and publications in physiological and psychological aesthetics, which he regarded as incomparably preferable to philosophical aesthetics. See James, "Notes for Philosophy 20a." See Barzun, *Stroll with William James*, and Leary, "William James and the Art of Human Understanding."

160. See also Bordogna, "Physiology and Psychology of Temperament."

161. Giovanni Papini [pseud. Gian Falco], "Morte e resurrezione della filosofia." Author's translation.

## CHAPTER SIX

1. Carter, "Critical Introduction" to Howells, *Hazard of New Fortunes*, xxxiv.

2. James to W. D. Howells, August 27, 1890, in *Correspondence*, 7: 87.

3. Howells, *Hazard of New Fortunes*, 263.

4. Ibid., 136–37, 353.

5. Ibid., 126, 390, 476.

6. Ibid., 296–97.

7. On the crisis of the autonomous self in late nineteenth-century America, see, e.g., Lears, *No Place of Grace*; Wiebe, *Self-Rule*; Sklansky, *Soul's Economy*. See also Bercovitch, "Rites of Assent"; McClay, *Masterless*; Ryan, *Vanishing Subject*; Cotkin, *William James*, 8. See also Trachtenberg, *Incorporation of America*, chap. 2, and Licht, *Industrializing America*, 130. For a period discussion, see H. D. Lloyd, *Wealth against Commonwealth*, 498. Historical accounts documenting a "collective crisis" of the self may exaggerate the diffusion of the phenomenon. Nevertheless many writers, social scientists, scientists, and philosophers perceived it as a real problem.

8. Lears, *No Place of Grace*, 60.

9. See, e.g., Huxley, "On the Hypothesis That Animals Are Automata."

10. On the centrality of occult practices to the modern reconfiguration of interiority, see Owen, *Place of Enchantment*, and "Occultism and the 'Modern' Self," esp. 80. In automatic writing, the hand of a person, unbeknownst to the mind, would write things of which the subject had no knowledge.

11. Sklansky, *Soul's Economy*, 142.

12. See Livingston, "Strange Career," and *Pragmatism, Feminism, and Democracy*, esp. 4–14.

13. McClay, *Masterless*, 150.

14. Among them was John Dewey. See Livingston, "Strange Career."

15. Bellamy, *Religion of Solidarity*, 14, 17–18, 24. See also Thomas, *Alternative America*, 87, and Guarneri, *Utopian Alternative*.

16. The literature on James's account of the self is rich. On the philosophical side, see, e.g., G. E. Myers, *William James*, chap. 12; Fontinell, *Self, God, and Immortality*; McDermott, "Promethean Self and Community"; Sprigge, *James and Bradley*; Gale, *Divided Self*; and Cooper, *Unity of William James's Thought*. On the more historical side, see Leary, "William James on the Self and Personality"; M. B. Smith, "William James and the Psychology of the Self"; Coon, "Salvaging the Self"; and Sklansky, *Soul's Economy*.

17. James, "Gospel of Relaxation," 124.

18. James, *Principles of Psychology*, 295. See also Gale, *Divided Self*, 18.

19. Most scholars associate James's early depression with his concerns about determinism. See, e.g., R. J. Richards, "Personal Equation in Science," and Seigfried, *William James's Radical Reconstruction*, 11. For a different point of view, see Cotkin, *William James*, 7, and chap. 2. See also Simon, *Genuine Reality*, chap. 6, and Feinstein, *Becoming William James*, 124–37. On James's neurasthenia, see Cotkin, *William James*, and Lutz, *American Nervousness*, 63–98.

20. Livingston, "Strange Career."

21. See, e.g., James to William M. Salter, April 8, 1898 (quoted in Coon, "Courtship with Anarchy," 125). See also Kloppenberg, *Uncertain Victory*, 168.

22. See Westbrook, *Democratic Hope*, 57.

23. Schirmer, "William James and the New Age."

24. James to Carl Schurz, March 16, 1900, in *Correspondence*, 9: 165. The Anti-Imperialist League also included Democrats, Republicans, labor leaders, and businessmen. According to some historians, however, the Mugwump section represented the spearhead of the movement. See Beisner, *Twelve against Empire*, xi.

25. James, "Philippine Tangle," 155. See also James to William Dean Howells, from Rome, November 16, 1900, in *Correspondence*, 9: 361, and Schirmer, "William James and the New Age," 439 (Schirmer quotes from James's address at the annual meeting of the New England Anti-Imperialist League, 1903).

26. See Beisner, *Twelve against Empire*, 35–52; Cotkin, *William James*, 129; and Westbrook, *Democratic Hope*, 54–58. James, for example, diverged from other Mugwumps on the momentous issue of federal monetary policy. See Coon, "Courtship with Anarchy," 142.

27. Kloppenberg, *Uncertain Victory*, 169.

28. See J. I. Miller, *Democratic Temperament*, esp. 25–32, and B. Lloyd, *Left Out*. For a discussion of these works, see Sklansky, *Soul's Economy*, 273–74.

29. James to William Dean Howells, Rome, November 16, 1900, in *Correspondence*, 9: 362. See Coon, "One Moment," esp. 71.

30. On James's "anarchism," see also Cotkin, *William James*.

31. James compared the "performance" of the United States in the Philippines to the "infernal adroitness of the great department store, which has reached perfect expertness in the art of killing silently and with no public . . . commotion the neighboring small concern" (James, "Philippine Tangle," 156). See also Beisner, *Twelve against Empire*, 46–47; William James to Henry James, February 20, 1899, and William James to Sarah Whitman, June 7, 1899, in *Correspondence*, 8: 545–46; Coon, "Courtship with Anarchy," 157ff.

32. See Westbrook, "Mumford, Dewey."

33. See Livingston, *Pragmatism, Feminism, and Democracy*, 4, and Sklar, *Corporate Reconstruction*.

34. Livingston, "Politics of Pragmatism," esp. 152. "When there was a socialist movement on the American scene, James did explicitly identify with it" (Livingston, *Pragmatism and the Political Economy of Cultural Revolution*, 166, 274–75). On James's socialism, see also Lentricchia, "On the Ideologies of Poetic Modernism."

35. See Browning, *Pluralism and Personality*; Kloppenberg, *Uncertain Victory*, 148–52; Cotkin, *William James*, chap. 7; Coon, "Courtship with Anarchy," and "One Moment"; Leary, "William

James on the Self and Personality," and "William James, the Psychologist's Dilemma"; J. I. Miller, *Democratic Temperament*; Seigfried, "James."

36. James, "On a Certain Blindness," 132. See also Gale, *Divided Self*, 247–48.

37. James, "On a Certain Blindness," 138. See also Seigfried, "James"; Gale, *Divided Self*, 251.

38. James, "What Makes Life Significant," 154–55. See also James, "On a Certain Blindness," 134, and Livingston, *Pragmatism and the Political Economy of Cultural Revolution*, 160ff.

39. Cotkin, *William James*, 111. See James, "What Makes Life Significant," 152.

40. See Westbrook, *Democratic Hope*, 59; Gale, *Divided Self*, 249.

41. James, preface to *Talks to Teachers*, 4–5; see also 149.

42. On this text, see also Seigfried, "James"; Coon, "Courtship with Anarchy"; and Gale, *Divided Self*.

43. James, "Philippine Question," 159. "'The Filipino mind, of course, was the absolutely vital feature in the situation: but this, being merely a psychological, and not a legal phenomenon, we disregarded it practically. . . . From the point of view of business . . . the only relations between man and man are legal" (James, "Diary of French Naval Officer"). See also Beisner, *Twelve against Empire*, 46.

44. See James, "Philippines Again," and "Governor Roosevelt's Oration," 164.

45. See Hollinger, *Cosmopolitanism and Solidarity*.

46. The tension was clearly formulated, for example, by James's friend Wincenty Lutoslawski. See Coon, "Courtship with Anarchy," 107.

47. See Hoopes, *Community Denied*, 54, 65.

48. For a full analysis of this text, see G. E. Myers, *William James*, chap. 12; Fontinell, *Self, God, and Immortality*; Leary, "William James on the Self and Personality"; and Gale, *Divided Self*, chap. 8.

49. James, *Principles of Psychology*, 280–82, 291.

50. Ibid., 282. James's discussion of the material self is heavily gendered.

51. Ibid., 286. See also James, *Essays in Radical Empiricism*, 19.

52. James, *Principles of Psychology*, 282, 295–96.

53. Ibid., 317.

54. Ibid., 320.

55. Ibid., 321–22. See Sklansky, *Soul's Economy*, 148–49.

56. See esp. E. I. Taylor, *William James on Consciousness*.

57. Pierre Janet, *L'automatisme psychologique*, 454. See also Pierre Janet, *Mental State of Hystericals*, 489–96.

58. James to Thomas Davidson, September 13, 1894, in *Correspondence*, 7: 540.

59. James, *Principles of Psychology*, 207, 222. See also James, "Hidden Self," and review of *État mental des hystériques* by Pierre Janet. For an insightful discussion of the latter text, see E. I. Taylor, *William James on Consciousness*, 52–54. James did not ascribe dissociation or hysteria exclusively to women.

60. For a description of the experiment, see James, "Notes on Automatic Writing."

61. In other cases of automatic writing, two consciousnesses could communicate but appeared not to be "on good terms." See James, "Notes on Automatic Writing," 48.

62. James, "Report of the Committee on Hypnotism." In one experiment two subjects were made blind to a "red patch laid on a piece of paper." While apparently insensitive to the red image, both reported perceiving what James knew must be its "after-image," a "bluish-green patch." This, James concluded, indicated that sensation of some sort did occur: the subject somehow indeed "felt" the sense impression. (See also James, *Principles of Psychology*, 208, 1206.) On these experiments, see E. I. Taylor, *William James on Consciousness*, 19–24.

63. For James's discussion of these experiments, see James, *Principles of Psychology*, 206, 1213. See also James, "Hidden Self," 268.

64. Gurney, "Peculiarities of Certain Post-Hypnotic States," 293, 311, 318; see also Oppenheim, *Other World*, 250. For James's indebtedness to Pierre Janet regarding this point, see Taves, "Fragmentation of Consciousness," 51.

65. See Taves, "Fragmentation of Consciousness," 69.

66. See Goldstein, "Advent of Psychological Modernism," 204–6.

67. Ribot, *Diseases of Personality*, 28ff.

68. James, *Manuscript Lectures*, 66. By then James was familiar with various works that looked at dissociation from the vantage point of the biology of colonial organisms. See, e.g., F. W. H. Myers, "Human Personality"; Prince, *Nature of Mind*; and Binet, *Alterations of Personality*.

69. The woman was the Boston medium Mrs. Leonora Piper.

70. James, "Hidden Self," 268.

71. Pierre Janet to James, March 23, 1890, in *Correspondence*, 7: 13–14.

72. James, *Essays in Psychical Research*, 230. James was cautious in inferring that conclusion. See, e.g., James, review of *Human Personality* by F. W. H. Myers, 205.

73. See, e.g., James, *Principles of Psychology*, 246ff. For a discussion of James's concept of the field of consciousness, see Lamberth, *William James*, and E. I. Taylor, *William James on Consciousness*.

74. See James, *Varieties of Religious Experience*, 162.

75. Ibid., 162–63, 189.

76. See H. James, *Portrait of a Lady*, 175.

77. See James, "Notes on Automatic Writing," 45.

78. F. W. H. Myers, *Human Personality*, 2: 568–71 (discussed in James, review of *Human Personality* by Myers, 206).

79. James, review of *Human Personality*, 209–11.

80. On Myers's spiritualism, see Oppenheim, *Other World*, 155.

81. James never committed himself to any of the many explanations of telepathy.

82. James, "Telepathy," 126, and "Possible Case."

83. James, "Suggestion about Mysticism," 161–62. See also Gale, *Divided Self*, 254ff.

84. See James, *Varieties of Religious Experience*, 307; James, *Will to Believe*, 217–21; and Simon, *Genuine Reality*, 141, 259.

85. James, *Varieties of Religious Experience*, 308. See also James, "Consciousness under Nitrous Oxide," and E. I. Taylor, *William James on Consciousness*, 91–92.

86. James, review of *Human Personality* by F. W. H. Myers, 206.

87. "The definitely closed nature of our personal consciousness is probably an average statistical resultant of many conditions, but not an elementary force or fact" (James, *Principles of Psychology*, 331; discussed in Leary, "William James on the Self and Personality," 115).

88. Royce, *World and the Individual*, vol. 2, 282.

89. On mental hygiene and mind cures in fin-de-siècle America, see, e.g., Caplan, *Mind Games*, and Pols, "Managing the Mind."

90. See Simon, *Genuine Reality*, 211–12.

91. On James's meeting with Vivekânanda, see E. I. Taylor, *William James on Consciousness*, 62–64.

92. James to Wincenty Lutoslawski, May 6, 1906, in *Correspondence*, 11: 220–22 (discussed in Cotkin, *William James*, 114).

93. Dresser, *Voices of Freedom*, 24, 33. The book, which James read in the summer of 1900, openly acknowledged James's influence. See also Dresser, *Perfect Whole*.

94. On James and meditation, see E. I. Taylor, *William James on Consciousness*, 64. Taylor also discusses James's relationships with Dresser (ibid., 94). On James and Dresser, see also James, *Pluralistic Universe*, 197.

95. On Fletcher and James, see Simon, *Genuine Reality*, 311.

96. His brother Henry, however, continued practicing the system for years. See Simon, *Genuine Reality*, 312.

97. See ibid., 343; and James, Diary (1907), entries for May.

98. James, *Principles of Psychology*, 116, 119.

99. Ibid., 127–28, 130. See also Leary, "William James on the Self and Personality," 113, and Cotkin, *William James*.

100. See Howe, *Unitarian Conscience*.

101. James, *Varieties of Religious Experience*, 134.

102. Ibid., 141–42.

103. Ibid., 162, 173.

104. Crary, *Suspensions of Perception*, 49.

105. See Vivekânanda, *Yoga Philosophy*, 7–8, 83. (James marked page 83 in his copy. See James, "Sources of William James.")

106. McDermott, "Promethean Self and Community," 53, 57.

107. The patient, Ansel Bourne, was an itinerant preacher who at the age of sixty-one had suddenly disappeared from his home. He found himself two months later in a small town close to Philadelphia, where he had opened up a "five cent" goods store and lived under the new name of "A. J. Brown." Brown knew nothing about Bourne nor could Bourne ever recall anything about Brown. James hypnotized Bourne several times between May 27 and June 7, 1890. He staged an encounter under hypnosis between "Brown" and the wife of Bourne, in Bourne's home. The encounter, however, was not a success, and neither of the two personalities ever acknowledged the existence of the other. See R. Hodgson, "Case of Double Consciousness," and James, *Principles of Psychology*, 371.

108. James, *Principles of Psychology*, 318.

109. As G. E. Myers put it, "the present self or act of thinking *both finds and fashions* the unity that causes us to think that we are the same person throughout successive experiences" (G. E. Myers, *William James*, 349). See also Gale, *Divided Self*, 130, 234–39.

110. Cotkin, *William James*, 114, and, more generally, chap. 5.

111. Puffer, "Loss of Personality."

112. "But man as man is essentially a weakling," James wrote to Lutoslawksi. A *"kräftige Seele* [strong soul] . . . has to be conquered every minute afresh by an act" (James to Lutoslawski, n.d., quoted in Cotkin, *William James*, 101).

113. On the (elitist) techniques for the cultivation of the unitary self in nineteenth-century France, see Goldstein, *Post-Revolutionary Self*.

114. Sklansky, *Soul's Economy*, 141–43.

115. James, "Horace Fletcher at Harvard," in James, *Essays, Comments*, 184–185.

116. James, *Pragmatism*, 139.

117. Flournoy, *Philosophy of William James*, 131.

118. James, "Gospel of Relaxation," 121, 123–24.

119. I borrow this term from Cotkin. See Cotkin, *William James*, 112.

120. Goldstein, *Post-Revolutionary Self*, 303.

121. James, *Human Immortality*, 92. On Fechner, see Dupéron, *G. T. Fechner*; and Heidelberger, *Nature*.

122. James, "Confidences of a 'Psychical Researcher,'" 374.

123. Ibid.

124. See Livingston, "Strange Career."

125. H. D. Lloyd, "Is Personal Development the Best Social Policy?" 190–91 (emphasis added). See also H. D. Lloyd, *Wealth against Commonwealth*, 527. On Lloyd, see Thomas, *Alternative America*.

126. Royce, "Self-Consciousness," 202.

127. Royce made self-consciousness derivative on "social consciousness," and egoistic impulses derivative on altruistic impulses. See Royce, "Self-Consciousness," 201. On Royce's vision of community, see R. J. Wilson, *In Quest of Community*, 144–170, and B. Kuklick, *Josiah Royce*. On "loyalty," see Royce, *Philosophy of Loyalty*. This text was based on lectures he delivered at Harvard and elsewhere in 1906–1908. Clendenning has argued that for Royce loyalty was an "intensely personal" virtue: it was "the only way that personality can be ethically expressed" (Clendenning, *The Life and Thought of Josiah Royce*, 299).

128. On that aspect of Royce's social theory, see Royce, *Race Questions, Provincialism, and Other American Problems*. For a period account of Royce's social philosophy, see S. G. Brown, *The Social Philosophy of Josiah Royce*.

129. James to Bradley, Jan. 22, 1905, in *Correspondence*, 10: 529.

130. The text where Bradley accomplished his "repudiation of the reality of the self" was *Appearance and Reality* (1893), which James carefully read.

131. Sprigge, *James and Bradley*, 511, 520. James was also well acquainted with Hugo Münsterberg's account of the self, one that painted the individual self as part of an "over-self," in the same way a drop was part of a stream. See Münsterberg, *Eternal Values*, 420. Münsterberg's account of the self was part of his efforts to promote a hierarchical type of society in which citizens would obey the "absolute Will" and the commands of the emperor. James was also familiar with C. S. Peirce's vision of the self and community. (On Peirce's account of the self, see R. J. Wilson, *In Quest of Community*; and Colapietro, *Peirce's Approach*.)

132. Howells, *Traveler from Altruria*, and *Through the Eye of the Needle*.

133. James, *Talks to Teachers*, 277. See also Livingston, *Pragmatism and the Political Economy of Cultural Revolution*, 163.

134. H. James Sr., *Society*, 196, 203, 285. Women were excluded from Henry James Sr.'s regenerate society. See Hoover, *Henry James, Sr.*

135. H. James Sr., *Society*, 406–7.

136. Trine, *In the Fire of the Heart*, 316–336. James owned a copy of this book. See ibid., chapters 5 and 6. See also Trine, *What All the World's A-Seeking*. The British Fabian socialist Annie Besant also proposed an account of the self which made the "individualized," "separated" self into a fragment of the "One," "great" self in whom "we live and move and have our being, open ever to Him, filled with His life." (Annie Besant, *A Study in Consciousness. A Contribution to the Science of Psychology*, Adyar, Madras, India: Theosophical Publishing House, 1938), 172.

137. James to Ernest Howard Crosby, Nov. 8, 1901, in *Correspondence*, 9: 557.

138. Barua, *Edward Carpenter*, chap. 3. On Carpenter, see also Tzuzuki, *Edward Carpenter*.

139. E. Carpenter, *Art of Creation*, 54–57.

140. Ibid., 57–59; see also Barua, *Edward Carpenter*, 155–56.

141. E. Carpenter, *Art of Creation*, 79, 90–91.

142. See Barua, *Edward Carpenter*, 95.

143. N. Mackenzie, ed., *Letters of Sidney and Beatrice Webb*, 2: 268, and Barua, *Edward Carpenter*, 158. Carpenter and James corresponded, and James once planned to visit Carpenter. James was also acquainted with the Webbs, whom he met in April 1898. See N. Mackenzie, ed., *Letters of Sidney and Beatrice Webb*, 2: 63.

144. Bucke, *Cosmic Consciousness*, 4–5. James read this book with great interest. See William James to Alice Howe Gibbens James, September 16, 1901, in *Correspondence*, 9: 542–43.

145. Bellamy, *Religion of Solidarity*, 17–18. On Bellamy, see Thomas, *Alternative America*, 83–88; Pittenger, *American Socialists*, chap. 4; and McClay, *Masterless*, 78–82. James most likely never read Bellamy's manuscript "The Religion of Solidarity," although he did read Bellamy's best-seller *Looking Backward* (1888). He also probably read other novels (some published posthumously by Howells) in which Bellamy sought to apply his religion of solidarity to the solution of

social problems. For other examples of Christian socialists linking communitarian visions to the "mystical bond of divine life," see Guarneri, *Utopian Alternative*, 55.

146. James, introduction to *The Literary Remains of Henry James*, 7. On this point, see Leary, "William James on the Self and Personality," 102.

147. James to Sydney Haldane Oliver, February 10, 1905, in *Correspondence*, 10: 547.

148. Gale observes that James "favored [Western] pluralistic mysticism . . . over its monistic Eastern version," a type of mysticism that allowed for unification without involving "complete numerical identity" among the terms unified (Gale, *Divided Self*, 14, 252, 271).

149. James to Sydney H. Oliver, February 10, 1905, in *Correspondence*, 10: 548.

150. Schiller, "Idealism," and James, "Mad Absolute." Sally Beauchamp was the name of one of the personalities of a "multiple" patient of Boston psychiatrist Morton Prince, a friend of James's.

151. The botanical language that James used in discussing the permeable "fence" separating the individual self from other selves closely echoed language used by Carpenter for similar purposes. See E. Carpenter, *Art of Creation*, 124.

152. G. E. Myers is an exception. See G. E. Myers, *William James*, 350. See also Gale, *Divided Self*.

153. James, *Manuscript Lectures*, 370. For an insightful philosophical discussion of this problem, see Sprigge, *James and Bradley*, chap. 4. See also chap. 2 above.

154. James, *Pluralistic Universe*, 83.

155. See Sprigge, *James and Bradley*, 177.

156. James explained that idealists assumed that the individual self "was" insofar as it was thought of by the absolute, yet also continued to *be* as it appeared to itself to be. But an individual's self-feeling must be very different from the way in which the absolute Self thinks of that individual. Given the idealistic equation between "to be" and "to be felt," James concluded that this implied a logical contradiction: How can I be at once what I take myself to be, and what the absolute mind thinks I am? (James, *Pluralistic Universe*, Lecture 5).

157. James, *Pluralistic Universe*, 112–22, 127. This is what Gale describes as "the mushing together of spatio-temporal neighbors" (see Gale, *Divided Self*, 253).

158. That solution allowed individual consciousness to combine yet to remain "each distinct from each other" (see Sprigge, *James and Bradley*, 180).

159. See, e.g., James, *Pluralistic Universe*, 131.

160. See Gale, *Divided Self*, 259; Sprigge, *James and Bradley*, 245. See also James to Bradley, January 22, 1905, in *Correspondence*, 10: 530.

161. James depicted many biological and cosmological theories of the time, rich in political implications, as counterparts of his technical metaphysical problem. See, e.g., Haeckel, "Zellseelen and Seelenzellen"; Royce, "'Mind-Stuff' and Reality"; Fechner, *Elemente der Psychophysik*, vol. 2, chap. 45; and Prince, *Nature of Mind* (all quoted in James, *Principles of Psychology*, 161–62 n. 15).

162. James, *Manuscript Lectures*, 366. See also James, *Pragmatism*, 282, 295, 298.

163. James to Bergson, July 28, 1908, in *Correspondence*, 11: 62. I discuss intimate international pragmatist communities in Bordogna, "Local Internationalism," and "L'Hotel Pragmatista."

164. See, e.g., James, *Talks to Teachers*, 151.

165. James, *Pragmatism*, 295, 276.

166. See Coon, "One Moment," esp. 83, 88.

167. "Damn great empires! Including that of the Absolute." James to Elizabeth Evans, February 15, 1901, in *Correspondence*, 9: 422.

168. James, *Pluralistic Universe*, 145.

169. James, *Manuscript Lectures*, 372, 415.

170. James to W. D. Howells, November 13, 1907, in *Correspondence*, 11: 478–79.

171. Wells, *Modern Utopia*, 20.

172. James, "Moral Equivalent of War," 170.

173. James believed that "stroke after stroke," people of genius could help the demolition of the "competitive régime." See William James to Henry James, December 19, 1908, in *Correspondence*, 3: 372.

174. The argument of "'The Moral Equivalent of War" was gendered. For a defense of James's views on women, see J. I. Miller, *Democratic Temperament*, chap. 3. On this issue, see also Seigfried, *Pragmatism and Feminism*. On "'The Moral Equivalent of War," see also McClay, *Masterless*, 33–34.

175. James, "Moral Equivalent of War," 171–72.

176. James, ibid., 172–73, quoting from Wells, *First and Last Things: A Confession of Faith and a Rule of Life* (New York: G. P. Putnam's Sons, 1908), 214–15.

177. James, "Moral Equivalent of War," 172, 173.

### CHAPTER SEVEN

1. James, introduction to the English translation of Fechner, *The Little Book of Life after Death*, 116.

2. See, e.g., Livingstone, *Putting Science in Its Place*, 7. On the role played by space in human interaction, see, e.g., Goffman, *Presentation of Self in Everyday Life*, and Cresswell, *In Place/Out of Place*.

3. The locus classicus for Latour's concept of "immutable mobiles" is Bruno Latour, *Science in Action*.

4. James, *Some Problems of Philosophy*, 19–20, and James, "Herbert Spencer Dead," 101.

5. H. Münsterberg, "Scientific Plan"; quotation on 92. On the St. Louis congress, see also Cahan and Rudd, *Science at the American Frontier*.

6. The president of the congress, Simon Newcomb, and the other vice president, the Chicago sociologist Albion Small, had each proposed plans for the conference, but Münsterberg's won the day, no doubt because of the zeal of its proponent. His plan, with a few revisions, was approved on February 23, 1903.

7. H. Münsterberg, "Scientific Plan," 95.

8. "Nothing [in the organization] could be left to chance methods and to casual contributions. The preparation needed the same administrative strictness which would be demanded for an encyclopedia" (H. Münsterberg, "Scientific Plan").

9. Ibid., 91.

10. Ibid., 95. "'The libraries of our specialistic work to-day form one big encyclopedia where one thing stands beside the other. This record [of the congress] would become at last a real system: the whole would be a real 'congress of the United Sciences'" (ibid.).

11. Charles A. Strong to William James, May 13, 1890, in *Correspondence*, 7: 26. Strong, who was a former student of William James, at the time was studying philosophy and science in Germany. In 1890 he attended Münsterberg's lectures "Einleitung in der Philosophie," where Münsterberg articulated his encyclopedic conception of philosophy.

12. H. Münsterberg, *Grundzüge der Psychologie*.

13. H. Münsterberg, "Position of Psychology."

14. The chart completed the partial classification previously sketched in the *Grundzüge* and represented a "graphic appendix" to that work. Münsterberg hoped it would enable readers to better follow the argument made in that bulky volume (H. Münsterberg, "Position of Psychology," 646).

15. In Münsterberg's philosophy, purposes and phenomena were the two main kinds of experiences that one might encounter in the world.

16. Münsterberg, "Position of Psychology," 652.

17. Ibid. See also H. Münsterberg, "St. Louis Congress."

18. H. Münsterberg, "Scientific Plan," 95.

19. Some of the differences no doubt reflected the complex negotiations between Münsterberg, Newcomb, and Small.

20. See, e.g., Dewey, "St. Louis Congress." For Münsterberg's response to Dewey, see H. Münsterberg, "International Congress," and "Scientific Plan."

21. Newcomb, in fact, presented the congress as a collective attempt to chart knowledge "by 300 leading scholars of all civilized nations." When the congress was over, he argued that, although the result obtained by that collaborative endeavor was perhaps not as complete as the organizers had wished, "nevertheless [we need to] accept it as we find it" (Newcomb to Münsterberg, January 19, 1905, Münsterberg Papers, Boston Public Library, MS acc. 1996, folder 2). Newcomb's vision of a community of scientists working together toward the unification of knowledge was shared by other scientists of the time, who agreed that the most significant unit for the production of knowledge was not the individual scientist but a scientific collective. Karl Pearson, for example, suggested that "an adequate classification could only be reached by a group of scientists having a wide appreciation of each other's fields, and a thorough knowledge of their own branches of learning," and "endowed with sympathy and patience" necessary for them to "work out a scheme in combination" (Pearson, *Grammar of Science*, 443).

22. H. Münsterberg, "Scientific Plan," 92–93.

23. Howison, "Philosophy: Its Fundamental Conceptions"; quotation on 177.

24. "May I finally ask at once your favor for a great undertaking of national and international extent, an undertaking which I think will be unique and will, I hope, not be unworthy of the national life under your Presidency. The St. Louis World's Fair plans a great exhibition of modern thought, with the participation of the leading scholars of the world and with the definite task of bringing out the Unity of Knowledge against the scattered specialistic work. . . . Above all, at the suggestion of a scientific committee, they last week accepted definitely the plans which I had drawn up for the whole International Congress. The tremendous responsibilities of the undertaking fall thus primarily on me, and it is therefore an almost instinctive desire on my part to ask from this first day for your favor and interest in this broad undertaking, which we all trust might become a blessing for American scholarship and public welfare. May I add privately that in working out my plans, I did not forget that this Congress will meet through the second half of September, 1904, a time when every far-reaching platform for the promulgation of sound views through your political friends and admirers might be of some importance in the great campaign. The plans of the Congress have indeed an abundance of room for the discussion of every question which refers to the national welfare" (Münsterberg to Roosevelt, September 20, 1901; unpublished letter, Münsterberg Papers, Boston Public Library, MS acc. 2394). Münsterberg's friendship with Roosevelt began in 1901, when Roosevelt was vice president.

25. Münsterberg sought to accomplish the same goal in other ways. For example, he was one of the promoters of the "Germanic Museum" at Harvard, which was to house German artifacts and casts of German "art treasures," a gift from the German emperor to Harvard College. In 1902 pictures of German monuments were donated to Harvard college by the emperor's brother, Prince Henry, at a dinner with two hundred guests at Münsterberg's home. During this carefully orchestrated diplomatic event, the prince expressed his "hope" that the gift would promote good feeling between the two nations.

26. Münsterberg expressed his hope that "as the whole plan is my personal work he [the emperor] may be willing to listen to my interpretation of it." See Münsterberg to Charlemagne-Tower, n.d. (Münsterberg Papers, Boston Public Library, Mss acc. 2414). See also Münsterberg to Roosevelt, September 20, 1901 (Münsterberg Papers). In August 1904, Münsterberg complained that Newcomb did not give him "credit."

27. Yeo, *Encyclopaedic Visions*, esp. 274–75. On maps of science and "science as a map of culture," see also Gieryn, *Cultural Boundaries*, x, 1–35.

28. Even those who had misgivings about the possibility of producing a satisfactory, permanent classification of the sciences, or the possibility of fully unifying them, could seldom resist the temptation of trying to accomplish those goals. For example, after acknowledging that any classificatory scheme was bound to remain temporary, Karl Pearson offered his own classification (Pearson, *Grammar of Science*). Similarly, St. George Mivart warned his readers that "all the sciences are connected by such a labyrinth of interrelations that the construction of a really satisfactory classification of them appears to be an insuperable task." Nevertheless, he provided his own "catalogue of the sciences," which arranged the sciences "in what seems, to our judgment, a not inconvenient order" (Mivart, *Groundwork of Science*, 26).

29. Comte, *Cours de philosophie positive*. The main critic of Comte's classification was Herbert Spencer. Spencer read Comte's classification as strictly linear, a veritable "échelle encyclopédique" (encyclopedic ladder), and argued that that linear arrangement failed to represent the genesis of the sciences from one another, through a branching process. Spencer also noted that Comte resorted to the metaphor of the "tree" of knowledge, describing the sciences as branches from a "tronc unique," and found that his use of that metaphor was inconsistent with the serial scheme. Spencer, "Genesis of Science," esp. 28, and "Classification of the Sciences."

30. See, e.g., Flint, *Philosophy as* Scientia Scientiarum; Durand de Gros, *Aperçus de Taxinomie Générale*.

31. Chambers, *Cyclopaedia: or, an Universal Dictionary of Arts and Sciences* (1728), and d'Alembert, *Preliminary Discourse*, discussed in Yeo, *Encyclopaedic Visions*, 141. See also Withers, "Encyclopaedism."

32. Serres, introduction to Comte, *Cours de philosophie positive*.

33. Darnton, "Philosophers Trim the Tree of Knowledge," 192.

34. L. Ward, "Place of Sociology."

35. Cattell, "Homo Scientificus Americanus."

36. Ostwald, "On the Theory of Science." See also Ostwald, "Das System der Wissenschaften."

37. On Ostwald's energeticism and its function as "an organizing principle of the sciences and of society," see Kim, "Practice and Representation," chap. 6.

38. Wundt, "Über die Einteilung der Wissenschaften." See also Wundt, *System der Philosophie*. Philosophical psychology, however, fell under the division of the philosophical sciences.

39. On Wundt's classification of psychology, see Hoorn and Verhave, "Wundt's Changing Conceptions"; Leary, "Wundt and After." More generally, see Ash, "Academic Politics," and "Wilhelm Wundt and Oswald Kuelpe"; Kusch, *Psychologism*; and Danziger, "Positivist Repudiation of Wundt."

40. H. Münsterberg, "Position of Psychology." See also Hale, *Human Science and Social Order*, 77.

41. It should be recalled that both Windelband and Rickert in 1913 signed a petition designed to prevent experimental psychologists from being appointed to philosophy chairs in German-speaking universities.

42. Windelband, "History and Natural Science," trans. Guy Oakes, *History and Theory* 19, no. 2 (1980): 165–68; Rickert, *Kulturwissenschaft und Naturwissenschaft*, 52. See also Rickert, *Die Grenzen der naturwissenschaftlichen Begriffsbildung*; and Oakes, *Weber and Rickert*, 69.

43. Ash, "Wilhelm Wundt and Oswald Kuelpe," 406–8. See also Kusch, *Psychologism*, 169.

44. See, e.g., Flint, *Philosophy as* Scientia Scientiarum.

45. The term "scientists" was introduced in the 1830s by the British philosopher William Whewell. He hoped that "scientists," understood as cultivators of science in general, would counteract the tendency to extreme specialization that, he feared, would lead the "commonwealth of science" to disintegrate "like a great empire falling to pieces" (Yeo, *Encyclopaedic Visions*, 248).

46. Spencer's taxonomy was thoroughly discussed both in Britain and abroad. For American discussions, see, e.g., Fiske, *Cosmic Philosophy*; Stanley, "On the Classification of the Sciences"; and Cogswell, "Classification of the Sciences."

47. Spencer, "Genesis of Science," 28–29.

48. The passage continues as follows: "As usage has defined it, Science consists of truths existing more or less separated and does not recognize these truths as entirely integrated" (Spencer, *First Principles*, 132).

49. Ibid., 133–34.

50. Ibid., 542. Spencer emphasized that the formula of the concomitant redistribution of matter and motion accounted "not only for nature," but also for human society and such "super-organic products" as "Language, Science, Art, and Literature" (ibid., 544).

51. Several British philosophers active in the late nineteenth century and at the turn of the twentieth were seduced by Spencer's conception of philosophy as "a system of completely unified knowledge," even though they sometimes mobilized it in an antipositivistic and anti-Spencerian vein. These included James's friend Henry Sidgwick, who complained that Spencer's definition of philosophy confined it to an overly restricted domain, and expanded Spencer's concept of philosophy so as to make philosophy into something greater than the "aggregate" of the sciences (H. Sidgwick, *Philosophy, Its Scope and Relations*, 28). Robert Flint, a former professor of Moral Philosophy and Political Economy at St. Andrews—a caricature of the turn-of-the-century figure of the unifier of knowledge—also appropriated, but revised, Spencer's conception. He gloomily announced that without an overarching knowledge encompassing all the sciences, the intellectual cosmos would fall apart, and become a "chaos," a collection of "*membra disjecta*." He contended that philosophy as *scientia scientiarum* would combine the sciences into a "harmonious" and "well-proportioned" cosmos by assigning each science "its proper place," and worked to reinstate the traditional image of philosophy as the "Queen of the Sciences." He believed that his philosophical unification of the sciences was sanctioned from above by the "Supreme Reason" that governed the universe. His pamphlet, *Philosophy as* Scientia Scientiarum, was completed in September 1904, a few days after the opening of the St. Louis Congress of Arts and Sciences, an event for which Flint had high praise (Flint, *Philosophy as* Scientia Scientiarum, esp. 6, 7, and 39).

52. He explained that this was a modification that Comte himself had introduced in his *Synthèse subjective* (Whittaker, "Compendious Classification," 21).

53. Janet, *Principes de métaphysique et de psychologie*. Janet's declared goal was to defend philosophy from the numerous attacks directed against it. On Janet's philosophy, see Fabiani, *Les philosophes de la République*; Ringer, *Fields of Knowledge*; Brooks, *Eclectic Legacy*; Barberis, "First Année sociologique"; and Goldstein, *Post-Revolutionary Self*.

54. See introduction to Galison and Stump, eds., *Disunity of Science*.

55. Rosmini divided the sciences into "ideological sciences," which dealt with the "ideal being"; "metaphysical sciences," which dealt with "real being"; and "deontological sciences," which dealt with moral being, right, and politics. In his scheme unity was guaranteed by the "idea of being," an innate idea of divine origin that in the end conferred on Catholicism an important role in the unification of knowledge. Rosmini's moral philosophy represented an attack against all attempts to extend the powers of the state, against which he posited the church. (See Morra, "Rosmini e lo spirito dell' 'Encyclopédie.'") Pope Pius IX initially favored Rosmini's political project, and promised him the "porpora cardinalizia"; later, however, he withdrew his favor. Another prominent Italian politician and philosopher, Vincenzo Gioberti, similarly advocated a confederation of Italian states under the pope, and proposed a tripartite classification. See Gioberti, *Degli errori filosofici di Antonio Rosmini*, and Abbagnano, *Storia della filosofia*, 220–21.

56. Likewise, prior to the founding of the German empire in 1871, a number of key scientists working in the German cultural sphere considered the unification of the sciences to be an important part of the project of unifying the German states. Liberal scientists engaged in a discourse of unity, piggybacking on the momentum toward political unification to harmonize the sciences and reconfigure the academy. See Jurkowitz, "Helmholtz and the Liberal Unification of Science." On broader debates about unity, see Ringer, *Decline of the German Mandarins*, and *Fields of Knowledge*; Weindling, "Theories of the Cell State," and "Dissecting German Social Darwinism."

57. Paulsen, *Autobiography*, 266, 339.

58. Paulsen, *German Universities*, 227–28.

59. Paulsen, *German Education*, 196.

60. Paulsen, *Introduction to Philosophy*. James wrote a preface for the English translation.

61. Harrington, *Reenchanted Science*, 24. See also Ringer, *Decline of the German Mandarins*.

62. For example, he wrote against the Subversion Bill, a bill devised by the emperor William II for the purpose of "crushing the Social Democrats" (Paulsen, *German Universities*). Paulsen also wrote against attempts to expel Social Democratic professors from the university. See Paulsen, *Autobiography*, esp. 229ff.

63. Paulsen, *Autobiography*, 248ff.

64. For uses of the metaphor of the state as an organism in Wilhelmine Germany, see, e.g., Weindling, "Theories of the Cell State."

65. Paulsen, *Introduction to Philosophy*, 209. Paulsen examined in detail the biologist Albert Schaefle's version of the analogy between organism and society, and between individuals and the cells of the organism (Schaefle, *Bau und Leben des sozialen Koerpers*). See Paulsen, *Introduction to Philosophy*, 191ff. He was also familiar with Haeckel's work.

66. Paulsen painted the "smallest and larger units" of the cosmos (individual living organisms and larger wholes) as similar, in this respect, to monads, "with windows and doors through which the 'influences' could enter" (Paulsen, *Introduction to Philosophy*, 215).

67. Ibid., 149.

68. Howison, "Philosophy: Its Fundamental Conceptions," 176.

69. Absolutists varied in allowing for more or less independence of the individual from the absolute.

70. For Howison, the political state was a harmonious unity in which citizens were invited to freely sacrifice their individual (or class) interests and goals for the good of the whole. James McLachlan argues that Howison's conception of God had radically democratic implications. See McLachlan, "George Holmes Howison: 'The City of God' and Personal Idealism," *Journal of Speculative Philosophy*, n.s., 20, no. 3 (2006): 224–42.

71. The social leadership of philosophers was the leitmotiv of Münsterberg's "Mandarin" social utopia. For German philosophers' struggles to retain some sort of social leadership, see Ringer, *Decline of the German Mandarins*.

72. H. Münsterberg, *American Traits*, 34–36, 194. Münsterberg wrote this book for an American public in order to reduce the widespread perception of friction between the United States and Germany. Münsterberg's public formulation of his political positions created a public scandal. At a dinner at the Cosmopolitan Club of Detroit, where he had been invited to present his new book *The Americans*—a book written for a German public—he observed that Roosevelt had never "in his speeches or writings . . . cited that socially equalizing Declaration of Independence." In the next couple of days, local and national newspapers reported that "Roosevelt had called the Declaration of Independence nothing but glittering generalities," and attributed the same opinion to Münsterberg. For Margaret Münsterberg's version of the affair, see M. Münsterberg, *Hugo Münsterberg*, 120–23. For an excellent discussion of Münsterberg's opinions

on American society and his cultural politics, see Hale, *Human Science and Social Order*, 56–65, 87–105.

73. The official itinerary for foreign speakers reads: "New York, Niagara Falls, Chicago, St. Louis, Washington, New York, Boston, New York" (Münsterberg Papers, Boston Public Library, MS acc. 2477). Thus the itinerary contemplated a significant detour for foreign speakers, from Washington, where they were to attend a dinner at the White House, to Cambridge, where they were to meet with the Harvard authorities and to attend a "Reception by Mr. and Mrs. Münsterberg at 7 Ware Street, Cambridge, at 9 o'clock" (Münsterberg Papers, Boston Public Library, MS acc. 2482).

74. H. Münsterberg to James, April 6, 1904, in James, *Correspondence*, 10: 392–93.

75. Münsterberg's wife, Margaret, stated that the Congress "was unique for its absence of cranks and its orderly cooperation of recognized leaders of thought" (M. Münsterberg, *Hugo Münsterberg*).

76. On this point, see William James to C. Stumpf, January 1, 1904, in *Correspondence*, 10: 356. Because of Münsterberg's political inclinations, he was suspected of being a German spy. On these rumors, and more generally Münsterberg's diplomatic efforts in preparing the congress, see Hale, *Human Science and Social Order*, 92–97.

77. Newcomb to Münsterberg, telegram, March 14, 1904; unpublished document, Münsterberg Papers, Boston Public Library.

78. Münsterberg Papers, Boston Public Library, MS acc. 2451.

79. James owned a copy of *The Philosophical System of Antonio Rosmini-Serbati*, translated by Thomas Davidson (London, 1882). (His copy is preserved at the Houghton Library.) He constantly referred to Spencer and Comte in his teaching lectures. James may have also been familiar with André-Marie Ampère's and Janet's taxonomies of the sciences (*Essai sur la philosophie des sciences*), since both were discussed in Paul Janet's *Principes*, which James read. James most likely also saw Whittaker's graph, which was published in an issue of *Mind* that included a book review authored by James.

80. James copiously commented on Spencer's *First Principles of a New System of Philosophy* (see James, "Spencer's First Principles"). James deployed the phrase "a system of completely unified knowledge" in *Some Problems of Philosophy: A Beginning of an Introduction to Philosophy*, a book that was published posthumously in 1911. The text was based on James's notes from his course Philosophy D.

81. James also read Paulsen's essay "Über das Verhältnis der Philosophie zum Wissenschaft," which dealt with the classification of the sciences.

82. Howison, "Philosophy and Science." James's copy is preserved at the Houghton Library. James also subscribed to journals such as *Mind*, *Revue Philosophique*, *Philosophical Review*, and *Vierteljahrsschrift für Wissenschaftliche Philosophie*, which often featured articles on the classification of the sciences and the unification of knowledge.

83. Münsterberg to the Visiting Committee of the Philosophical Division, Harvard, March 20, 1901, Münsterberg Papers, Boston Public Library, RB XH.901.M94T. Münsterberg wrote the letter in his capacity as chair of the department of philosophy.

84. Ibid.

85. Münsterberg in a published article listed William James among those to be invited to the International Congress of Arts and Sciences. In a letter to A. O. Lovejoy, James wrote: "I must hasten to correct the error into which Münsterberg's printing of my name has led you. I told him unequivocally that I should n't go next September—probably the article which named me was written ere he had asked me, but I hope that his other American names are not similar creatures of hope" (James to Lovejoy, February 7, 1904, in *Correspondence*, 10: 377).

86. James to Münsterberg, August 10, 1904, in James, *Correspondence*, 10.

87. James to C. Stumpf, January 1, 1904, in *Correspondence*, 10: 355.

88. James to T. Flournoy, June 14, 1904, in *Correspondence*, 10: 415.

89. Ibid.

90. James to George Herbert Palmer, April 3, 1903, in *Correspondence*, 10: 227.

91. James to Münsterberg, April 11, 1901, in *Correspondence*, 9: 463–64.

92. "Surely it isn't too late for both architect and site to be reconsidered? Longfellow is capable of any atrocity" (James to George Herbert Palmer, April 3, 1903, in *Correspondence*, 10: 227–28).

93. "Must the building go on Quincy St.? Is n't the Holmes field region, with 'power' for the laboratory etc., accessible from outside, better? I think this question ought to be thoroughly threshed out, before any irrevocable step is taken" (ibid).

94. "Emerson Memorial Hall." The article had, no doubt, Münsterberg's approval. Münsterberg himself, in an article that he published a few months later, described the new hall as a "Greek, brick building with brick columns and rich limestone trimmings" (H. Münsterberg, "Emerson Hall," 6). He described the new building as a "noble monumental building," emphasizing that at Harvard "now at last [philosophy] has unity and dignity and its imposing quarters have quickly become a new centre for Harvard's intellectual life" (Münsterberg, *Science and Idealism*).

95. James, "Notes for Philosophy 20c," 26. Elsewhere, commenting on the St. Louis congress, James dismissed Münsterberg's philosophy as "thoroughly artificial" (James to C. Stumpf, July 17, 1904, in *Correspondence*, 10: 435).

96. H. Münsterberg, "Emerson Hall," 13–14.

97. James to Münsterberg, August 3, 1901, in *Correspondence*, 9: 525.

98. James, "Ph.D. Octopus"; quotation on 70. On this article, see Coon, "Courtship with Anarchy," 188–90, and Simon, *Genuine Reality*, 327.

99. A few years later, Münsterberg and James were to have a serious falling-out on the matter of the PhD title. A Harvard student, Buck, who had failed to receive his PhD, had been refused a job as a teacher of English literature at Bryn Mawr. After a disagreeable discussion, James wrote Münsterberg: "What I had in mind about Buck was this, that if we establish the *principle* that PhDs must take precedence, we run the risk (in Buck's case we have it before us) of sacrificing the interests of an individual who has studied here, and whose fate to a certain degree is in our hands, to the interests of our own PhD machine. This is one of those injustices which every organized machine brings with it, and against which it seems to me we must be most strenuously on our guard." This "would gradually bring in an atmosphere of academic politics, and the individual cases with their interests would easily become mere counters in our academic game. I think we must guard against such degeneration as this as against a pestilence. . . . I am very sure you understand me . . . and sympathize with my principle, even though you venerate the PhD institution more than I do." In a postscript, he added: "For Heavens [sic] sake don't reply, if you agree. Still less if you disagree." James to Münsterberg, August 15, 1906, in *Correspondence*, 11: 255.

100. James to T. Flournoy, June 14, 1904, in *Correspondence*, 10: 415, and James to James Ward, July 31, 1904, in *Correspondence*, 10: 441. Similarly, in a letter to Schiller, James stated: "It seems to me little less than an insane exhibition of the schematizing impulse run mad. My wonder is that M. should have talked his co-committee men over successfully to his ideas. I wouldn't touch it for 100 dollars" (James to F. C. S. Schiller, June 12, 1904, in *Correspondence*, 10: 412).

101. James, "Notes for Philosophy 20c," 326.

102. He asked: "What, on pragmatist terms, does 'Nature itself' signify? To my mind it sig-
nifies the non-artificial, the artificial having certain definite aesthetic characteristics which I
dislike," the latter instances of "bad taste." "All neat schematisms with permanent & absolute
distinctions, classifications with absolute pretensions, systems with pigeon holes etc., have this
character." Ibid.

103. Ibid.

104. James to G. H. Howison, July 4, 1903, in *Correspondence*, 10: 176.

105. James, *Pragmatism*, 18.

106. "A philosophy that breathes out nothing but refinement will never satisfy the empiricist
temper of mind; it will seem rather a monument of artificiality" (ibid.).

107. James to Münsterberg, June 28, 1906, in *Correspondence*, 11: 241–42.

108. Münsterberg to James, July 1, 1906, in *Correspondence*, 11: 245.

109. James, "World of Pure Experience," 42. James extensively dealt with ontological "novelty" in *Some Problems of Philosophy*.

110. On this, see Conant, "James/Royce Dispute," 201–2.

111. See "Notes for Philosophy 1," esp. 186–87; "Notes for Philosophy 9," esp. 355;
"Syllabus in Philosophy D," esp. 378–80. On unity, see "Notes for Philosophy 10," esp. 204;
"Notes for Philosophy 20b," esp. 218; "Notes on 'Conclusions of Lotze Course,'" esp. 260; "Syl-
labus of Philosophy 3," esp. 268–70.

112. James, *Some Problems of Philosophy*.

113. For a similar discussion, see also James, "Notes for Philosophy 1," esp. 186–87; "Notes
for Philosophy 9," esp. 355; "Syllabus for Philosophy D," esp. 378–80.

114. See, e.g., Lewes, *Biographical History of Philosophy*, xv, and *History of Philosophy from
Thales to Comte*, 1:xxiv–xxvii and 2: 749–50.

115. James, *Some Problems of Philosophy*, 112.

116. See J. Ward, "Progress of Philosophy," esp. 215. Ward argued that, as soon as
a body of problems was solved, it split from philosophy and was given the title of a science. For example, "most of the old Natural Philosophy has become the Natural Sciences,"
and "much of the old Moral Philosophy has become the Moral Sciences." Philosophy
proper, Ward added, "still remains as the 'leader' or main growing-point of the whole tree
of knowledge, and so regarded seems as inchoate and nascent as in the days of Thales or
Pythagoras" (Ward, "Progress of Philosophy," 225). See also Paul Janet, *Principes*. Shadworth Hodgson portrayed philosophy as "the parent stem of knowledge," from which "all
the sciences have come, as it were, by fission" (S. H. Hodgson, "The Practical Bearing of
Speculative Philosophy," 21). Philosophy alone could unify the sciences weaving them into
a "living and ever growing web" that would embrace "the whole body of experience, the
whole of the phenomena of the universe" (S. H. Hodgson, "The Relation of Philosophy to Science," 7).

117. See also James, *Manuscript Lectures*, 355.

118. James, *Some Problems of Philosophy*, 11–13. James praised Thomas Aquinas, who had
written on almost everything, "from God down to matter, with angels, men, and demons taken
in on the way" (ibid., 12).

119. Ibid., 13–14.

120. James, introduction to Fechner, *Little Book of Life after Death*.

121. Fechner "was not a professional philosopher at all" (James, *Pluralistic Universe*, 8).

122. James, *Some Problems of Philosophy*, 19. James was referring to the section of Spencer's
*First Principles of a New System of Philosophy* that dealt with the unifying function of philosophy.
James heavily underlined those pages in his own copy of the book. See also James, "Herbert
Spencer Dead": Spencer "brought us back to the old ideal of philosophy, which since Locke's

time had well nigh taken flight, the ideal, namely, of a 'completely unified knowledge,' into which the physical and mental worlds should enter on equal terms. This was the original Greek ideal of philosophy, to which men surely must return" (101).

123. James, "Syllabus for Philosophy D," 380.

124. Ibid.

125. Spencer, *First Principles*, 541–44.

126. James, "Importance to Philosophy."

127. H. Münsterberg, *Eternal Values*, 80.

128. James, *Some Problems of Philosophy*, 55. For a very enlightening discussion of James's metaphors of leaving doors and windows open, see Gale, *Divided Self*, 4.

129. Boutroux, address delivered before the Academy of Moral and Political Sciences, Paris, April 23, 1910 (quoted in Perry, *Thought and Character*, 2: 568). Boutroux described James's Cambridge house as a "charming" home, "a dwelling-place marvelously suited to study and meditation," where "a most amiable sociability [reigned]" (ibid.).

130. James to Alice James (quoted from Gale, *Divided Self*, 4).

131. I borrow this phrase from Bjork, *William James*.

132. James, introduction to Fechner, *Little Book of Life after Death*.

133. H. Münsterberg, *Science and Idealism*, 3.

134. H. Münsterberg, "Emerson Hall," 13.

135. James, *Pragmatism*, 33.

136. Papini, "Il pragmatismo messo in ordine," *Leonardo*, anno III (April 1905): 45–48. I discuss this metaphor in my article "L'Hotel pragmatista."

137. James, "Papini"; quotation on 146. On the "hotel" as a threshold space, see, e.g., Clifford, *Routes*, and de Certeau, *Practice of Everyday Life*.

138. For a similar discussion of the notion of unity, see also James, "Notes on 'Conclusions of Lotze Course,'" esp. 260; "Syllabus of Philosophy 3," esp. 268–70; and *Pragmatism*, Lecture 4.

139. See, e.g., James, "Sentiment of Rationality," in *Will to Believe*.

140. James, *Pragmatism*, Lecture IV, esp. 77.

141. Ibid.

## CHAPTER EIGHT

1. Schiller, "William James," 145. See also Schiller to James, February 22, 1904, in *Correspondence*, 10: 380.

2. On this point, see especially David A. Hollinger, "Inquiry and Uplift."

3. James, *Principles of Psychology*, 125–26.

4. Philosophers pursued "reasoned" truths, while common men were content with suppositions. James, *Pluralistic Universe*, 11; James to H. Höffding, Cambridge, January 3, 1910, in *Correspondence*, 12: 405.

5. William James to Henry James, May 4, 1907, in *Correspondence*, 3: 339.

6. Santayana, "William James," 105; Schiller, "William James," 146.

7. See C. S. Peirce to William James, March 9, 1909, *Correspondence*, 12: 171–72; Peirce, preface to *Collected Papers*, 5: 13 (discussed in D. J. Wilson, *Science, Community*, 127); and Peirce, "Ethos of Terminology," in *Collected Papers*, 2: 226.

8. Schiller, "William James," 146. The abridged version of *Principles—Psychology, Briefer Course* (1894) raised similar criticism; James Ward, for example, found that its style was inappropriately "picturesque." See Sokal, introduction to *Psychology, Briefer Course*.

9. Creighton, "Nature and Criterion of Truth," 593. Creighton contended that "philosophy, like all other genuine sciences" had "passed beyond the stage of the merely striking or suggestive treatment of problems," and aimed no longer "at interesting or picturesque results." Its "object

[was] . . . not to make the world interesting, but to satisfy the mind's demands for intelligibility." Ibid.

10. H. Münsterberg, *Eternal Values*, ix–x.

11. A. E. Taylor, "Philosophy," 48.

12. The full passage reads: "Popular science, popular art, popular theology; only one thing was lacking—popular philosophy. And now [the pragmatists] give that to us. What a triumph for a weak cause!" Schinz believed that, because of its popular tone and availability to the masses, pragmatism would have destabilizing political effects, and that it deserved "to be smothered in its cradle." Schinz, *Anti-Pragmatism*, xi, xv–xvi.

13. "If by chance anyone writes popularly and about results only [rather than about other philosophers' arguments]," he complained, "it is reckoned *oberflächliches zeug*, and *ganz unwissenschaftlich*" (James, *Pluralistic Universe*).

14. James, ibid., 12–14.

15. James, ibid., and Paulsen, *Autobiography*, 339.

16. The abstract program of pragmatism, James wrote Papini, "*must* be sketched extravagantly" (James to Papini, May 15, 1906, in *Essays in Philosophy*, 272).

17. James, *Will to Believe*, 7. For a discussion of this point, see Hollinger, *American Province*, chap. 1.

18. William James to Henry James, May 4, 1907, in *Correspondence*, 3: 339.

19. James to Wilhelm Jerusalem, September 15, 1907, in *Correspondence*, 11: 448.

20. Quoted in Perry, *Thought and Character*, 2: 467. In a letter to his German ally Wilhelm Jerusalem, who was then preparing a German translation of *Pragmatism*, James predicted that the book's "untechnicality of style" would "make the German *Gelehrtes Publikum*, as well as the professors, consider it *oberflächlichles Zeug* [a superficial tool]" (James to Jerusalem, September 15, 1907, in *Correspondence*, 11: 448).

In other writings, especially the essays on radical empiricism, James deployed a technical style. He also realized that many of the misunderstandings and disputes surrounding pragmatism stemmed from the style he had adopted, and endeavored to formulate his theories in more technical ways. On other occasions, he tried to produce more popular statements of technical philosophical doctrines that, he feared, only few could understand. For example, in a letter to Giovanni Papini, who had agreed to translate several of James's articles on radical empiricism into Italian, James wrote that those articles were "highly technical, polemical, abstract, and unnatural" writings, addressed to an audience of specialists. James declined Papini's offer, and stated that he would prefer instead to wait and publish first "a digestible and popular volume" on radical empiricism, which he was planning to write. See James to Papini, May 21, 1906, in *Correspondence*, 11: 226.

21. James, *Principles of Psychology*, 111, 114, 117, 126.

22. James, "Thomas Davidson," 91–92.

23. Ibid., 90. Also: "Our undisciplinables are our proudest product" (James, "True Harvard," 77).

24. James, "Teaching of Philosophy," 4.

25. James, *Some Problems of Philosophy*, 10–11. See also "Notes from Philosophy 1a."

26. James, *Some Problems of Philosophy*, 11.

27. Anderson, "Philosophy as Teaching," 241.

28. James, "Energies of Men," presidential address, 131.

29. Ibid., 136, 139, and 143.

30. Ibid., 145–46.

31. Ibid., 132 and 140.

32. Ibid., 131.

33. Ibid.

34. Ibid., 135.

35. Ibid., 136.

36. Ibid., 132 and 146.

37. "Our life is contracted like the field of vision of an hysteric subject, but with less excuse, for the poor hysteric is diseased, while in the rest of us it is only an inveterate habit—the habit of inferiority to our full self—that is bad" (James, "Energies of Men," 144).

38. Star and Griesemer, "Institutional Ecology"; see also Fujimura, "Crafting Science."

## CONCLUSIONS

1. See, e.g., Schiller, "William James"; James, "Social Value of the College-Bred," esp. 109; and Kittelstrom, "Against Elitism." Also, as Hollinger noted, James's intended "popular" audiences were often limited to college graduates and an intelligent, educated readership. Those were the kinds of individuals who, James hoped, would be able to choose good leaders for the country, and make democracy safe for America (Hollinger, *American Province*, chap. 1).

2. See, e.g., Rorty, *Philosophical Papers*, vol. 1, *Objectivity, Relativism, and Truth*; Latour, *Making Things Public*, 28; Latour, *Reassembling the Social*; Latour, "What Is the Style of Matters of Concern?" Menand, "Undisciplined"; Menand, "The Marketplace of Ideas." For a rejoinder to Menand and to the use of pragmatism for projects of postdisciplinarity, see Haskell, "Menand's Postdisciplinary Project." On postdisciplinarity and pragmatism, see also Kimball, *The Condition of American Liberal Education*. Rorty deployed Dewey's pragmatism and pragmatist "anti-representationalism" in order to challenge the existence of "epistemological differences" between such "disciplinary matrices as theoretical physics and literary criticism" (Rorty, *Objectivity, Relativism, and Truth*, 1). Latour has resorted to James's pragmatist account of truth and metaphysics of radical empiricism in his attempt to articulate a program that transgresses the boundaries between science studies, sociology, ecology, politics, and metaphysics. The ultimate goal of his program is to break down accounts of truth, nature, objectivity, and society, which, he contends, scientists and philosophers have long used to validate their claims to their exclusive access to truth and exclusive right to power.

3. For a few examples, see Pandora, *Rebels within the Ranks*, 15, and Bordogna, "Scientific Personae." See also Hollinger, "Defense of Democracy."

4. See, e.g., Daston, "The Moralized Objectivities of Science," and Daston and Galison, "The Image of Objectivity."

# BIBLIOGRAPHY

## ARCHIVES

Archives of the Society for Psychical Research, Cambridge University Library, Cambridge, UK

Archivio Papini, Fondazione Primo Conti, Fiesole, Italy

Ernst Mach Nachlass, Deutsches Museum, Munich

F. W. H. Myers Papers, Trinity College, Cambridge University, Cambridge, UK

Harvard University Archives, Pusey Library, Harvard University, Cambridge, Mass.

Henri Bergson Papers, Bibliothèque littéraire Jacques Doucet, Paris

Henry Sidgwick Papers, Trinity College, Cambridge, UK

Hugo Münsterberg Papers, Boston Public Library, Boston

William James Papers,1803–1941 (inclusive), 1862–1910 (bulk). bMS Am 1092.9–11, Houghton Library, Harvard University, Cambridge, Mass.

## WORKS BY WILLIAM JAMES

### Standard Editions

*The Correspondence of William James*. 12 vols. Edited by Ignas K. Skrupskelis and Elizabeth M. Berkeley. Charlottesville, Va.: University of Virginia Press, 1992–2004.

*Essays, Comments, and Reviews*. The Works of William James. Gen. ed. Frederick Burkhardt. Cambridge: Harvard University Press, 1987.

*Essays in Philosophy*. The Works of William James. Gen. ed. Frederick Burkhardt. Cambridge, Mass.: Harvard University Press, 1978.

*Essays in Psychical Research*. The Works of William James. Gen. ed. Frederick Burkhardt. Cambridge, Mass.: Harvard University Press, 1986.

*Essays in Psychology.* The Works of William James. Gen. ed. Frederick Burkhardt. Cambridge, Mass.: Harvard University Press, 1983.

*Essays in Radical Empiricism.* The Works of William James. Gen. ed. Frederick Burkhardt. Cambridge, Mass.: Harvard University Press, 1976.

*Essays in Religion and Morality.* The Works of William James. Gen. ed. Frederick Burkhardt. Cambridge, Mass.: Harvard University Press, 1982.

*The Letters of William James.* Edited by Henry James. 2 vols. Boston: Atlantic Monthly Press, 1920. Reprint, London: Longmans, Green, 1920.

*Manuscript Essays and Notes.* The Works of William James. Gen. ed. Frederick Burkhardt. Cambridge, Mass.: Harvard University Press, 1988.

*Manuscript Lectures.* The Works of William James. Gen. ed. Frederick Burkhardt. Cambridge, Mass.: Harvard University Press, 1988.

*The Meaning of Truth.* The Works of William James. Gen. ed. Frederick Burkhardt. Cambridge, Mass.: Harvard University Press, 1975.

*A Pluralistic Universe.* The Works of William James. Gen. ed. Frederick Burkhardt. Cambridge, Mass.: Harvard University Press, 1977.

*Pragmatism.* The Works of William James. Gen. ed. Frederick Burkhardt. Cambridge, Mass.: Harvard University Press, 1975.

*The Principles of Psychology.* The Works of William James. Gen. ed. Frederick Burkhardt. Cambridge, Mass.: Harvard University Press, 1981.

*Psychology, Briefer Course.* The Works of William James. Gen. ed. Frederick Burkhardt. Cambridge, Mass.: Harvard University Press, 1984.

*Some Problems of Philosophy.* The Works of William James. Gen. ed. Frederick Burkhardt. Cambridge, Mass.: Harvard University Press, 1979.

*Talks to Teachers on Psychology, and to Students on Some of Life's Ideals.* The Works of William James. Gen. ed. Frederick Burkhardt. Cambridge, Mass.: Harvard University Press, 1983.

*The Varieties of Religious Experience.* The Works of William James. Gen. ed. Frederick Burkhardt. Cambridge, Mass.: Harvard University Press, 1985.

*The Will to Believe and Other Essays in Popular Psychology.* Gen. ed. Frederick Burkhardt. Cambridge, Mass.: Harvard University Press, 1979.

## Other Works by William James

"Absolutism and Empiricism." *Mind* 33 (1884): 281–86.

"Abstractionism and 'Relativismus'" (1909). In *The Meaning of Truth,* 134–45.

"Address of the President before the Society for Psychical Research" (1896). In *Essays in Psychical Research,* 127–37.

"Address on the Medical Registration Bill" (1898). In *Essays, Comments, and Reviews,* 56–62.

"Aesthetics" (1878–85). bMS Am 1092.9 4393, William James Papers, Houghton Library, Harvard University. In *Manuscript Lectures,* 123–26.

"Answer to Roosevelt on the Venezuelan Crisis" (1896). In *Essays, Comments, and Reviews,* 152–53.

"Confidences of a 'Psychical Researcher'" (1909). In *Essays in Psychical Research,* 361–75.

"Consciousness under Nitrous Oxide" (1898). In *Essays in Psychology,* 322–25.

Diary 1 (1868–1873). Unpublished MS, William James Papers, Houghton Library, Harvard University.

Diary (1902–1903). Unpublished MS, William James Papers, Houghton Library, Harvard University.

Diary (1905). Unpublished MS, William James Papers, Houghton Library, Harvard University.

Diary (1907). Unpublished MS, William James Papers, Houghton Library, Harvard University.

"Diary of French Naval Officer: Observations at Manila" (1900). In *Essays, Comments, and Reviews*, 167–68.

"Discussion: The Meaning and Criterion of Truth," in "Proceedings of the 1907 Meeting of the American Philosophical Association." *Philosophical Review* 17 (1908): 180–86. (James's fellow panelists are J. E. Creighton, Charles M. Bakewell, John Grier Hibben, and C. A. Strong.)

"Does Consciousness Exist?" (1904). In *Essays in Radical Empiricism*, 3–20.

"The Energies of Men." Presidential Address before the American Philosophical Association, December 28, 1906. In *Essays in Religion and Morality*, 129–46.

"The Existence of Julius Caesar" (1908). In *The Meaning of Truth*, 120–22.

"The Experience of Activity" (1905). President's Address before the American Psychological Association, Philadelphia Meeting, December 1904. In *Essays in Radical Empiricism*, 79–95.

"Frederic Myers's Service to Psychology" (1901). In *Essays in Psychical Research*, 192–202.

"The Function of Cognition" (1885). In *The Meaning of Truth*, 13–32.

"The Gospel of Relaxation" (1899). In *Talks to Teachers on Psychology*, 117–31.

"Governor Roosevelt's Oration" (1899). In *Essays, Comments, and Reviews*, 162–69.

"Herbert Spencer Dead" (1903). In *Essays in Philosophy*, 96–101.

"The Hidden Self" (1890). In *Essays in Psychology*, 247–68.

"Horace Fletcher at Harvard" (1905). In *Essays, Comments, and Reviews*, 184–85.

*Human Immortality: Two Supposed Objections to the Doctrine* (1898). In *Essays in Religion and Morality*, 74–101.

"Humanism and Truth" (1904). In *The Meaning of Truth*, 37–60. Originally published in *Mind*, n.s., 13 (1904): 457–75.

"Humanism and Truth Once More" (1905). *Mind*, n.s., 14 (1905): 190–99.

"The Importance to Philosophy of the Appointment of Professors Wundt and Hitzig at Zürich" (1875). In *Essays, Comments, and Reviews*, 14.

"International Congress of Experimental Psychology: Instructions to the Person Undertaking to Collect Answers to the Question on the Other Side" (1889). In *Essays in Psychical Research*, 57–58.

Introduction to *The Literary Remains of the Late Henry James* (1884). In *Essays in Religion and Morality*, 3–63.

Introduction to *The Little Book of Life after Death*, by G. T. Fechner. In James, *Essays in Religion and Morality*, 116–19.

"Johns Hopkins Lectures on 'The Senses and the Brain and Their Relation to Thought'" (1878). In *Manuscript Lectures*, 3–16.

"The Knowing of Things Together" (1895). President's Address before the American Psychological Association, December 1894. In *Essays in Philosophy*, 71–89.

Lecture, May 18, 1906, at the Psychology Club, Harvard. In *Essays in Religion and Morality*, 199–200.

"Lowell Lectures on 'The Brain and the Mind'" (1878). In *Manuscript Lectures*, 16–43.

"The Mad Absolute" (1906). In *Essays in Philosophy*, 149–50.

"The Medical Registration Act" (1894). In *Essays, Comments, and Reviews*, 145–49.

"The Medical Registration Bill" (1894). In *Essays, Comments, and Reviews*, 149–50.

"Messrs. Lehman and Hansen on Telepathy" (1899). In *Essays in Psychical Research*, 172.

"The Miller-Bode Objections, 1905–1908." In *Manuscript Essays and Notes*, 65–129.

[Ignoramus, pseud.]. "The Mood of Science and the Mood of Faith" (1874). In *Essays, Comments, and Reviews*, 115–17.

"The Moral Equivalent of War" (1910). In *Essays in Religion and Morality*, 162–73.

"Myers's Service to Psychology" (1901). In *Essays in Psychical Research*, 192–202.

"Notes for Natural History 2: Physiological Psychology, 1876–1877." In *Manuscript Lectures*, 16–129.

"Notes for Philosophy 1: General Introduction to Philosophy (1890–1901)." In *Manuscript Lectures*, 186–98.

"Notes for Philosophy 1a, 1905–1906." William James Papers, bMS Am 1092.9 4516, Houghton Library, Harvard University.

"Notes for Philosophy 9: Metaphysics (1905–1906)." In *Manuscript Lectures*, 347–74.

"Notes for Philosophy 10: Descartes, Spinoza, and Leibniz (1890–1891)." In *Manuscript Lectures*, 198–205.

"Notes for Philosophy 20a: Psychological Seminary. Aesthetics (1891–1892)." bMS Am 1092.9 4392, William James Papers, Houghton Library, Harvard University. In *Manuscript Lectures*, 206–11.

"Notes for Philosophy 20b: Psychological Seminary—The Feelings (1895–1896)." In *Manuscript Lectures*, 212–30.

"Notes for Philosophy 20c: Metaphysical Seminary (1903–1904)." In *Manuscript Lectures*, 319–27.

"Notes on Automatic Writing" (1889). In *Essays in Psychical Research*, 37–55.

"Notes on 'Conclusions of Lotze Course' (1897–1898)." In *Manuscript Lectures*, 259–65.

"On a Certain Blindness in Human Beings" (ca. 1898). In *Talks to Teachers on Psychology*, 132–49.

"On Some Hegelisms" (1882). In *The Will to Believe*, 263–98.

"Papini and the Pragmatist Movement in Italy" (1906). In *Essays in Philosophy*, 144–48.

"Person and Personality" (1895). In *Essays in Psychology*, 315–21.

"The Ph.D. Octopus" (1903). In *Essays, Comments, and Reviews*, 67–74.

"The Philippine Question" (1899). In *Essays, Comments, and Reviews*, 159–60.

"The Philippine Tangle" (1899). In *Essays, Comments, and Reviews*, 154–58.

"The Philippines Again" (1899). In *Essays, Comments, and Reviews*, 160–62.

"Philosophical Conceptions and Practical Results" (1898). In *Pragmatism*, 255–74.

"Philosophies Paint Pictures." MS 4439, William James Papers, Houghton Library, Harvard University. In "The Many and the One" (1903–04), in *Manuscript Essays and Notes*, 3–6.

Photographic Album. MS Am 1092.9 4568. William James Papers, Houghton Library, Harvard University.

"Plea for Psychology as a Natural Science" (1892). In *Essays in Psychology*, 270–77.

"A Possible Case of Projections of the Double" (1909). In *Essays in Psychical Research*, 376–77.

"The Powers of Men" (1907). In *Essays in Religion and Morality*, 147–61.

"The Pragmatist Account of Truth and Its Misunderstanders" (1908). In *The Meaning of Truth*, 99–116.

Preface to *Introduction to Philosophy*, by F. Paulsen (1895). In *Essays in Philosophy*, 90–93.

"Professor Hébert on Pragmatism" (1908). In *The Meaning of Truth*, 126–33.

"Professor Pratt on Truth" (1907). In *The Meaning of Truth*, 90–98.

"Psychical Research" (1896). In *Essays in Psychical Research*, 138–42.

"Psychology at Harvard University" (1890). In *Essays, Comments, and Reviews*, 31.

"The Psychology of Belief." *Mind* 55 (1889): 321–52.

"Quelques considérations sur la méthode subjective" (1878). In *Essays in Philosophy*, 23–31. Translation in the same volume, 331–38.

"Reaction-Time in the Hypnotic Trance" (1887). In *Essays in Psychology*, 200–203.

"Reflex Action and Theism" (1882). In *The Will to Believe*, 90–113.

"Remarks at the Peace Banquet" (1904). In James, *Essays in Religion and Morality*, 120–23.

"Remarks on Spencer's 'Definition of Mind as Correspondence'" (1878). In *Essays in Philosophy*, 7–22.

"Report of a Discussion in Philosophy 20e: Seminary in the Theory of Knowledge 1909." In *Manuscript Lectures*, 429–43.

"Report of the Committee on Hypnotism" (1886). In *Essays in Psychology*, 191–92.

"Report on Mrs. Piper's Hodgson-Control" (1909). In *Essays in Psychical Research*, 253–360.

Review of *The Data of Ethics*, by Herbert Spencer (1879). In *Essays, Comments, and Reviews*, 347–53.

Review of *État mental des hystériques*, by Pierre Janet, vol. 1 (1892) and vol. 2 (1894). In *Essays, Comments, and Reviews*, 470–74.

Review of *The Functions of the Brain*, by David Ferrier; *The Physiology and Pathology of Mind*, by Henry Maudsley; and *Le Cerveau et ses fonctions*, by Jules Luys (1877). In *Essays, Comments, and Reviews*, 332–37.

Review of *Grundzüge der Physiologischen Psychologie*, by Wilhelm Wundt (1875). In *Essays, Comments, and Reviews*, 296–303.

Review of *Human Personality and Its Survival of Bodily Death*, by F. W. H. Myers (1903). In *Essays in Psychical Research*, 203–15.

Review of *Pain, Pleasure, and Aesthetics*, by Henry Rutgers Marshall (1894). In *Essays, Comments, and Reviews*, 489–93.

Review of *The Physical Basis of Mind*, by George Henry Lewes (1877). In *Essays, Comments, and Reviews*, 342–45.

[Two] reviews of *Principles of Mental Physiology, with Their Applications to the Training and Discipline of the Mind and the Study of Its Morbid Conditions*, by W. B. Carpenter (1874). In *Essays, Comments, and Reviews*, 269–75.

Review of *Psychology: Descriptive and Explanatory*, by G. T. Ladd (1894). In *Essays, Comments, and Reviews*, 476–85.

Review of *The Relations of Mind and Brain*, by H. Calderwood (1879). In *Essays, Comments, and Reviews*, 361–63.

Review of the "Report of the Census of Hallucinations" (1895). In *Essays in Psychical Research*, 66–73.

Review of *Revue Philosophique de la France et de l'Étranger*, by T. Ribot (1876). In *Essays, Comments, and Reviews*, 319–20.

Review of *Über das Gedächtnis*, by Hermann Ebbinghaus (1885). In *Essays, Comments, and Reviews*, 388–90.

Seminar on "The Feelings" (1895–96). In *Manuscript Lectures*, 220.

"The Sentiment of Rationality" (1879). In *Essays in Philosophy*, 32–64.

"The Sentiment of Rationality" (1897). In *The Will to Believe*, 57–89.

"The Social Value of the College-Bred" (1907). In *Essays, Comments, and Reviews*, 106–12.

"Spencer's First Principles." bMS Am 1092.9 4492, folders 1 and 2. William James Papers, Houghton Library, Harvard University.

"A Suggestion about Mysticism" (1909). In *Essays in Philosophy*, 157–65.

"Syllabus in Philosophy D: General Problems of Philosophy (1906–1907)." In *Manuscript Lectures*, 378–428.

"Syllabus of Philosophy 3: The Philosophy of Nature (1902–1903)." In *Manuscript Lectures*, 257–73.

"The Teaching of Philosophy in Our Colleges" (1876). In *Essays in Philosophy*, 3–6.

"Telepathy" (1895). In *Essays in Psychical Research*, 119–26.

"Thomas Davidson: Individualist" (1903). In *Essays, Comments, and Reviews*, 86–98.

"The Tigers in India" (1895). In *The Meaning of Truth*, 33–36.

"The True Harvard" (1903). In *Essays, Comments, and Reviews*, 74–77.

"Truth and Reality, etc., 1907–1909." bMS Am 1092.9 4549, William James Papers, Houghton Library, Harvard University. In *Manuscript Essays and Notes*, 234–46.

"What Is an Emotion?" (1884). In *Essays in Psychology*, 168–87.

"What Makes Life Significant" (n.d.). In *Talks to Teachers on Psychology*, 150–67.

"What Psychical Research Has Accomplished" (1892). In *Essays in Psychical Research*, 89–106.
"What Psychical Research Has Accomplished" (1896). In *The Will to Believe*, 222–41.
"What the Will Effects" (1888). In *Essays in Psychology*, 216–34.
"William James's Sources." bMS Am 1092.9 (4578), folders 1 and 2. William James Papers, Houghton Library, Harvard University.
"The Will to Believe" (1896). In *The Will to Believe*, 13–33.
"A Word More about Truth" (1907). In *The Meaning of Truth*, 78–89.
"A World of Pure Experience" (1904). In *Essays in Radical Empiricism*, 21–44.

## PRIMARY SOURCES

American Society for Psychical Research. *Proceedings of the American Society for Psychical Research*. Vol. 1. Boston: Rand, Avery, 1885–89.

Angell, James R. "The Province of Functional Psychology." President's Annual Address before the American Psychological Association, New York, 1906. *Psychological Review*, n.s., 14, no. 2 (1907): 61–91.

———. "The Relations of Structural and Functional Psychology to Philosophy." *Philosophical Review* 12 (1903): 243–71.

———. *The Relations of Structural and Functional Psychology to Philosophy*. Vol. 3. The Decennial Publications of the University of Chicago. Chicago: University of Chicago Press, 1903.

Arnold, Matthew. *Culture and Anarchy*. London: Smith, Elder, 1869.

———. "Culture and Its Enemies." *Cornhill Magazine* 16 (1867): 36–53.

———. "Literature and Science." In *The Portable Matthew Arnold*, edited by Lionel Trilling, 405–29. New York: Viking Press, 1949.

Bain, Alexander. *The Emotions and the Will*. 1855. 3rd ed., New York: Appleton, 1888.

———. *Mental Science: A Compendium of Psychology*. New York: Appleton, 1868.

———. *The Senses and the Intellect*. 3rd rev. ed. London: Longmans, Green, 1868.

Bakewell, Charles M. "On the Meaning of Truth." *Philosophical Review* 17 (1908): 579–91.

Baldwin, James Mark, ed. *Dictionary of Philosophy and Psychology*. 3 vols. New York: Macmillan, 1901–05.

———. "The Limits of Pragmatism." *Psychological Review* 11 (1904): 30–60.

———. "Psychology: Past and Present." *Psychological Review* 1 (1894): 363–91.

Beard, G. M. "Experiments with Living Human Beings." *Popular Science Monthly* 14 (1879): 611–21, 751–57.

———. "The Psychology of Spiritism." *North American Review* 129 (1879): 65–80.

———. "The Scientific Study of Human Testimony." *Popular Science Monthly* 13 (1878): 53–54, 173–83, 328–38.

Bellamy, Edward. *Looking Backward*. 1888. Cambridge, Mass.: Belknap Press, 1967.

———. *The Religion of Solidarity*. 1874. Edited by Arthur E. Morgan. Yellow Springs, Ohio: Antioch Bookplate, 1940.

Bergson, Henri. *The Creative Mind*. Translated by M. Andison. New York: Philosophical Library, 1946.

———. *An Introduction to Metaphysics*. Translated by T. E. Hulme. New York: Putnam's Sons, 1912.

———. "Les etudes gréco-latines et la démocratie." In *L'éducation de la démocratie*, by Ernest Lavisse et al. Paris: F. Alcan, 1903.

———. " 'Phantasms of the Living' and 'Psychical Research': Presidential Address to the Society for Psychical Research, London, May 28, 1913." In *Mind-Energy: Lectures and Essays*. New York: Henry Holt, 1920.

Binet, Alfred. *Alterations of Personality*. Translated by Helen Green Baldwin, with notes and a preface by J. Mark Baldwin. New York: Appleton, 1896. French edition, 1891.

Boring, Edwin G. "Edward Bradford Titchener, 1867–1927." *American Journal of Psychology* (1927): 489–506.

Boutroux, Émile. Address delivered before the Academy of Moral and Political Sciences, Paris, April 23, 1910.

———. *De la contingence des lois de la nature.* Paris: F. Alcan, 1895.

———. "La philosophie en France depuis 1867." *Revue de métaphysique et de morale* (1908): 683–716.

———. *Science et religion dans la philosophie contemporaine.* Paris: Flammarion, 1908.

———. *William James.* Paris: A. Colin, 1911.

Bowditch, H. "Letter to the Editor." *Boston Medical Surgical Journal* 82 (1870): 307.

Bowen, Francis. *Principles of Metaphysics and Ethical Science Applied to the Evidences of Religion.* Boston: Brewer and Tileston, 1855.

———. *The Principles of Political Economy.* 1856. Reprint, Boston: Little, Brown, 1863.

Bradley, Francis H. *Appearance and Reality.* Oxford, 1893.

———. "On Truth and Practice." *Mind* 13 (1904): 309–35.

Brown, S. G. *The Social Philosophy of Josiah Royce.* New York: Syracuse University Press, 1905.

Buchner, E. F. "A Quarter Century of Psychology in America: 1878–1903." *American Journal of Psychology* 14 (1903): 402–16.

Bucke, Richard M. *Cosmic Consciousness: A Study in the Evolution of the Human Mind.* 1901. New York: E. P. Dutton, 1969.

Call, Annie Payson. *As a Matter of Course.* Boston: Roberts Brothers, 1894.

———. *Power through Repose.* Boston: Roberts Brothers, 1891.

Carpenter, Edward. *The Art of Creation: Essays on the Self and Its Powers.* 1901. New and enlarged edition, London: George Allen, Ruskin House, 1907.

———. *Towards Democracy.* London: J. Heywood, 1883.

Carpenter, William Benjamin. [unsigned] "Electro-Biology and Mesmerism." *Quarterly Review* 93 (1853): 501–57.

———. "Letter from Dr. Carpenter, Dec. 24, 1869." In *Report on Spiritualism of the London Dialectical Society,* 266–77. London: Longmans and Green, 1871.

———. *Mesmerism, Spiritualism, Table-Turning and Odylism.* New York: Appleton, 1877.

———. *Principles of Human Physiology.* 5th ed. London: Churchill, 1855.

———. *Principles of Mental Physiology, with Their Applications to the Training and Discipline of the Mind and the Study of Its Morbid Conditions.* New York: Appleton, 1874.

———. "The Radiometer and Its Lessons." *Nineteenth Century* (1877): 242–56.

[———]. "Spiritualism and Its Recent Converts." *Quarterly Review* 131 (1871): 301–53.

Carus, Paul. *Philosophy as a Science.* Chicago: Open Court, 1909.

Cattell, James McKeen. "Address of the President before the American Psychological Association, 1895." *Psychological Review* 3 (1896): 134–48.

———. "Homo Scientificus Americanus." In Poffenberger, *James McKeen Cattell,* 2:185–96.

———. "Homo Scientificus Americanus: Address of the President of the American Society of Naturalists." *Science* 17 (1903): 561–70.

———. "Our Psychological Association and Research." *Science,* n.s., 45 (1917): 275–84.

———. "Psychical Research: Address by the President before the Society for Psychical Research, William James." *Psychological Review* 3 (September 1896): 582–83.

Chambers, Ephraim. *Cyclopaedia: or, an Universal Dictionary of Arts and Sciences.* 2 vols. London: 1728.

Clendenning, John, ed. *The Letters of Josiah Royce.* Chicago: University of Chicago Press, 1970.

Clifford, W. K. "The Ethics of Belief" (1877). In W. K. Clifford, *Lectures and Essays,* 1:339–63. London: Macmillan, 1886.

———. *Lectures and Essays.* London: Macmillan, 1879.

Cogswell, G. A. "The Classification of the Sciences." *Philosophical Review* 8 (1899): 494–512.

Cohen, Morris R. "The Conception of Philosophy in Recent Discussion." *Journal of Philosophy, Psychology and Scientific Methods* 7 (1910): 401–10.

Comte, Auguste. *Cours de philosophie positive.* Edited by Michel Serres, François Dagognet, and Allal Sianceur. Paris: Hermann, 1999.

*Congress of Arts and Science, Universal Exposition, St. Louis, 1904.* Edited by Howard J. Rogers. 8 vols. Boston: Houghton, Mifflin, 1905–1907.

Courtier, Jules. "Rapport sur les séances d'Eusapia Palladino à l'Institut général psychologique en 1905, 1906, 1907 et 1908." *Bulletin de l'Institut général psychologique* (1908): 407–546.

Creighton, J. E. "The Idea of a Philosophical Platform." *Journal of Philosophy* 6 (1909): 141–45.

———. "The Nature and Criterion of Truth." *Philosophical Review* 17 (1908): 592–605.

———. "Purpose as a Logical Category." Paper presented at the American Philosophical Association annual meeting, Princeton, N.J., December 1903. *Philosophical Review* 13 (1904): 284–97.

———. "The Purposes of a Philosophical Association." *Philosophical Review* 11 (1902): 219–37.

Crookes, William. "Experimental Investigation of a New Force." *Quarterly Journal of Science* (July 1871): 339–49.

———. "Some Further Experiments on Psychic Force." *Quarterly Journal of Science* (October 1871): 471–93.

d'Alembert, Jean Le Rond. *Preliminary Discourse to the Encyclopedia of Diderot.* Translated by R. N. Schwab. 1751. Chicago: University of Chicago Press, 1995.

Davies, William Harper. "Proceedings of the Fifteenth Annual Meeting of the American Psychological Association, New York City, Dec. 27–29, 1906. Report of the Secretary." *Psychological Bulletin* 4 (July 1907): 201–5.

Dewey, John. *The Early Works, 1882–1898.* 5 vols. Carbondale, Ill.: Southern Illinois University Press, 1967–72.

———. *Psychology.* 1887. 3rd ed., New York: Harper, 1893.

———. "Psychology as a Philosophic Method." In Dewey, *Early Works,* 1:144–67.

———. "The Reflex-Arc Concept in Psychology." In Dewey, *Early Works,* 5:96–109.

———. "The St. Louis Congress of the Arts and Sciences." *Science,* n.s., 18 (1903): 275–78.

———. *Studies in Logical Theory.* Chicago: University of Chicago Press, 1903.

Dresser, Horatio W. *The Perfect Whole: An Essay on the Conduct and Meaning of Life.* Boston: G. H. Ellis, 1896.

———. *Voices of Freedom and Studies in the Philosophy of Individuality.* New York: G. P. Putnam's Sons; Knickerbocker Press, 1899.

Durand de Gros, Joseph Pierre. *Aperçus de Taxinomie Générale.* Paris: Alcan, 1899.

Eliot, Charles W. *Charles W. Eliot: The Man and His Beliefs.* Edited by William Allan Neilson. New York: Harper & Brothers, 1926.

———. "Inaugural Address of Dr. Eliot. Delivered on October 19, 1869, when Charles William Eliot Became President of Harvard College." In Neilson, *Charles W. Eliot.*

"Emerson Memorial Hall." *Harvard Graduates Magazine,* December 1905, 249–50. Harvard University Archives, Pusey Library, Harvard University.

Fechner, G. T. *Elemente der Psychophysik.* 2 vols. in 1. Leipzig: Breitkopf und Härtel, 1860.

———. *The Little Book of Life after Death.* With an introduction by William James. Translated by Mary C. Wadsworth. Boston: Little, Brown, 1904.

Fernberger, S. W. "The American Psychological Association: A Historical Summary, 1892–1930." *Psychological Bulletin* 29 (1932): 1–89.

Ferrier, David. *The Functions of the Brain.* New York: Putnam's Sons, 1876.

Ferrier, James Frederick. *Institutes of Metaphysic: The Theory of Knowing and Being.* Edinburgh: Wm. Blackwood, 1854.

Fiske, John. *Outlines of Cosmic Philosophy*. New York: Houghton, Mifflin, 1874.

Fletcher, Horace. *Happiness as Found in Forethought Minus Fearthought*. Chicago: H. S. Stone, 1897.

Flint, Robert. *Philosophy as Scientia Scientiarum; and A History of the Classification of the Sciences*. Edinburgh: William Blackwood and Sons, 1904.

Flournoy, Theodore. *The Philosophy of William James*. New York: Henry Holt, 1917. First French edition, 1911.

Foá, Pio, A. Herlitzka, C. Foá, and A. Aggazzotti. "Ce que le prof. P. Foá, de l'Université de Turin, et trois Docteurs, assistants du professeur Mosso, ont constaté avec Eusapia Palladino." *Annales des sciences psychiques* 4 (1907).

Frege, Gottlob. *The Basic Laws of Arithmetics*. Edited by M. Furth. Berkeley: University of California Press, 1964.

———. *The Foundations of Arithmetic: A Logico-Mathematical Inquiry into the Concept of Number*. Translated by J. L. Austin. 1884. Evanston, Ill: Northwestern University Press, 1980.

Frisbie, Levi. "Notices of Professor Frisbie." In *A Collection of the Miscellaneous Writings of Professor Frisbie*, edited by A. Norton. Boston: Cummings, Hilliard, 1823.

Fullerton, G. S. "The Psychological Standpoint." *Psychological Review* 1 (1894): 113–33.

Gardiner, H. N. "The First Twenty-Five Years of the American Philosophical Association." *Philosophical Review* 35 (1926): 135–58.

———. "The Problem of Truth." *Philosophical Review* 17 (1908): 113–37.

Gioberti, Vincenzo. *Degli errori filosofici di Antonio Rosmini*. Capolago: Tipografia elvetica, 1846.

Gore, Willard C. "The Mad Absolute of a Pluralist." *Journal of Philosophy, Psychology, and Scientific Methods* 3 (1906): 575–77.

Green, T. H. "Can There Be a Natural Science of Man?" *Mind* 7 (1882): 1–29, 161–85, 321–48.

———. "Evidence to the Royal Commission on the University of Oxford" (1877). Reprinted in Green, *Miscellaneous Writings, Speeches and Letters*, 199–215.

———. *Miscellaneous Writings, Speeches and Letters*. With an introduction by Peter Nicholson. Bristol: Thoemmes Press, 2003.

———. *Prolegomena to Ethics*. Edited by A. C. Bradley. Oxford: Clarendon Press, 1883.

———. "The Province of Ethics" (1878). Reprinted in Green, *Miscellaneous Writings, Speeches and Letters*, 193–94.

Gronlund, Laurence. *Cooperative Commonwealth*. Boston: Lee and Shepard, 1884.

Gurney, Edmund. "Peculiarities of Certain Post-Hypnotic States." *Proceedings of the Society for Psychical Research* 4 (1886–87): 268–323.

Gurney, Edmund, Frederic W. H. Myers, and Frank Podmore. *Phantasms of the Living*. 2 vols. London: Rooms of the Society for Psychical Research; Trübner, 1886.

Haeckel, Ernst. *The Wonders of Life: A Popular Study of Biological Philosophy*. Translated by Joseph McCabe. New York: Harper and Brothers, 1905. First German edition, 1899.

———. "Zellseelen und Seelenzellen." Vortrag, gehalten am 22. März 1878 in der "Concordia" zu Wien. In *Gesammelte populäre Vorträge aus dem Gebiete der Entwickelungslehre*, 1:143–92. Bonn: Verlag von Emil Strauss, 1878.

Hall, G. Stanley. "The Affiliation of Psychology with Philosophy and with the Natural Sciences." *Science*, n.s., 23 (1906): 297–301.

———. *Aspects of German Culture*. Boston: James R. Osgood, 1881.

———. "College Instruction in Philosophy." *Nation* 23 (September 21, 1876): 178–79.

———. *Life and Confessions of a Psychologist*. New York: Appleton, 1923.

———. "On the History of American College Text Books and Teaching in Logic, Ethics, Psychology, and Allied Subjects." *Proceedings of the American Antiquarian Society* 9 (1894): 137–74.

———. "Philosophy in the United States." *Popular Science Monthly*, suppl. 1 (1879): 57–68.

————. Review of *The Principles of Psychology*, by William James. *American Journal of Psychology* 3 (1891): 578–91.

————. Review of *Proceedings of the English Society for Psychical Research*, July 1882–May 1887, and of *Phantasms of the Living*, by E. Gurney, F. H. Myers, and F. Podmore. *American Journal of Psychology* 1 (1887): 128–48.

Hamilton, William. "Advertisement by the Editor." In *The Collected Works of Stewart Dugald*, 2:viii–ix. Edinburgh: Thomas Constable, 1854.

————. *Lectures on Metaphysics and Logic*. 4 vols. Edinburgh: W. Blackwood and Sons, 1859–60.

Hammond, William A. "The Relations of Logic to Allied Disciplines." *Psychological Review* 13 (1906): 1–22.

Hansen, F. C. C., and A. G. L. Lehmann. "Ueber unwillkürliches Flüsten. Eine kritische und experimentelle Untersuchung der sogenannten Gedankenübertragung." *Philosophisce Studien* 11 (1895): 471–530.

Hébert, Marcel. *Le Pragmatisme: Étude de ses diverses formes Anglo-Américaines, Françaises, Italiennes et de sa valeur religieuse*. Paris: Libraire Critique Émile Nourry, 1908.

Helmholtz, Hermann von. "The Facts in Perception." In Helmholtz, *Science and Culture: Popular and Philosophical Essays*, edited by David Cahan, 342–66. Chicago: University of Chicago Press, 1995.

Hibben, John Grier. "The Test of Pragmatism." *Philosophical Review* 17 (1908): 365–82.

Hocking, William Ernest. Review of "Die Wiedergeburt der Philosophie," by C. Stumpf (Leipzig: Barth, 1908). *Journal of Philosophy, Psychology, and Scientific Methods* (1908): 612–13.

Hodgson, Richard. "A Case of Double Consciousness." In *Proceedings of the Society for Psychical Research* 7 (1891–92): 221–57.

Hodgson, Shadworth H. "Inter-Relation of the Academical Sciences." In *Proceedings of the British Academy* (1905–1906): 219–34.

————. "The Method of Philosophy: An Address Delivered before the Aristotelian Society, Oct. 9, 1882." London: Williams and Norgate, 1882. William James Papers, Houghton Library, Harvard University.

————. "Philosophy and Science: I, II, and III." *Mind* 1 (1876): 67–81, 223–35, 351–62.

————. "Philosophy in Relation to Its History: An Address Delivered before the Aristotelian Society, Oct. 11, 1880." London: Williams and Norgate, 1880. William James Papers, Houghton Library, Harvard University.

————. "The Practical Bearing of Speculative Philosophy: An Address Delivered before the Aristotelian Society, Oct. 10, 1881." London: Williams and Norgate, 1881.

————. "The Relation of Philosophy to Science, Physical and Psychological: An Address Delivered before the Aristotelian Society on October 20, 1884." London: Williams and Norgate, 1884. William James Papers, Houghton Library, Harvard University.

Hoernlé, R. F. Alfred. "Pragmatism v. Absolutism." *Mind*, n.s., 14 (July 1905): 297–334.

Höffding, Harald. *Outlines of Psychology*. Translated from the German edition by Mary E. Lowndes. London: Macmillan, 1893. First English edition, 1891.

Howells, W. D. *A Hazard of New Fortunes*. 1889. New York: Modern Library, 2002.

————. *Through the Eye of the Needle: A Romance*. New York: Harper & Brothers, 1907.

————. *A Traveler from Altruria: Romance*. New York: Harper & Brothers, 1894.

Howison, George Holmes. "Philosophy: Its Fundamental Conceptions and Its Methods." In *Congress of Arts and Science, Universal Exposition, St. Louis, 1904*, edited by Howard J. Rogers, 1:173–94. Boston: Houghton, Mifflin, 1905.

————. "Philosophy and Science: An Address before the Science Association of the University of California, September 10, 1903." Reprinted from the *University Chronicle*, October 1903. William James Papers, Houghton Library, Harvard University.

————. "Psychology and Logic: Further Views." *Psychological Review* 3 (1896): 652–57.

Husserl, Edmund. *Logical Investigations*. Translated by J. N. Findlay. 2 vols. London: Routledge and Kegan Paul, 1900.

————. "Philosophie als strenge Wissenschaft." *Logos* 1 (1910–11): 289–341.

————. *Psychologische Studien zur Elementaren Logik* (1894). Reprinted in *Edmund Husserl: Aufsätze und Rezensionen (1890–1910)*, edited by Bernhard Rang, 92–123. The Hague: Martinus Nijhoff, 1979.

Huxley, T. H. *Collected Essays*. Vol. 3, *Science and Education*. London: Macmillan, 1893.

————. "Contribution to 'A Modern Symposium on the Influence upon Morality of a Decline in Religious Belief.'" *Nineteenth Century* 1 (April 1877): 331–58.

————. *Man's Place in Nature and Other Anthropological Essays*. 1863. Reprint, New York: Appleton, 1902.

————. *Methods and Results*. 1896. New York: Appleton, 1911.

————. "On the Advisableness of Improving Natural Knowledge" (1866). Reprinted in Huxley, *Methods and Results*, 18–41.

————. "On the Hypothesis That Animals Are Automata and Its History" (1874). Reprinted in Huxley, *Methods and Results*, 199–250.

Ilbert, Courtenay Peregrine. "Evidence." *Encyclopedia Britannica*, 11th ed. (1910–11), 11–21.

James, Henry. *Portrait of a Lady*. 1881. Reprint, New York: W. W. Norton, 1995.

————. *The Wings of the Dove*. 1909. Reprint, New York: W. W. Norton, 2003.

James, Henry, Sr. *Lectures and Miscellanies*. New York: Redfield, Clinton Hall, 1852.

————. *Society: The Redeemed Form of Man*. Boston: Houghton, Osgood, 1879.

Janet, Paul, *Principes de métaphysique et de psychologie; leçons professées à la Faculté des lettres de Paris, 1888–1894*. Paris: C. Delagrave, 1897.

Janet, Pierre. *L'automatisme psychologique*. Paris: Alcan, 1889.

————. *The Mental State of Hystericals*. New York: G. P. Putnam's Sons, 1901. First French edition, 1893–94.

Jastrow, Joseph. "The Problems of 'Psychic Research.'" *Harper's New Monthly Magazine* (1889): 76–82.

————. "The Reconstruction of Psychology." *Psychological Review* 34 (1927): 169–95.

Jerusalem, Wilhelm. *Der kritische Idealismus und die reine Logik: Ein Ruf im Streite*. Vienna: Wilhelm Braumüller, 1905.

————. *Einleitung in die Philosophie*. 5. und 6. aufl. Wien: Wilhelm Braumüller, 1913.

————. *Introduction to Philosophy*. Translated by Charles F. Sanders. New York: Macmillan, 1910.

————. *Urtheilsfunction: eine psychologische und Erkenntniskritische Untersuchung*. Wien: Wilhelm Braumüller, 1895.

Johnson, W. E. "On Feeling as Indifference." *Mind* 13 (1888): 80–83.

Joseph, Horace W. B. "Professor James on 'Humanism and Truth.'" *Mind*, n.s., 14 (1905): 28–41.

Kant, Immanuel. *Anthropology from a Pragmatic Point of View*. Translated by Victor Lyle Dowdell. Revised and edited by Hans H. Rudnick. Carbondale: Southern Illinois University Press, 1978.

————. *The Critique of Pure Reason*. 2nd ed. Translated by Norman Kemp Smith. London: Macmillan, 1950.

————. *Logic*. Translated by R. S. Hartman and W. Schwarz. New York: Dover, 1988.

————. *Logik*. Edited by G. B. Jäsche. Vol. 8, pt. 1 of *Werke*, edited by P. Hartenstein. Leipzig: L. Voss, 1867–68.

Kronecker, H., and G. S. Hall. "Die willkürliche Muskelaction." *Archiv für Physiologie. Physiologische Abtheilung des Archives für Anatomie und Physiologie*, Supplement-Band (1879): 11–47.

Külpe, Oswald. *Introduction to Philosophy*. Translated by W. B. Pillsbury and E. B. Titchener. London: Swan Sonnenschein; New York: Macmillan, 1897. German edition, 1895.

Ladd, George Trumbull. "Is Psychology a Science?" *Psychological Review* 1 (July 1894): 392–95.

———. *Outlines of Descriptive Psychology*. New York: Scribner's Sons, 1898.

———. "President's Address before the New York Meeting of the American Psychological Association." *Psychological Review* 1 (1894): 1–21.

———. "Psychology as So-Called 'Natural Science.'" *Philosophical Review* 1 (1892): 24–53.

Lange, F. A. *Geschichte des Materialismus*. Iserlohn: J. Baedeker, 1873.

Laycock, Thomas. "On the Reflex Function of the Brain." *British and Foreign Medical Review* 19 (1845): 298–311.

Leighton, J. A. "Pragmatism." *Journal of Philosophy, Psychology, and Scientific Methods* 1 (1904): 148–56.

Lewes, George Henry. *The Biographical History of Philosophy from Its Origin in Greece down to the Present Day*. Enlarged edition. London: John W. Parker and Son, 1857.

———. *History of Philosophy from Thales to Comte*. 2 vols. 4th ed. London: Longmans, Green, 1867–1871.

———. *Problems of Life and Mind*. 2 vols. London: Trübner, 1877.

Lipps, T. "Die Aufgabe der Erkenntnistheorie und die Wundt'sche Logik." *Philosophische Monatshefte* 16 (Leipzig, 1880): 529–39.

Lloyd, Henry Demarest. "Is Personal Development the Best Social Policy?" (1902). Reprinted in H. D. Lloyd, *Mazzini and Other Essays*, 190–200. New York: G. P. Putnam's Sons; Knickerbrocker Press, 1910.

———. *Wealth against Commonwealth*. New York: Harper & Brothers, 1894.

Lotze, Hermann. *Medicinische Psychologie, order Physiologie der Seele*. Leipzig: Weidmann, 1852.

———. *Microcosmus: An Essay Concerning Man and His Relation to the World*. 2 vols. Translated by Elizabeth Hamilton and E. E. Constance Jones. New York: Scribner & Welford, 1886.

Lovejoy, Arthur O. "The Thirteen Pragmatisms." *Journal of Philosophy, Psychology, and Scientific Methods* 5, no. 1 (1908): 5–12 (pt. 1); and 5, no. 2 (1908): 29–39 (pt. 2). Reprinted in Lovejoy, *Thirteen Pragmatisms*, 1–29.

———. *The Thirteen Pragmatisms and Other Essays*. Baltimore, Md.: Johns Hopkins University Press, 1963.

Lutoslawski, Vincenti. "Un peuple individualiste." *Bibliothèque Universelle* 68 (1895): 575–92.

Mackenzie, Norman, ed. *The Letters of Sidney and Beatrice Webb*. Vol. 2. Cambridge: Cambridge University Press, 1978.

Mantegazza, Paolo. *Fisiologia del piacere*. 1854. Reprint, Milano: tipografia Bernardoni, 1890.

———. *Physiologie de la douleur*. Paris: Librairie illustrée, 1888.

Marey, E.-J. "Mesure à prendre pour l'uniformisation des méthodes et le contrôle des instruments employés en Physiologie." *Comptes rendus des séances de l'Académie des Sciences* 127 (1898): 375–81.

———. *La méthode graphique dans les sciences expérimentales et principalement en physiologie et en medicine*. Paris: Masson, 1878.

Marshall, Henry Rutgers. *Aesthetic Principles*. 1895. 3rd ed., New York: Macmillan, 1901.

———. *Pain, Pleasure, and Aesthetics*. London: Macmillan, 1894.

Maudsley, Henry. "The Double Brain." *Mind* 14 (1889): 161–87.

———. *The Physiology and Pathology of the Mind*. New York: Appleton, 1876.

"Meeting of the New York Society for Neurology and Electrology, Jan. 18, 1875." *Medical Record* 10 (1875): 132.

Mill, John Stuart. *Auguste Comte and Positivism*. London: Kegan Paul, 1891.

————. *On Liberty*. New York: Norton, 1975.

Miller, Dickinson S. "The Conditions of Greatest Progress in American Philosophy." *Journal of Philosophy, Psychology, and Scientific Methods* 3 (1906): 72.

————. "The Conditions of Greatest Progress in American Philosophy." In "Proceedings of the American Philosophical Association." *Philosophical Review* 15 (1906): 160.

Minutes of the Harvard Natural History Society. MS. Harvard University Archives, HUD.

Mivart, St. George. *Groundwork of Science: The Study of Epistemology*. New York: G. P. Putnam's Sons, 1898.

Montague, W. P. "May a Realist Be a Pragmatist?" *Journal of Philosophy, Psychology, and Scientific Methods* 6 (1909): pt. 1, "The Two Doctrines Defined," 460–63; pt. 2, "The Implications of Instrumentalism," 485–90; pt. 3, "The Implications of Psychological Pragmatism," 543–48; pt. 4, "The Implications of Humanism and of the Pragmatic Criterion," 561–71.

Moore, Addison Webster. "Pragmatism and Its Critics." *Philosophical Review* 14 (1905): 322–43.

————. *Pragmatism and Its Critics*. Chicago: University of Chicago Press, 1910.

Moore, G. E. *Philosophical Papers*. London: Allen and Unwin, 1959.

————. *Philosophical Studies*. London: Routledge and Kegan Paul, 1922.

————. "Professor James's 'Pragmatism'" (1907–08). Reprinted as "William James's 'Pragmatism,'" in Moore, *Philosophical Studies*, 97–146.

Morselli, E. *Psicologia e spiritismo: Impressioni e note critiche sui fenomeni di Eusapia Paladino*. Torino: Fratelli Bocca editori, 1908.

Münsterberg, Hugo. "The Affiliation of Psychology with Philosophy and the Natural Sciences." In "Proceedings of the American Philosophical Association: The Fifth Annual Meeting, Emerson Hall, Harvard University, Cambridge, Mass., Dec. 27–29, 1905." *Philosophical Review* 15 (1906): 175.

————. *American Traits from the Point of View of a German*. Boston: Houghton, Mifflin, 1901.

————. "The Congress of Arts and Sciences in St. Louis." MS acc. 2479. Münsterberg Papers, Boston Public Library.

————. "Emerson Hall." *Harvard Psychological Studies* 2 (1906): 2–39.

————. *The Eternal Values*. Boston: Houghton, Mifflin, 1909.

————. *Grundzüge der Psychologie*. Leipzig: Verlag von Johann Ambrosius Barth, 1900.

————. "The International Congress of Arts and Science." *Science*, n.s., 18 (1903): 559–63.

————. "The Need of Emerson Hall." Letter addressed to the Visiting Committee of the Overseers of Harvard University, March 20, 1901. Münsterberg Papers, Boston Public Library, RB XH.901.M94T.

————. "The Position of Psychology in the System of Knowledge." *Harvard Psychological Studies* 1 (1903): 641–54.

————. *Psychology and Life*. Boston: Houghton, Mifflin, 1899.

————. "Psychology and Mysticism." *Atlantic Monthly* 83 (1898): 67–85.

————. "Science and Education." *Educational Review* 16 (1898): 105–32.

————. *Science and Idealism*. Boston: Houghton, Mifflin, 1906.

————. "The Scientific Plan of the Congress." In *Congress of Arts and Science, Universal Exposition, St. Louis, 1904*, edited by Howard J. Rogers, 1:85–134. Boston: Houghton, Mifflin, 1905.

————. "The St. Louis Congress of Arts and Sciences." *Atlantic Monthly* 91 (1903): 671–84.

Münsterberg, Margaret. *Hugo Münsterberg: His Life and Work*. New York: Appleton, 1922.

*Museum of Comparative Zoology Bulletin* 3 (1871–76). Harvard University Archives, Pusey Library, Harvard University.

Myers, F. W. H. "Automatic Writing, II." *Proceedings of the Society for Psychical Research* 3 (1885): 1–63.

————. "Human Personality." *Proceedings of the Society for Psychical Research* 4 (1886–87): 1–24.

————. *Human Personality and Its Survival of Bodily Death.* 2 vols. New York: Longmans, Green, 1903.

Ostwald, Wilhelm. "Das System der Wissenschaften." *Annalen der Naturphilosophie* 8 (1909): 266–72.

————. *Individuality and Immortality.* Boston: Houghton, Mifflin, 1906.

————. "On the Theory of Science." In *Congress of Arts and Science. Universal Exposition, St. Louis, 1904,* edited by Howard J. Rogers, 1:333–52. Boston: Houghton, Mifflin, 1905.

Papini, Giovanni [pseud. Gian Falco]. "Morte e resurrezione della filosofia." *Leonardo* 1, no. 11 (December 20, 1903): 2.

————. *Sul pragmatismo: Saggi e ricerche, 1903–1911.* Milan: Libreria Editrice Milanese, 1913.

Paulhan, Frédéric. "L'invention." *Revue Philosophique* 45 (1898): 225–58.

Paulsen, Friedrich. *An Autobiography.* Translated and edited by Theodor Lorenz. New York: Columbia University Press, 1938.

————. *German Education: Past and Present.* Translated by Theodor Lorenz. New York: Scribner's Sons, 1908.

————. *German Universities: Their Character and Historical Development.* Translated by E. D. Perry. New York: Macmillan, 1895.

————. *Introduction to Philosophy.* Translated by Frank Thilly. With a preface by William James. New York: Henry Holt, 1895.

————. "Über das Verhältnis der Philosophie zum Wissenschaft." *Vierteljahrsschrift für Wissenschaftliche Philosophie* 1 (1876): 15–50.

Pearson, Karl. *Grammar of Science.* 1892. Reprint, Bristol: Toemmes, 1991.

Peirce, Charles S. "The Architectonic Character of Philosophy" (ca. 1896). In Peirce, *Collected Papers,* 1:176–79.

————. "Backward State of Metaphysics" (1898). In Peirce, *Collected Papers,* 6:1–5.

————. *Collected Papers of Charles Sanders Peirce.* Edited by Charles Hartshorne and Paul Weiss. 8 vols. Cambridge, Mass.: Belknap Press of Harvard University Press, 1960–66.

————. "Criticism of 'Phantasms of the Living.'" *Proceedings of the American Society for Psychical Research* 1 (1885–89): 150–57.

————. "A Detailed Classification of the Sciences" (1902). In Pierce, *Collected Papers,* 1:203–83.

————. "The Ethos of Terminology" (1903). In Pierce, *Collected Papers,* 2:219–26.

————. "The Fixation of Belief." *Popular Science Monthly* 12 (1877): 1–15.

————. "Grounds of Validity" (1869). In Peirce, *Collected Papers,* 5:318–57.

————. "How to Make Our Ideas Clear" (1878). In Peirce, *Collected Papers,* 5:388–410.

————. "Mr. Peirce's Rejoinder." *Proceedings of the American Society for Psychical Research* 1 (1885–89): 180–215.

————. "Notes on Scientific Philosophy" (1905). In Peirce, *Collected Papers,* 1:126–75.

————. Preface to Peirce, *Collected Papers,* vol. 1 (ca. 1897). In Peirce, *Collected Papers,* 1:2–14.

————. Preface to Peirce, *Collected Papers,* vol. 5 (1902). In Peirce, *Collected Papers,* 5:1–13.

————. "Questions Concerning Certain Faculties Claimed for Man." *Journal of Speculative Philosophy* 2 (1868): 103–14.

————. "Scientific Method" (1902). In Peirce, *Collected Papers,* 7:49–138.

————. "Some Consequences of Four Incapacities" (1868). In Peirce, *Collected Papers,* 5:264–317.

————. "Vitally Important Topics" (1898). In Peirce, *Collected Papers,* 1:616–76.

————. "What Pragmatism Is" (1905). In Peirce, *Collected Papers,* 5:411–37.

————. *Writings of Charles S. Peirce: A Chronological Edition.* 6 vols. Editorial director, Christian J. W. Kloesel. Bloomington: Indiana University Press, 1982–2000.

Perry, Ralph Barton. "A Review of Pragmatism as a Philosophical Generalization." *Journal of Philosophy, Psychology, and Scientific Methods* 4 (1907): 420–28.

―――. "A Review of Pragmatism as a Theory of Knowledge." *Journal of Philosophy, Psychology, and Scientific Methods* 4 (1907): 365–74.

―――. *The Thought and Character of William James.* 2 vols. Boston: Little, Brown, 1935.

―――. *The Thought and Character of William James. Briefer Version.* Cambridge: Harvard University Press, 1948.

Poffenberger, A. T., ed. *James McKeen Cattell: Man of Science.* 2 vols. Lancaster, Pa.: Science Press, 1947.

Porter, Noah. *The Human Intellect, with an Introduction upon Psychology and the Soul.* 4th ed. New York: Scribner, 1899.

―――. "Professor Tyndall's Last Deliverance" (*New Englander*, 1878). In Porter, *Science and Sentiment*, 192–221.

―――. *Science and Sentiment.* New York: Scribner, 1882.

―――. "The Sciences of Nature versus the Science of Man." Address delivered before the Phi Beta Kappa Societies at Harvard and Trinity Colleges, June and July 1871. Reprinted in Porter, *Science and Sentiment*, 38–76.

Pratt, James Bisset. "Truth and Its Verification." *Journal of Philosophy, Psychology, and Scientific Methods* 4 (1907): 320–24.

―――. *What Is Pragmatism?* New York: Macmillan, 1909.

Prichard, Harold Arthur. "A Criticism of the Psychologists' Treatment of Knowledge." *Mind*, n.s., 16 (1907): 27–53.

Prince, Morton. *The Dissociation of a Personality: A Biographical Study in Abnormal Psychology.* New York: Longmans, Green, 1906.

―――. *The Nature of Mind and Human Automatism.* Philadelphia: J. B. Lippincott, 1885.

"Proceedings of the American Philosophical Association: The Fifth Annual Meeting, Emerson Hall, Harvard University, Cambridge Mass., Dec. 27–29, 1905." *Philosophical Review* 15 (1906): 157–81.

"Proceedings of the 1907 Meeting of the American Philosophical Association." *Philosophical Review* 17 (1908): 167–90.

Puffer, Ethel Dench. "The Loss of Personality." *Atlantic Monthly* 85 (1900): 185–204.

Putnam, J. J. "Contributions to the Physiology of the Cortex Cerebri." *Boston Medical and Surgical Journal* 91 (July 16, 1874): 49–52.

Ribot, Théodule. *Creative Imagination.* Chicago: Open Court, 1906.

―――. *The Diseases of Personality.* Chicago: Open Court, 1891.

―――. *English Psychology.* New York: Appleton, 1874.

―――. "Philosophy in France." *Mind* 2 (1877): 366–86.

Rickert, Heinrich. *Der Gegenstand der Erkenntnis: Einführung in die tranzendentale Philosophie.* 1892. 2nd ed., revised and expanded, Tübingen: Mohr, 1904.

―――. *Die Grenzen der naturwissenschaftlichen Begriffsbildung.* 1902. 2nd rev. ed., Tübingen: Mohr: 1913.

―――. *Kulturwissenschaft und Naturwissenschaft.* Freiburg: Mohr, 1899.

―――. *Zwei Wegen der Erkenntnistheorie.* Halle: C. A. Kraemmer, 1909.

Ritchie, David G. "The Relation of Logic to Psychology, I and II." *Philosophical Review* 5 (1896): 585–600, and 6 (1897): 1–17.

Robertson, George Croom. "Prefatory Words." *Mind* 1 (1876): 1–6.

Rogers, A. K. "Professor James' Theory of Knowledge." *Philosophical Review* 15 (1906): 577–96.

Royce, Josiah. "The Eternal and the Practical." Presidential address before the American Philosophical Association, third annual meeting, Princeton, December 30, 1903. *Philosophical Review* 13 (March 1904): 113–42.

―――. "'Mind-Stuff' and Reality." *Mind* 6 (1881): 365–77.

―――. *Philosophy of Loyalty.* 1908. New York: Macmillan, 1924.

————. *Race Questions, Provincialism, and Other American Problems.* New York: Macmillan, 1908.

————. "Self-Consciousness, Social Conciousness and Nature." In Royce, *Studies of Good and Evil*, 198–248.

————. "Some Observations on the Anomalies of Self-Consciousness" (1894). Reprinted in Royce, *Studies of Good and Evil*, chap. 7.

————. *Studies of Good and Evil: A Series of Essays upon Problems of Philosophy and Life.* New York: Appleton, 1898.

————. *William James and Other Essays on the Philosophy of Life.* New York: Macmillan, 1911.

————. *The World and the Individual.* 2 vols. Gifford Lectures delivered before the University of Aberdeen. 2nd series, "Nature, Man, and the Moral Order." New York: Macmillan, 1901.

Russell, Bertrand. *Philosophical Essays.* London: Longmans, Green, 1910.

————. "Pragmatism" (1909). Reprinted in Russell, *Philosophical Essays*, 104.

————. "Transatlantic Truth." *Albany Review* (January 1908): 393–410.

Russell, J. E. "Pragmatism as the Salvation from Philosophic Doubt." *Journal of Philosophy* 4 (1907): 47–64.

————. "Some Difficulties with the Epistemology of Pragmatism and Radical Empiricism." *Philosophical Review* 14 (1906): 406–13.

Sabine, George H. "Discussion: Radical Empiricism as a Logical Method" (1905). Reprinted in *Pure Experience: The Response to William James*, edited by E. I. Taylor and. Robert H. Wozniak, 79–89. Bristol, UK: Thoemmes Press, 1996.

————. "Review of J. M. Baldwin, 'The Limits of Pragmatism.'" *Philosophical Review* 13 (1904): 236.

Santayana, George. "William James." In Simon, *William James Remembered*, 89–105. Also published in Santayana, *Character and Opinion in the United States* (New York: Doubleday Anchor, 1920).

Schiller, F. C. S. "Can Logic Abstract from the Psychological Conditions of Thinking?" Symposium by F. C. S. Schiller, B. Bosanquet, and H. Rashdall. *Proceedings of the Aristotelian Society* 6 (1906): 224–37.

————. *Formal Logic: A Scientific and Social Problem.* New York: Macmillan, 1912.

————. *Humanism: Philosophical Essays.* London: Macmillan, 1903.

————. "Idealism and the Dissociation of Personality." *Journal of Philosophy, Psychology, and Scientific Methods* 3 (1906): 477–82.

————. "Logic or Psychology?" *Mind*, n.s., 18 (1909): 400–406.

————. "Psychology and Knowledge." *Mind*, n.s., 16 (1907): 244–48.

————. "Psychology and Psychical Research: A Reply to Professor Münsterberg." *Proceedings of the Society for Psychical Research* 14 (June 1899): 348–65.

————. *Studies in Humanism.* London: Macmillan, 1907.

————. "William James." In Simon, *William James Remembered*, 139–52.

Schinz, Albert. *Anti-Pragmatism: An Examination into the Respective Rights of Intellectual Aristocracy and Social Democracy.* Boston: Small, Maynard, 1909.

Schmidt, George P. "The Old Time College President." PhD diss., Columbia University, 1930.

Schurman, Jacob Gould. "Prefatory Note." *Philosophical Review* 1, no. 1 (1892).

Scripture, Edward W. "Accurate Work in Psychology." *American Journal of Psychology* 6 (June 1894): 427–30.

————. *Thinking, Feeling, Doing.* New York: Chautauqua Century Press, 1895.

Sidgwick, Arthur, and Eleanor M. Sidgwick. *Henry Sidgwick: A Memoir.* London: Macmillan, 1906.

Sidgwick, Henry. "Is Philosophy the Germ or the Crown of Science?" (n.d.). Sidgwick Papers, Trinity College, Wren Library, Cambridge.

————. "Philosophy at Cambridge." *Mind* 1 (1876): 235–46.

———. *Philosophy, Its Scope and Relations: An Introductory Course of Lectures.* London: Macmillan, 1902.

———. "President's Address," Society for Psychical Research, first general meeting, July 17, 1882. *Proceedings of the Society for Psychical Research* 1 (1882): 7–12.

———. "Review of *The History of Philosophy from Thales to Comte,* by G. H. Lewes (4th. ed.)" (1871). Reprinted in Henry Sidgwick, *Miscellaneous Essays 1870–1899,* edited by Eleanor Mildred Sidgwick and Arthur Sidgwick. New York: Macmillan, 1904.

Sidgwick, Henry, Mrs. H. Sidgwick, and G. A. Smith. "Experiments in Thought-Transference." *Proceedings of the Society for Psychical Research* 6 (1889–90): 128–70.

Sigwart, Christoph. *Handbuch zu Vorlesungen über die Logik.* Tübingen: C. F. Oslander, 1835.

———. *Logic.* Vol. 1. Translated by Helen Dendy. London: Swan Sonnenschein, 1895.

———. *Logik.* 2 vols. 3rd ed. Tübingen: J. C. B. Mohr, 1904.

———. "Logische Fragen." *Vierteiljahresschrift für wissenschaftliche Philosophie* 4 (1880): 454–55.

Simon, Linda, ed. *William James Remembered.* Lincoln: University of Nebraska Press, 1996.

Small, Albion. "The Era of Sociology." *American Journal of Sociology* 1 (1895): 1–27.

———. "The Relation between Sociology and the Other Social Sciences." *American Journal of Sociology* 12 (1906): 11–31.

———. "The Scope of Sociology." *American Journal of Sociology* 5 (1899–1900): 506–26.

Smith, Hannah Whitall. *The Christian's Secret of a Happy Life.* 1875. Chicago: Fleming H. Revell, 1883.

———. *The Unselfishness of God and How I Discovered It: A Spiritual Autobiography.* New York: Fleming H. Revell, 1883.

"The Society for Psychical Research: Objects of the Society." *Proceedings of the Society for Psychical Research* 1 (1882–83): 3–6.

Spencer, Herbert. *An Autobiography.* Vols. 20–21 of *The Works of Herbert Spencer.* Osnabrück: Otto Zeller, 1966.

———. "Classification of the Sciences" (1864). In Spencer, *Essays: Scientific, Political, and Speculative,* 2:74–117.

———. *Data of Ethics.* New York: Appleton, 1879.

———. *Essays: Scientific, Political, and Speculative.* 3 vols. New York: D. Appleton, 1892.

———. *First Principles of a New System of Philosophy.* Vol. 1 of *The Works of Herbert Spencer.* Osnabrück: Otto Zeller, 1966.

———. "The Genesis of Science" (1854). In Spencer, *Essays: Scientific, Political, and Speculative,* 2:1–73.

———. *The Principles of Psychology.* Vols. 4–5 of *The Works of Herbert Spencer.* Osnabrück: Otto Zeller, 1966.

———. "What Knowledge Is of Most Worth" (1859). In Spencer, *Education: Intellectual, Moral, and Physical,* 1–87. Vol. 16 of *The Works of Herbert Spencer.* Osnabrück: Otto Zeller, 1966.

Stanley, Hiram M. "On the Classification of the Sciences." *Mind* 9 (1884): 265–74.

———. "Relation of Feeling to Pleasure and Pain." *Mind* 14 (1889): 537–44.

Stephen, J. Fitzjames. *Digest of the Law of Evidence.* 4th ed. London: Macmillan, 1893.

Stephen, Leslie. *A Life of Sir James Fitzjames Stephen, a Judge of the High Court of Justice, by His Brother.* New York: G. P. Putnam, 1895.

Stewart, B. "Mr. Crookes on the 'Psychic' Force." *Nature* (July 27, 1871): 237.

Stratton, George. "The Relation between Psychology and Logic." *Psychological Review* 3 (1896): 313–20.

Strong, C. A. "A Naturalistic Theory of the Reference of Thought to Reality." *Journal of Philosophy* 1 (1904): 253–60.

———. "Pragmatism and Its Definition of Truth." *Journal of Philosophy* 5 (1908): 256–64.

Stumpf, Carl von. "Antrittsrede." *Sitzungsberichte der Preussichen Akademie der Wissenschaften* (1895): 736–37.

———. "Die Wiedergeburt der Philosophie" (1908). In Stumpf, *Philosophische Reden und Vorträge*, 161–96. Leipzig: Barth, 1910.

———. "Psychologie und Erkenntnistheorie." *Abhandlungen der philosophischen Klasse der königlich Bayerischen Akademie der Wissenschaften*, I cl., vol. 19 (1892): 464–516.

———. *William James nach Seinen Briefen. Leben. Charakter. Lehre.* Berlin: Pan-Verlag Rolf Heise, 1928.

Sully, James. *Outlines of Psychology, with Special Reference to the Theory of Education.* 1884. Reprint, New York: Appleton, 1910.

———. "Review of William James's *Principles of Psychology*." *Mind* 16 (July 1891): 393–404.

Taylor, A. E. "Philosophy." In *Recent Developments in European Thought*, edited by F. S. Marvin., 25–63. London: Oxford University Press, 1920.

———. "The Place of Psychology in the Classification of the Sciences." *Philosophical Review* 15 (1906): 380–86.

———. *Plato.* London: Archibald Constable, 1908.

———. "Some Side Lights of Pragmatism." *McGill University Magazine* 3 (1903): 44–66.

———. "Truth and Consequences." *Mind* 15 (1906): 81–93.

———. "Truth and Practice." *Philosophical Review* 14 (1905): 264–89.

Thilly, Frank. "Psychology, Natural Science, and Philosophy." *Philosophical Review* 15 (1906): 130–44.

Titchener, Edward B. *A Beginner's Psychology.* 2nd ed. London: Macmillan, 1918.

———. "Discussion and Correspondence: Applied Psychology." *Science* 49 (1919): 169–70.

———. "The Feeling of 'Being Stared At.'" *Science* 8 (1898): 897.

———. "On 'Psychology as the Behaviorist Views It.'" *Proceedings of the American Philosophical Society* 53 (1914): 1–17.

———. "The Postulates of a Structural Psychology." *Philosophical Review* 7 no. 5 (September 1898): 449–65.

———. "Psychology: Science or Technology?" *Popular Science Monthly* 84 (1914): 39–51.

———. Reply to James. *Science*, May 12, 1899. Reprinted in James, *Essays in Psychical Research*, 176.

———. *Systematic Psychology: Prolegomena.* With a foreword by R. B. Evans and R. B. MacLeod. Ithaca: Cornell University Press, 1972.

Trine, Ralph Waldo. *In the Fire of the Heart.* New York: McClure, Phillips, 1906.

———. *In Tune with the Infinite: or Fullness of Peace, Power, and Plenty.* New York: Crowell, ca. 1897.

———. *On the Open Road: Being Some Thoughts and a Little Creed of Wholesome Living.* New York: Crowell, ca. 1908.

———. *What All the World's A-Seeking.* New York: T. Y. Crowell, 1896.

Tyndall, John. "An Address to Students" (1868–69). In Tyndall, *Fragments of Science*, 1:91–101. New York: D. Appleton, 1897.

Vivekânanda, Swâmi. *Yoga Philosophy: Lectures Delivered in New York, Winter of 1895–6.* New York: Longmans, Green, 1896.

Ward, James. "The Progress of Philosophy." *Mind* 15 (1890): 213–33.

———. "Psychological Principles." *Mind* 8 (1883): 465–86.

———. "Psychology." In *Encyclopedia Britannica*, 11th ed. (1910), 547–98. Article first published in 1886 edition.

Ward, Lester. "The Place of Sociology among the Sciences." *American Journal of Sociology* 1 (1895–96): 16–27.

Washburn, M. F. "Margaret Floy Washburn." In *A History of Psychology in Autobiography*, edited by A. Murchison., 2:340, 344. Worcester, Mass.: Clark University Press, 1932.

Wells, H. G. *First and Last Things: A Confession of Faith and a Rule of Life*. New York: G. P. Putnam's Sons, 1908.

———. *A Modern Utopia*. London: Thomas Nelson and Sons, 1905.

Whittaker, Thomas. "A Compendious Classification of the Sciences." *Mind*, n.s., 12 (1903): 21–34.

Windelband, Wilhelm. *Die Philosophie im deutschen Geistesleben des XIX. Jahrunderts. Funf Vorlesungen*. Tübingen: Mohr, 1909.

———. *Präluden: Aufsätze und Reden zur Einleitung in die Philosophie*. Freiburg: Mohr, 1884.

Woodbridge, Frederick J. E. "The Field of Logic." *Science*, n.s., 20, no. 18 (1904): 587–600.

———. *Nature and Mind*. New York: Columbia University Press, 1937.

———. "Notes." *Journal of Philosophy, Psychology, and Scientific Methods* 1 (1904): 27.

———. "The Problem of Metaphysics." *Philosophical Review* 12 (1903): 367–85.

Woodward, William R. "James's Evolutionary Epistemology: 'Necessary Truths and the Effects of Experience.'" In *Reinterpreting the Legacy of William James*, edited by M. Donnelly., 153–69. Washington, D.C.: American Psychological Association, 1992.

Wundt, Wilhelm. "Philosophy in Germany." *Mind* 2 (1877): 493–518.

———. *System der Philosophie*. Leipzig: W. Engelmann, 1889.

———. "Über die Einteilung der Wissenschaften." *Philosophische Studien* 5 (1899): 1–55.

———. "Über empirische und metaphysische Psychologie: Eine kritische Betrachtung." *Archiv für die gesamte Psychologie* 2 (1904): 333–61.

Youmans, Edward L., ed. *The Culture Demanded by Modern Life: A Series of Addresses and Arguments on the Claims of Scientific Education*. 1867. Reprint, New York: Appleton, 1897.

———. "Introduction—On Mental Discipline in Education." In Youmans, *Culture Demanded by Modern Life*, 1–56.

———. "Observations on the Scientific Study of Human Nature: A Lecture Delivered before the London College of Preceptors." In Youmans, *Culture Demanded by Modern Life*, 371–414.

———. "The Progress of Rational Education." *Popular Science Monthly* (1897): 111–13.

## SECONDARY SOURCES

Abbagnano, Nicola. *Storia della filosofia*. Torino: Utet, 1950.

Abbott, Andrew. *Chaos of Disciplines*. Chicago: University of Chicago Press, 2001.

———. *Department and Discipline: Chicago Sociology at One Hundred*. Chicago: University of Chicago Press, 1999.

———. "The Emergence of American Psychiatry, 1880–1930." PhD diss., University of Chicago, 1982.

———. *The System of Professions: An Essay on the Division of Expert Labor*. Chicago: University of Chicago Press, 1988.

———. "Things of Boundaries." *Social Research* 62 (1995): 857–82.

Allen, Gay Wilson. *William James: A Biography*. New York: Viking Press, 1967.

Anderson, Douglas R. "Philosophy as Teaching: James's 'Knight Errant,' Thomas Davidson." *Journal of Speculative Philosophy* 18 (2004): 239–47.

Annan, Noel. *Leslie Stephen: The Godless Victorian*. London: Random House, 1984.

Antliff, Mark. *Inventing Bergson*. Princeton: Princeton University Press, 1993.

Appel, Toby A. "A Scientific Career in the Age of Character: Jeffries Wyman and Natural History at Harvard." In *Science at Harvard University: Historical Perspectives*, edited by Clark A. Elliott and Margaret W. Rossiter, 96–120. Bethlehem, Pa.: Lehigh University Press, 1992.

Ash, Mitchell G. "Academic Politics in the History of Science: Experimental Psychology in Germany, 1879–1941." *Central European History* 13 (1980): 255–86.

——. *Gestalt Psychology in German Culture, 1890–1967: Holism and the Quest for Objectivity.* Cambridge: Cambridge University Press, 1995.

——. "The Self-Presentation of a Discipline: History of Psychology in the United States between Pedagogy and Scholarship." In *Functions and Uses of Disciplinary Histories,* edited by Loren Graham, Wolf Lepenies, and Peter Weingart, 143–89. Dordrecht: D. Reidel, 1983.

——. "Wilhelm Wundt and Oswald Kuelpe on the Institutional Status of Psychology: Academic Controversy in Historical Context." In Bringmann and Tweney, *Wundt Studies,* 396–432.

Ash, Mitchell G., and William R. Woodward, eds. *Psychology in Twentieth Century Thought and Society.* Cambridge: Cambridge University Press, 1987.

Ayer, A. J. *The Origins of Pragmatism: Studies in the Philosophy of Charles Sanders Peirce and William James.* London: Macmillan, 1968.

Baker, G. P., and P. M. S Hacker. "Frege's Anti-Psychologism." In *Perspectives on Psychologism,* edited by Mark A. Notturno. Leiden: E. J. Brill, 1989.

Bannan, John F. "Emotions and Biology: Remarks on the Contemporary Trend." *Review of Metaphysics* 58 (2004).

Bannister, Robert C. *Social Darwinism: Science and Myth in Anglo-American Social Thought.* Philadelphia: Temple University Press, 1979.

Barberis, Daniela S. "The First *Année sociologique* and Neo-Kantian Philosophy in France." PhD diss., University of Chicago, 2001.

Barua, Dilip Kumar. *Edward Carpenter, 1844–1929: An Apostle of Freedom.* Burdwan, India: University of Burdwan, 1991.

Barzun, Jacques. *A Stroll with William James.* New York: Harper and Row, 1983.

——. "William James: The Mind as Artist." In *A Century of Psychology as Science,* edited by Sigmund Koch and David E. Leary, 904–10. New York: McGraw-Hill, 1985.

Becher, Tony. *Academic Tribes and Territories: Intellectual Enquiry and the Cultures of Disciplines.* Buckingham: Society for Research into Higher Education and Open University Press, 1989.

Bechtel, William, ed. *Integrating Scientific Disciplines.* Dordrecht: Martinus Nijhoff, 1986.

Beisner, Robert L. *Twelve against Empire: The Anti-Imperialists, 1898–1900.* New York: McGraw-Hill, 1968.

Ben-David, J., and R. Collins. "Social Factors in the Origins of a New Science: The Case of Psychology." *American Sociological Review* 31 (1966): 451–65.

Benschop, Ruth G., and Douwe Draaisma. "In Pursuit of Precision: The Calibration of Minds and Machines in Late Nineteenth-Century Psychology." *Annals of Science* 57 (2000): 1–25.

Bercovitch, Sacvan. "The Rites of Assent: Rhetoric, Ritual, and the Ideology of American Consensus." In *The American Self: Myth, Ideology, and Popular Culture,* edited by S. B. Girgus, 5–42. Albuquerque: University of New Mexico Press, 1981.

Bird, Graham. *William James: The Arguments of the Philosophers.* New York: Routledge and Kegan Paul, 1986.

Bjork, Daniel W. *The Compromised Scientist: William James in the Development of American Psychology.* New York: Columbia University Press, 1983.

——. *William James: The Center of His Vision.* New York: Columbia University Press, 1988.

Bledstein, Burton J. *The Culture of Professionalism: The Middle Class and the Development of Higher Education in America.* New York: W. W. Norton, 1976.

Blondel, Christine. "Eusapia Palladino: La méthode expérimentale et la 'diva des savants.'" In *Des Savantes face à l'occulte, 1870–1940,* edited by Bernadette Bensaude-Vincent and Christine Blondel, 143–71. Paris: Éditions La Découverte, 2002.

Boller, Paul F., Jr. *American Thought in Transition: The Impact of Evolutionary Naturalism, 1865–1900.* Boston: University Press of America, 1981.

Bordogna, Francesca. "A Geography of Evidence: The Case of Psychical Research from the 1870s to the Early 1900s." Paper presented at the conference "Test, Proof, and Demonstration," Max Planck Institute for the History of Science, Berlin, July 1998.

———. "L'Hotel pragmatista: Viaggi, scienza, e filosofia." In *Studi sul 900 toscano offerti a Giorgio Luti*, edited by E. Ghidetti and A. Nozzoli, 1–26. Firenze: Università di Firenze, 2006.

———. "Local Internationalism: A Turn-of-the-Twentieth-Century Pragmatist Network." Paper presented at the History of Science Society Meeting, Minneapolis, 2005.

———. "Physiology and Psychology of Temperament." *Journal of the History of the Behavioral Sciences* 37 (2001): 3–25.

———. "Pragmatism and Objectivity." In "The Scientific Contexts of William James's Pragmatist Epistemology," 235–60. PhD diss., University of Chicago, 1998.

———. "Pragmatism in Context: Physiology and Psychology of Temperament." *Journal of the History of the Behavioral Sciences* 37 (2001): 3–26.

———. "Pragmatismo e oggettività in William James." *Rivista di filosofia* 86 (December 1995): 387–412.

———. "The Scientific Contexts of William James's Pragmatist Epistemology." PhD diss., University of Chicago, 1998.

———. "Scientific Personae in American Psychology: Three Case Studies." *Studies in the History and Philosophy of the Biological and Biomedical Sciences* (2005): 95–134.

Bourdieu, Pierre. "The Genesis of the Concepts of *Habitus* and of Field." *Sociocriticism* 2 (1985): 11–24.

———. *Homo academicus*. Paris: Editions de Minuit, 1984.

———. "Intellectual Field and Creative Project." *Social Science Information* 8 (1969): 89–119.

———. *Outline of a Theory of Practice*. Cambridge: Cambridge University Press, 1977.

Bourguet, Marie-Noëlle, Christian Licoppe, and H. Otto Sibum. *Instruments, Travel, and Science: Itineraries of Precision from the Seventeenth to the Twentieth Century*. London: Routledge, 2002.

Brain, Robert. "The Graphic Method: Inscription, Visualization, and Measurement in Nineteenth-Century Science and Culture." PhD diss., University of California, Los Angeles, 1995.

Braun, Marta. *Picturing Time: The Work of Etienne-Jules Marey*. Chicago: University of Chicago Press, 1992.

Bringmann, Wolfgang G., and Ryan D. Tweney, ed. *Wundt Studies: A Centennial Collection*. Toronto: C. J. Hogrefe, 1980.

Brooks, John I. *The Eclectic Legacy: Academic Philosophy and the Human Sciences in Nineteenth-Century France*. Newark: University of Delaware Press, 1998.

Brown, Alan. *The Metaphysical Society: Victorian Minds in Crisis, 1869–1880*. New York: Octagon Books, 1973.

Brown, Edward M. "Neurology and Spiritualism in the 1870s." *Bulletin of the History of Medicine* 57 (1983): 563–77.

Browning, Don S. *Pluralism and Personality: William James and Some Contemporary Cultures of Psychology*. Lewisburg: Bucknell University Press, 1980.

———. "William James's Philosophy of the Person: The Concept of the Strenuous Life." *Zygon* 10 (June 1975): 162–74.

Bryson, Gladys. "The Emergence of the Social Sciences from Moral Philosophy." *International Journal of Ethics* 42 (1932): 304–23.

Buck, Paul H., ed. *Social Sciences at Harvard, 1860–1920*. Cambridge, Mass.: Harvard University Press, 1965.

Bunn, G. C., A. D. Lovie, and G. D. Richards, eds. *Psychology in Britain: Historical Essays and Personal Reflections*. Leicester: British Psychological Society, 2001.

Cahan, David, and M. Eugene Rudd. *Science at the American Frontier: A Biography of DeWitt Bristol Brace.* Lincoln: University of Nebraska Press, 2000.

Camfield, Thomas M. "The Professionalization of American Psychology, 1870–1917." *Journal of the History of Behavioral Sciences* 9 (1973): 66–75.

Caplan, Eric. *Mind Games: American Culture and the Birth of Psychotherapy.* Berkeley: University of California Press, 1998.

Capshew, James H. "Psychologists on Site: Reconnaissance of the Historiography of the Laboratory." *American Psychologist* 47 (February 1992): 132–42.

———. *Psychologists on the March: Science, Practice, and Professional Identity in America, 1929–1969.* Cambridge: Cambridge University Press, 1999.

Carl, Wolfgang. *Frege's Theory of Sense and Reference: Its Origins and Scope.* Cambridge: Cambridge University Press, 1994.

Carroy, J., and R. Plas. "The Origin of French Experimental Psychology." *History of the Human Sciences* 9 (1996): 73–84.

Carter, Everett. "Critical Introduction." In *A Hazard of New Fortunes,* by W. D. Howells. New York: Modern Library, 2002.

Chandler, James K., Arnold Ira Davidson, and Harry D. Harootunian, eds. *Questions of Evidence: Proof, Practice and Persuasion across the Disciplines.* Chicago: University of Chicago Press, 1994.

Clarke, Edwin, and L. S. Jacyna, eds. *Nineteenth-Century Origins of Neuroscientific Concepts.* Berkeley: University of California Press, 1987.

Clendenning, John. *The Life and Thought of Josiah Royce.* Nashville: Vanderbilt University Press, 1999.

Clifford, James. *Routes: Travel and Translation in the Late Twentieth Century.* Cambridge, Mass.: Harvard University Press, 1997.

Coady, C. A. J. *Testimony: Philosophical Study.* Oxford: Clarendon Press, 1992.

Colapietro, Vincent. *Peirce's Approach to the Self.* New York: SUNY Press, 1989.

Collini, Stefan. "The Ordinary Experience of Civilized Life: Sidgwick's Politics and the Method of Reflective Analysis." In *Essays on Henry Sidgwick,* edited by Bart Schultz, 333–67. Cambridge: Cambridge University Press, 2002.

———. *Public Moralists: Political Thought and Intellectual Life in Britain, 1850–1930.* Oxford: Clarendon Press, 1991.

Conant, James. "The James/Royce Dispute and the Development of James's 'Solution.'" In R. A. Putnam, *Cambridge Companion to William James,* 186–213.

Conn, Steven. *Museums and American Intellectual Life, 1876–1926.* Chicago: University of Chicago Press, 1998.

Connell, W. F. *The Educational Thought and Influence of Matthew Arnold.* London: Routledge and Kegan Paul, 1950.

Coon, Deborah J. "Courtship with Anarchy: The Socio-Political Foundations of William James's Pragmatism." PhD diss., Harvard University, 1988.

———. "One Moment in the World's Salvation: Anarchism and the Radicalization of William James." *Journal of American History* 83 (1996): 70–99.

———. "Salvaging the Self in a World without Soul: William James's *The Principles of Psychology*." *History of Psychology* 3 (2000): 81–183.

———. "Standardizing the Subject: Experimental Psychologists, Introspection, and the Quest for a Technoscientific Ideal." *Technology and Culture* 34 (October 1993): 757–83.

———. "Testing the Limits of Sense and Science: American Experimental Psychologists Combat Spiritualism, 1889–1920." *American Psychologist* 47, no. 2 (1992): 143–51.

Cooper, Wesley. *The Unity of William James's Thought.* Nashville: Vanderbilt University Press, 2002.

Cooter, Roger. *The Cultural Meaning of Popular Science: Phrenology and the Organization of Consent in Nineteenth-Century Britain*. Cambridge: Cambridge University Press, 1984.

Cotkin, George. *William James, Public Philosopher*. Baltimore: Johns Hopkins University Press, 1990.

Crary, Jonathan. *Suspensions of Perception: Attention, Spectacle, and Modern Culture*. Cambridge: MIT Press, 2000.

———. *Techniques of the Observer: On Vision and Modernity in the Nineteenth Century*. Cambridge, Mass.: MIT Press, 1990.

Cresswell, Tim. *In Place/Out of Place: Geography, Ideology, and Transgression*. Minneapolis: University of Minnesota Press, 1996.

Croce, Paul J. *Science and Religion in the Era of William James*. Vol. 1, *The Eclipse of Certainty, 1820–1880*. Chapel Hill: University of North Carolina Press, 1995.

Danziger, Kurt. *Constructing the Subject: Historical Origins of Psychological Research*. Cambridge: Cambridge University Press, 1990.

———. "History of Introspection Reconsidered." *Journal of the History of the Behavioral Sciences* 16 (1980): 241–62.

———. "History, Practice and Psychological Objects: Reply to Commentators." *Annals of Theoretical Psychology* 8 (1993): 71–84.

———. "Mid-Nineteenth-Century British Psycho-Physiology: A Neglected Chapter in the History of Psychology." In Woodward and Ash, *Problematic Science*, 119–46.

———. "The Positivist Repudiation of Wundt." *Journal of the History of Behavioral Sciences* 15 (1979): 205–30.

———. "The Unknown Wundt: Drive, Apperception, and Volition." In *Wilhelm Wundt in History: The Making of a Scientific Psychology*, edited by Robert W. Rieber and David K. Robinson, 95–120. New York: Kluwer Academic/Plenum, 2001.

———. "Wilhelm Wundt and the Emergence of Experimental Psychology." In *Companion to the History of Modern Science*, edited by R. C. Olby, G. N. Cantor, J. R. R. Christie, and J. S. Hodge, 396–409. London: Routledge, 1990.

Darnton, Robert. "Philosophers Trim the Tree of Knowledge: The Epistemological Strategy of the *Encyclopédie*." In Darnton, *The Great Cat Massacre and Other Episodes in French Cultural History*, 191–213. New York: Basic Books, 1984.

Daston, Lorraine. "The Moral Economy of Science." *Osiris* 10 (1995): 3–24.

———. "The Moralized Objectivities of Science." In *Warheit und Geschichte*, edited by Wolfgang Carl and Lorraine Daston, 78–100. Göttingen: Vandenhoeck and Ruprecht, 1999.

———. "Theory of Will versus the Science of Mind." In Woodward and Ash, *Problematic Science*, 88–115.

Daston, Lorraine, and Peter Galison. "The Image of Objectivity." *Representations* 40 (1992): 81–128.

de Certeau, Michel. *The Practice of Everyday Life*. Berkeley: University of California Press, 1984.

de Chadarevian, Soraya. "Graphical Method and Discipline: Self-Recording Instruments in Nineteenth-Century Physiology." *Studies in the History and Philosophy of Science* 24, no. 2 (1993): 267–91.

Decleva, Fernanda Caizzi. *Immagini del filosofo*. Milano: Unicopli, 1984.

Delabarre, Edmund B., Edwin D. Starbuck, and Roswell P. Angier. "Students' Impression of James." *Psychological Review* 50 (1943): 125–34.

Deleuze, Gilles. *L'image-mouvement*. Paris: Editions de Minuit, 1983.

Desmond, Adrian. *Huxley: From Devil's Advocate to Evolution's High Priest*. Reading, Mass.: Addison-Wesley, 1997.

———. "Redefining the X Axis: 'Professionals,' 'Amateurs,' and the Making of Mid-Victorian

Biology—A Progress Report." *Journal of the History of Biology* 34 (2001): 3–50.

Diamond, Solomon. "Wundt before Leipzig." In *Wilhelm Wundt in History: The Making of a Scientific Psychology*, edited by Robert W. Rieber and David K. Robinson, 3–68. New York: Kluwer Academic/Plenum, 2001.

Diggins, Patrick. *The Promise of Pragmatism: Modernism and the Crisis of Knowledge and Authority.* Chicago: University of Chicago Press, 1994.

Donnelly, Margaret E., ed. *Reinterpreting the Legacy of William James.* Washington, D.C.: American Psychological Association, 1992.

Draaisma, Douwe, and Sarah de Rijcke. "The Graphic Strategy: The Uses and Functions of Illustrations in Wundt's *Grundzüge*." *History of the Human Sciences* 14, no. 1 (2001): 1–24.

Dupéron, Isabelle. *G. T. Fechner, le parallélisme psychophysiologique.* Paris: PUF, 2000.

Ellenberger, Henri. *The Discovery of the Unconscious: The History and Evolution of Dynamic Psychiatry.* New York: Basic Books, 1970.

Evans, Rand B. "Introduction. The Historical Context." In James, *Principles of Psychology*, xli–lxviii.

———. "Origins of American Academic Psychology." In *Wilhelm Wundt and the Making of Scientific Psychology*, edited by Robert W. Rieber, 17–60. New York: Plenum Press, 1980.

———. "The Scientific and Psychological Positions of E. B. Titchener." In Evans and Leys, *Defining American Psychology*, 1–38.

———. "William James and His *Principles*." In Johnson and Henley, *Reflections on "The Principles of Psychology,"* 11–32.

———. "William James, 'The Principles of Psychology' and Experimental Psychology." *American Journal of Psychology* 103, no. 4 (1990): 433–47.

Evans, Rand B., and Ruth Leys, eds. *Defining American Psychology: The Correspondence between Adolf Meyer and Edward Bradford Titchener.* Baltimore: Johns Hopkins University Press, 1990.

Fabiani, Jean-Louis. "Enjeux et usages de la 'crise' dans la philosophie universitaire en France au tournant du siecle." *Annales: Economies, Sociétés, Civilisations* (March–April 1985): 377–409.

———. *Les Philosophes de la République.* Paris: Editions de Minuit, 1988.

Feffer, Andrew. *The Chicago Pragmatists and American Progressivism.* Ithaca: Cornell University Press, 1993.

Feinstein, Howard M. *Becoming William James.* Ithaca: Cornell University Press, 1984.

Fisch, M. H. "Alexander Bain and the Genealogy of Pragmatism." *Journal of the History of Ideas* (1954): 413–44.

———. "Was There a Metaphysical Club in Cambridge?: A Postscript." *Transactions of the Charles S. Peirce Society* (Spring 1981): 128–30.

Flower, Elizabeth, and Murray G. Murphy. *A History of Philosophy in America.* 2 vols. New York: G. P. Putnam's Sons, Capricorn Books, 1977.

Fontinell, Eugene. *Self, God, and Immortality: A Jamesian Investigation.* Philadelphia: Temple University Press, 1986.

Forgan, Sophie. "The Architecture of Display: Museums, Universities and Objects in Nineteenth-Century Britain." *History of Science* 32 (1994): 139–62.

———. "The Architecture of Science and the Idea of a University." *Studies in the History and Philosophy of Science* 20 (1986): 405–34.

Frank, Robert G. "American Physiologists in German Laboratories, 1865–1914." In Geison, *Physiology in the American Context*, 33–38.

Franzese, Sergio. *L'uomo indeterminato: Saggio su William James.* Rome: D'Anselmi, 2000.

Fujimura, Joan H. "Crafting Science: Standardized Packages, Boundary Objects, and 'Translation.'" In *Science as Practice and Culture*, edited by Andrew Pickering, 168–211. Chicago: University of Chicago Press, 1992.

Fuller, Steve. "Disciplinary Boundaries and the Rhetoric of the Social Sciences." In Messer-Davidow, Shumway, and Sylvan, *Knowledges*, 125–49.

Fullerton, G. S. "The Psychological Standpoint." *Psychological Review* 1 (1894): 113–33.

Furumoto, L. "Shared Knowledge: The Experimentalists." In Morawski, *Rise of Experimentation in American Psychology*, 94–113.

Fye, W. Bruce. *The Development of American Physiology: Scientific Medicine in the Nineteenth Century*. Baltimore: Johns Hopkins University Press, 1987.

Gale, Richard. *The Divided Self of William James*. Cambridge: Cambridge University Press, 1999.

Galison, Peter, and Lorraine Daston. "Wissenschaftliche Koordinations als Ethos und Epistemologie." In *Instrumente in Kunst und Wissenschaft*, edited by H. Schramm, L. Schwarte, and J. Lazardig, 319–361. Berlin: Walter de Gruyter, 2006.

Galison, Peter, and David J. Stump, eds. *The Disunity of Science: Boundaries, Contexts, and Power*. Stanford: Stanford University Press, 1996.

Galison, Peter, and Emily Thompson, eds. *The Architecture of Science*. Cambridge, Mass.: MIT Press, 1999.

Gauld, Alan. *The Founders of Psychical Research*. New York: Schocken Books, 1968.

———. *A History of Hypnotism*. Cambridge: Cambridge University Press, 1992.

Geiger, Roger. *To Advance Knowledge: The Growth of American Research Universities, 1900–1940*. New York: Oxford University Press, 1986.

Geison, Gerald L., ed. *Physiology in the American Context, 1850–1940*. Bethesda, Md.: American Physiological Society, 1987.

Gieryn, Thomas F. "Boundaries of Science." In *Handbook of Science and Technology Studies*, edited by Sheila Jasanoff et al., 393–443. Thousands Oaks, Calif.: Sage, 1995.

———. "Boundary Work and the Demarcation of Science from Non-Science: Strains and Interests in the Professional Ideologies of Scientists." *American Sociological Review* 48 (1983): 781–95.

———. *Cultural Boundaries of Science: Credibility on the Line*. Chicago: University of Chicago Press, 1999.

———. "Three Truth-Spots." *Journal of the History of the Behavioral Sciences* 38, no. 2 (Spring 2002): 113–32.

Ginzburg, Carlo. "Checking the Evidence: The Historian and the Judge." *Critical Inquiry* 18 (1991): 79–92.

———. "Clues: Roots of an Evidential Paradigm." In Ginzburg, *Clues, Myths, and the Historical Method*, 96–125. Baltimore, Md.: Johns Hopkins University Press, 1989.

Giri, Ananta Kumar. "Transcending Disciplinary Boundaries: Creative Experiments and the Critiques of Modernity." *Critique of Anthropology* 18 (1998): 379–404.

Goetz, Christopher G., Michel Bonduelle, and Toby Gelfand. *Constructing Neurology: Jean-Martin Charcot, 1825–1893*. New York: Oxford University Press, 1995.

Goffman, Erving. *The Presentation of Self in Everyday Life*. London: Allen Lane, 1969.

Goldstein, Jan E. "The Advent of Psychological Modernism in France: An Alternate Narrative." In Ross, *Modernist Impulses in the Human Sciences*, 190–209.

———. "Foucault and the Post-Revolutionary Self: The Uses of Cousinian Pedagogy in Nineteenth-Century France." In *Foucault and the Writing of History*, edited by Goldstein, 99–115. Cambridge: Blackwell, 1994.

———. "Foucault's Technologies of the Self and the Cultural History of Identity." *Arcadia* 33 (1998): 46–63.

———. *The Post-Revolutionary Self: Politics and Psyche in France, 1750–1850*. Cambridge, Mass.: Harvard University Press, 2005.

Good, James M. M. "Disciplining Social Psychology: A Case Study of Boundary Relations in the History of the Human Sciences." *Journal of the History of Behavioral Sciences* 36 (2000): 383–403.

———. "Quests for Interdisciplinarity: The Rhetorical Constitution of Social Psychology." In Roberts and Good, *Recovery of Rhetoric*, 239–62.

Gregory, Frederick. *Scientific Materialism in Nineteenth Century Germany*. Dordrecht, Holland: Reidel, 1977.

Guarneri, Carl J. *The Utopian Alternative: Fourierism in Nineteenth-Century America*. Ithaca: Cornell University Press, 1991.

Haakonssen, Lisbeth. *Medicine and Morals in the Enlightenment: John Gregory, Thomas Percival and Benjamin Rush*. Amsterdam: Rodopi, 1997.

Hacking, Ian. *Rewriting the Soul: Multiple Personality and the Sciences of Memory*. Princeton: Princeton University Press, 1994.

Hadot, Pierre. *Exercices spirituels et philosophie antique*. Paris: Institut d'études augustiniennes, 1993.

———. *La philosophie comme manière de vivre: Entretiens avec Jeannie Carlier et Arnold I. Davidson*. Paris: Albin Michel, 2001.

———. *Philosophy as a Way of Life: Spiritual Exercises from Socrates to Foucault*. Edited by Arnold Ira Davidson. Oxford: Blackwell, 1995.

Hagner, Michael, ed. *Ecce Cortex: Beitraege zur Geschichte des modernen Gehirns*. Göttingen: Wallstein, 1999.

———. *Homo cerebralis: Der Wandel vom Seelenorgan zum Gehirn*. Berlin: Berlin verlag, 1997.

Hagner, Michael, and Bettina Wahrig-Schmidt, eds. *Johannes Müller und die Philosophie*. Berlin: Akademie verlag, 1992.

Hale, Matthew, Jr. *Human Science and Social Order: Hugo Münsterberg and the Origins of Applied Psychology*. Philadelphia: Temple University Press, 1980.

Haraway, Donna. "Situated Knowledges: The Science Question in Feminism and the Privilege of Partial Perspective." *Feminist Studies* 14 (1988): 575–99.

Harper, R. S. "The Laboratory of William James." *Harvard Alumni Bulletin* 51 (1949): 169–73.

Harrington, Anne. *Medicine, Mind, and the Double Brain: A Study in Nineteenth-Century Thought*. Princeton: Princeton University Press, 1987.

———. *Reenchanted Science: Holism in German Culture from Wilhelm II to Hitler*. Princeton: Princeton University Press, 1996.

Haskell, Thomas L., ed. *The Authority of Experts: Studies in History and Theory*. Bloomington: Indiana University Press, 1984.

———. *The Emergence of Professional Social Science*. Urbana: University of Illinois Press, 1977.

———. "Menand's Postdisciplinary Project." *Intellectual History Newsletter* 24 (2002): 107–119.

———. "Professionalism versus Capitalism: Tawney, Durkheim, and C. S. Peirce on the Disinterestedness of Professional Communities." In Haskell, *Objectivity Is Not Neutrality: Explanatory Schemes in History*, chap. 4. Baltimore: Johns Hopkins University Press, 1998.

———. "Two Observations on the Context of Charles S. Peirce's Community of Inquiry." *Krisis* 1 (Summer 1983): 27–38.

Heidelberger, Michael. *Nature from Within: Gustave Theodor Fechner and His Psychophysical Worldview*. Pittsburgh: University of Pittsburgh Press, 2004.

Herzog, Max. "William James and the Development of Phenomenological Psychology in Europe." *History of the Human Sciences* 8 (1995): 29–46.

Heyck, T. W. *Transformation of Intellectual Life in Victorian England*. New York: St. Martin's Press, 1982.

Hilgard, Ernest R. "The Trilogy of Mind: Cognition, Affection, and Conation." *Journal of the History of the Behavioral Sciences* 16 (1980): 107–17.

Hinsley, Curtis M. "The Museum Origins of Harvard Anthropology, 1866–1915." In *Science at Harvard University: Historical Perspectives*, edited by Clark A. Elliott and Margaret W. Rossiter, 121–45. Bethlehem, Pa.: Lehigh University Press, 1992.

Hollinger, David A. *Cosmopolitanism and Solidarity*. Madison: University of Wisconsin Press, 2006.

———. "'Damned for God's Glory': William James and the Scientific Vindication of Protestant Culture." In *William James and a Science of Religions: Reexperiencing "The Varieties of Religious Experience,"* edited by Wayne Proudfoot, 9–30. New York: Columbia University Press, 2004.

———. "The Defense of Democracy and Robert K. Merton's Formulation of the Scientific Ethos." *Knowledge and Society* 4 (1983): 1–15.

———. *In the American Province: Studies in the History and Historiography of Ideas*. Bloomington: Indiana University Press, 1985.

———. "Inquiry and Uplift: Late Nineteenth-Century American Academics and the Moral Efficacy of Scientific Practice." In Haskell, *The Authority of Experts*, 142–56.

———. "James, Clifford, and the Scientific Conscience." In R. A. Putnam, *Cambridge Companion to William James*, 69–89.

———. *Morris R. Cohen and the Scientific Ideal*. Cambridge: MIT Press, 1975.

———. *Science, Jews, and Secular Culture*. Princeton: Princeton University Press, 1996.

Hoopes, James. *Community Denied*. Ithaca: Cornell University Press, 1998.

Hoorn, Willem van, and Thom Verhave. "Wundt's Changing Conceptions of a General and Theoretical Psychology." In Bringmann and Tweney, *Wundt Studies*, 71–113.

Hoover, Dwight W. *Henry James, Sr. and the Religion of Community*. Grand Rapids, Mich.: William B. Eerdmans, 1969.

Howe, Daniel Walker. *The Unitarian Conscience: Harvard Moral Philosophy, 1805–1861*. 1970. Reprint, Middletown, Conn.: Wesleyan University Press, 1988.

Hull, David. *Science as a Process: An Evolutionary Account of the Social and Conceptual Development of Science*. Chicago: University of Chicago Press, 1988.

Jasanoff, Sheila. "Contested Boundaries in Policy-Relevant Science." *Social Studies of Science* 17 (1990): 195–230.

Johnson, Jim [pseud.]. See Latour, Bruno, "Mixing Humans and Nonhumans Together."

Johnson, Michael G., and Tracy B. Henley, eds. *Reflections on "The Principles of Psychology": William James after a Century*. Hillsdale, N.J.: Lawrence Erlbaum Associates, 1990.

Jurkowitz, Edward. "Helmholtz and the Liberal Unification of Science." *Historical Studies in the Physical and Biological Sciences* 32 (Spring 2002): 291–317.

———. *Liberal Pursuits: Hermann von Helmholtz, Ernst Mach and the Framing of Physics and the Human Mind*. Forthcoming.

Kim, Mimi. "Practice and Representation: Investigative Programs of Chemical Affinity in the Nineteenth Century." PhD diss., UCLA, 1990.

Kimball, Bruce A. *The Condition of American Liberal Education: Pragmatism and a Changing Tradition, with Commentaries by Noted Scholars*, edited by Robert Orrill. New York: College Entrance Examination Board, 1995.

Kittelstrom, Amy. "Against Elitism: Studying William James in the Academic Age of the Underdog." *William James Studies* 1, no. 1 (2006): 1–18.

Klein, Julie T. *Crossing Boundaries: Knowledge, Disciplinarities, Interdisciplinarities*. Charlottesville: University of Virginia Press, 1996.

Kloppenberg, James T. "Pragmatism: An Old Name for Some New Ways of Thinking?" In *A Pragmatist's Progress: Richard Rorty and American Intellectual History*, edited by John Pettegrew, 19–60. Lanham, Md.: Rowman and Littlefield, 2000.

——— *Uncertain Victory: Social Democracy and Progressivism in European and American Thought, 1870–1920*. New York: Oxford University Press, 1986.

Kraushaar, Otto F. "Lotze's Influence on the Psychology of William James." *Psychological Review* 43 (1936): 235–57.

Kuklick, Bruce. *Churchmen and Philosophers: From Jonathan Edwards to John Dewey.* New Haven: Yale University Press, 1985.

———. *Josiah Royce: An Intellectual Biography.* Indianapolis, Ind.: Bobbs-Merrill, 1972.

———. *Philosophy in America, 1720–2000.* Oxford: Clarendon Press, 2002.

———. *The Rise of American Philosophy: Cambridge, Massachusetts, 1860–1930.* New Haven: Yale University Press, 1977.

Kuklick, Henrika. "Boundary Maintenance in American Sociology: Limitations to Academic 'Professionalization.'" *Journal of the History of the Behavioral Sciences* 16 (1980): 201–19.

———. "Professionalization and the Moral Order." In *Disciplinarity at the Fin de Siècle,* edited by Amanda Anderson and Joseph Valente, 126–52. Princeton: Princeton University Press, 2002.

Kusch, Martin. "The Criticism of Husserl's Arguments against 'Psychologism' in German Philosophy, 1901–1920." In *Mind, Meaning, and Mathematics,* edited by L. Haaparanta, 51–85. Boston: Kluwer Academic Publishers, 1994.

———. *Psychological Knowledge: A Social History and Philosophy.* London: Routledge, 1999.

———. *Psychologism: A Case Study in the Sociology of Philosophical Knowledge.* New York: Routledge, 1995.

———. "Recluse, Interlocutor, Interrogator: Natural and Social Order in Turn-of-the-Century Psychological Research Schools." *Isis* 86 (1995): 419–39.

———. "The Sociology of Philosophical Knowledge: A Case Study and a Defense." In *The Sociology of Philosophical Knowledge,* edited by Martin Kusch, 15–38. Dordrecht: Kluwer Academic Publishers, 2000.

Lamberth, David C. "Squaring Logic with Life: Metaphysics, Experience, and Religion in William James's Philosophical Worldview." PhD diss., Harvard University, 1996.

———. *William James and the Metaphysics of Experience.* Cambridge: Cambridge University Press, 1999.

Larson Sarfatti, Magali. "The Production of Expertise and the Constitution of Expert Power." In Haskell, *Authority of Experts,* 28–80.

———. *The Rise of Professionalism: A Sociological Analysis.* Berkeley: University of California Press, 1977.

Latour, Bruno, ed. *Making Things Public.* Cambridge, Mass.: MIT Press, 2005.

——— [Jim Johnson pseud.] "Mixing Humans and Nonhumans Together: The Sociology of a Door-Closer." *Social Problems* 35, no. 2 (April 1988): 298–310.

———. "What is the Style of Matters of Concern? Two Lectures in Empirical Philosophy." Two lectures given in Amsterdam in April and May 20005.

———. *Reassembling the Social: An Introduction to Actor-Network Theory.* Oxford: Oxford University Press, 2005.

———. *Science in Action: How to Follow Scientists and Engineers through Society.* Chicago: University of Chicago Press, 1987.

Lears, T. J. Jackson. *No Place of Grace: Anti-Modernism and the Transformation of American Culture, 1880–1920.* New York: Pantheon Books, 1981.

Leary, David E. "'Authentic Tidings': What Wordsworth Gave to William James." Paper presented at the annual meeting of Cheiron, Berkeley, June 2005.

———. "The Influence of Literature in the Life and Work of William James." Paper presented at the Fishbein Center for the History of Science, University of Chicago, May 2005.

———. *Metaphors in the History of Psychology.* Cambridge: Cambridge University Press, 1990.

———. "A Profound and Radical Change: How William James Inspired the Reshaping of American Psychology." In *The Anatomy of Impact: What Makes the Great Works of Psychology Great,*

edited by Robert J. Sternberg, 19–42. Washington, D.C.: American Psychological Associa-
tion, 2003.
————. "Telling Likely Stories: The Rhetoric of the New Psychology, 1880–1920." *Journal of the
History of the Behavioral Sciences* 23 (1987): 315–31.
————. "William James and the Art of Human Understanding." *American Psychologist* 47 (1992):
152–60.
————. "William James on the Self and Personality: Clearing the Ground for Subsequent Theo-
rists, Researchers, and Practitioners." In Johnson and Henley, *Reflections on "The Principles of
Psychology*," 101–37.
————. "William James, the Psychologist's Dilemma and the Historiography of Psychology:
Cautionary Tales." *History of the Human Sciences* 8 (1995): 91–105.
————. "Wundt and After: Psychology's Shifting Relations with the Natural Sciences, Social
Sciences, and Philosophy." *Journal of the History of the Behavioral Sciences* 15 (1979): 231–41.
Lenoir, Tim. "Discipline of Nature / Nature of Disciplines." In Messer-Davidow, Shumway, and
Sylvan, *Knowledges*, 70–102.
————. *Instituting Science: The Cultural Production of Scientific Disciplines*. Stanford: Stanford
University Press, 1997.
Lentricchia, Frank. *Ariel and the Police: Michel Foucault, William James, Wallace Stevens*. Madison:
University of Wisconsin Press, 1988.
————. "On the Ideologies of Poetic Modernism, 1890–1913: The Example of William James."
In *Reconstructing American Literary History*, edited by Sacvan Bercovitch, 220–49. Cambridge,
Mass.: Harvard University Press, 1986.
Levinson, R. B. "Gertrude Stein, William James, and Grammar." *American Journal of Psychology*
54 (1941): 124–28.
Leys, Ruth. "The Correspondence between Meyer and Titchener." In Evans and Leys, *Defining
American Psychology*, 58–114.
Licht, Walter. *Industrializing America: The Nineteenth Century*. Baltimore: Johns Hopkins Univer-
sity Press, 1995.
Lipton, Peter. "The Epistemology of Testimony." *Studies in the History and Philosophy of Science*
29 (1998): 1–31.
Livingston, James. "The Politics of Pragmatism." *Social Text* 49, no. 4 (Winter 1996): 149–72.
————. *Pragmatism and the Political Economy of Cultural Revolution, 1850–1940*. 2nd ed. Chapel
Hill: University of North Carolina Press, 1997.
————. *Pragmatism, Feminism, and Democracy: Rethinking the Politics of American History*.
New York: Routledge, 2001.
————. "The Strange Career of the 'Social Self.'" *Radical History Review* 76 (2000): 53–79.
Livingstone, David N. *Putting Science in Its Place: Geographies of Scientific Knowledge*. Chicago:
University of Chicago Press, 2003.
————. "The Spaces of Knowledge: Contributions towards the Historical Geography of Sci-
ence." *Environment and Planning D: Society and Space* 13, no. 1 (1995): 5–34.
Lloyd, Brian. *Left Out: Pragmatism, Exceptionalism, and the Poverty of American Marxism, 1890–1920*.
Baltimore: Johns Hopkins University Press, 1997.
Lutz, Tom. *American Nervousness, 1903: An Anecdotal History*. Ithaca: Cornell University Press,
1991.
Mackenzie, Lynne. "William James and the Problem of Interest." *Journal of the History of the
Behavioral Sciences* 16 (1980): 175–85.
Madden, E. H. Introduction to *The Will to Believe*. The Works of William James.
McClay, Wilfred M. *The Masterless: Self and Society in Modern America*. Chapel Hill: University of
North Carolina Press, 1994.
McDermott, John J. Introduction to *Essays in Radical Empiricism*. The Works of William James.

————, ed. Introduction to *The Writings of William James*. New York: Modern Library, 1968.

————. "The Promethean Self and Community in the Philosophy of William James." In Mc-Dermott, *Streams of Experience: Reflections on the History and Philosophy of American Culture*, 44–58. Amherst: University of Massachusetts Press, 1986.

McFarland, Gerald W., ed. *Moralists or Pragmatists? The Mugwumps, 1884–1900*. New York: Simon and Schuster, 1975.

————. *Mugwumps, Morals, and Politics, 1884–1920*. Amherst: University of Massachusetts Press, 1975.

McLachlan, James. "George Holmes Howison: 'The City of God' and Personal Idealism." *Journal of Speculative Philosophy*, n.s., 20, no. 3 (2006): 224–42.

Menand, Louis. "The Marketplace of Ideas." *ACLS (American Council of Learned Societies) Occasional Paper No. 49* (2001).

————. *The Metaphysical Club: A Story of Ideas in America*. New York: Farrar, Straus and Giroux, 2001.

————. "Undisciplined." *Wilson Quarterly* 57 (2001): 51–59.

Messer-Davidow, Ellen, David R. Shumway, and David J. Sylvan, eds. *Knowledges: Historical and Critical Studies in Disciplinarity*. Charlottesville: University Press of Virginia, 1993.

Miller, Joshua I. *Democratic Temperament: The Legacy of William James*. Lawrence: University Press of Kansas, 1997.

Morawski, Jill C. "Assessing Psychology's Moral Heritage through Our Neglected Utopias." *American Psychologist* 37 (1982): 1082–95.

————, ed. *The Rise of Experimentation in American Psychology*. New Haven: Yale University Press, 1988.

————. "Self-Regard and Other-Regard: Reflexive Practices in American Psychology, 1890–1940." *Science in Context* 5 (1992): 281–308.

————. "There Is More to Our History of Giving: The Place of Introductory Textbooks in American Psychology." *American Psychologist* 47, no. 2 (1992): 161–69.

Morra, Gianfranco. "Rosmini e lo spirito dell' 'Encyclopédie.'" In *Rosmini e l'enciclopedia delle scienze: Atti del congresso internazionale diretto da Maria Adelaide Raschini*. Firenze: Olschki, 1998.

Müller, Detlef K., Fritz Ringer, and Brian Simon, eds. *The Rise of the Modern Educational System: Structural Change and Social Reproduction, 1870–1920*. Cambridge: Cambridge University Press, 1987.

Mumford, Lewis. *The Golden Day: A Study in American Literature and Culture*. New York: Boni and Liveright, 1926.

Murchison, C. A., E. G. Boring, and L. Gardner. *A History of Psychology in Autobiography*. Vol. 4. Worcester, Mass.: Clark University Press, 1952.

Murphy, Gardner, and Robert O. Ballou, eds. *William James on Psychical Research*. London: Chatto and Windus, 1961.

Myers, Gerald E. *William James: His Life and Thought*. New Haven: Yale University Press, 1986.

Neary, Francis. "A Question of 'Peculiar Importance': George Croom Robertson, Mind and the Changing Relationship between British Psychology and Philosophy." In Bunn, Lovie, and Richards, *Psychology in Britain*, 54–71.

Noakes, Richard. "The 'Bridge Which Is between Physical and Psychical Research': William Fletcher Barrett, Sensitive Flames, and Spiritualism." *History of Science* 42 (2004): 419–64.

Oakes, Guy. *Weber and Rickert: Concept Formation in the Cultural Sciences*. Cambridge, Mass.: MIT Press, 1988.

O'Donnell, John M. *The Origins of Behaviorism: American Psychology, 1870–1920*. New York: New York University Press, 1985.

Olesko, Kathryn M. "The Meaning of Precision: The Exact Sensibility in Early Nineteenth-Century Germany." In Wise, *Values of Precision*, 103–34.

———. *Physics as a Calling: Discipline and Practice in the Königsberg Seminar for Physics*. Ithaca: Cornell University Press, 1991.

Ophir, Adi, and Steven Shapin. "The Place of Knowledge: A Methodological Survey." *Science in Context* 4 (1991): 3–21.

Oppenheim, Janet. *The Other World: Spiritualism and Psychical Research in England, 1850–1914*. Cambridge: Cambridge University Press, 1985.

Outram, Dorinda. "New Spaces in Natural History." In *Cultures of Natural History*, edited by N. Jardine, E. Spary, and A. Secord, 249–65. Cambridge: Cambridge University Press, 1996.

Owen, Alex. *The Darkened Room: Women, Power, and Spiritualism in Late Victorian England*. 1990. Reprint, Chicago: University of Chicago Press, 2004.

———. "Occultism and the 'Modern' Self in Fin-de-Siècle Britain." In *Meanings of Modernity: Britain from the Late-Victorian Era to World War II*, edited by Martin Daunton and Bernhard Rieger, 71–96. New York: Oxford University Press, 2001.

———. *The Place of Enchantment: British Occultism and the Culture of the Modern*. Chicago: University of Chicago Press, 2004.

Pandora, Katherine. *Rebels within the Ranks: Psychologists' Critique of Scientific Authority and Democratic Realities in New Deal America*. Cambridge: Cambridge University, 1997.

Parshall, Karen Hunger, and David E. Rowe. *The Emergence of the American Mathematical Research Community, 1876–1900: J. J. Sylvester, Felix Klein, and E. H. Moore*. Providence, R.I.: American Mathematical Society, 1994.

Passmore, John. *A Hundred Years of Philosophy*. London: Gerald Duckworth, 1957.

Paulston, Rolland G., ed., *Social Cartography: Mapping Ways of Seeing Social and Educational Change*. New York: Garland, 1996.

Pireddu, Nicoletta. *Antropologi alla corte della bellezza: Decadenza ed economia simbolica nell'Europa fin de siècle*. Verona: Fiorini, 2002.

Pittenger, Mark. *American Socialists and Evolutionary Thought, 1870–1920*. Madison: University of Wisconsin Press, 1993.

Pols, Hans. "Managing the Mind: The Culture of American Mental Hygiene, 1910–1950." PhD diss., University of Pennsylvania, 1997.

Porter, Theodore. *Karl Pearson: The Scientific Life in a Statistical Age*. Princeton: Princeton University Press, 2004.

———. *The Rise of Statistical Thinking, 1820–1900*. Princeton: Princeton University Press, 1986.

Posnock, Ross. *The Trial of Curiosity: Henry James, William James, and the Challenge of Modernity*. New York: Oxford University Press, 1991.

Putnam, Hilary. "James's Theory of Truth." In R. A. Putnam, *Cambridge Companion to William James*, 166–85.

Putnam, Hilary, and Ruth Anna Putnam. "William James's Ideas." In *Realism with a Human Face*, edited by James Conant, 217–31. Cambridge, Mass.: Harvard University Press, 1990.

Putnam, Ruth Anna, ed. *The Cambridge Companion to William James*. Cambridge: Cambridge University Press, 1997.

Rabinbach, Anson. *The Human Motor: Energy, Fatigue, and the Origins of Modernity*. New York: Basic Books, 1990.

Reed, Edward S. "The Psychologist's Fallacy as a Persistent Framework in William James's Psychological Theorizing." *History of the Human Sciences* 8, no. 1 (1995): 61–72.

———. "Space Perception and the Psychologist's Fallacy in James's Principles." In Johnson and Henley, *Reflections on "The Principles of Psychology,"* 231–48.

Reingold, Nathan, and Ida H. Reingold, eds. *Science in America*. Chicago: University of Chicago Press, 1981.

Reuben, Julie. *The Making of the Modern University.* Chicago: University of Chicago Press, 1996.

Rice, Charles E. "Uncertain Genesis: The Academic Institutionalization of American Psychology in 1900." *American Psychologist* 55 (2000): 488–91.

Richards, Graham. "Edward Cox, the Psychological Society of Great Britain (1875–1879), and the Meanings of an Institutional Failure." In Bunn, Lovie, and Richards, *Psychology in Britain,* 33–53.

———. *Putting Psychology in Its Place: An Introduction from a Critical Historical Perspective.* London: Routledge, 1996.

Richards, Robert J. *Darwin and the Emergence of Evolutionary Theories of Mind and Behavior.* Chicago: University of Chicago Press, 1987.

———. "'The Personal Equation in Science: William James's Psychological and Moral Uses of the Darwinian Theory." In *A William James Renascence: Four Essays by Young Scholars,* edited by Mark R. Schwhen. *Harvard Library Bulletin* 30 (October 1982): 387–425.

Richter, Melvin. *Politics of Conscience: T. H. Green and His Age.* Cambridge, Mass.: Harvard University Press, 1964.

Rieber, Robert W., ed. *Wilhelm Wundt and the Making of a Scientific Psychology.* New York: Plenum Press, 1980.

Rieber, Robert W., and David Robinson, eds. *Wilhelm Wundt in History: The Making of a Scientific Psychology.* New York: Kluwer Academic/Plenum, 2001.

Ringer, Fritz K. *The Decline of the German Mandarins: The German Academic Community, 1890–1933.* Cambridge: Cambridge University Press, 1969. Reprint, Hanover, N.H.: University Press of New England, 1990.

———. *Education and Society in Modern Europe.* Bloomington: Indiana University Press, 1979.

———. *Fields of Knowledge: French Academic Culture in Comparative Perspective, 1890–1920.* Cambridge: Cambridge University Press, 1992.

Roberts, Richard H., and James M. M. Good. "Introduction: Persuasive Discourse in and between Disciplines in the Human Sciences." In Roberts and Good, *Recovery of Rhetoric,* 1–21.

———, eds. *The Recovery of Rhetoric: Persuasive Discourse and Disciplinarity in the Human Sciences.* Charlottesville: University Press of Virginia, 1993.

Rodgers, Daniel T. *Atlantic Crossings: Social Politics in a Progressive Age.* Cambridge, Mass.: Belknap Press of Harvard University Press, 1998.

Rorty, Richard. "Faith, Responsibility and Romance." In R. A. Putnam, *Cambridge Companion to William James,* 19–36.

———. *Philosophical Papers.* Vol. 1, *Objectivity, Relativism, and Truth.* Cambridge: Cambridge University Press, 1991.

———. "Putnam and the Relativist Menace." *Journal of Philosophy* 90, no. 9 (September 1993): 443–61.

———. "Texts and Lumps." In Rorty, *Philosophical Papers,* 1:78–92.

Rorty, Richard, J. B. Schneewind, and Quentin Skinner, eds. *Philosophy in History: Essays on the Historiography of Philosophy.* Cambridge: Cambridge University Press, 1984.

Ross, Dorothy. "The Development of the Social Sciences." In *The Organization of Knowledge in Modern America, 1860–1920,* edited by Alexandra Oleson and John Voss, 107–38. Baltimore: Johns Hopkins University Press, 1979.

———. *G. Stanley Hall: The Psychologist as a Prophet.* Chicago: University of Chicago Press, 1972.

———, ed. *Modernist Impulses in the Human Sciences, 1870–1930.* Baltimore: Johns Hopkins University Press, 1994.

———. *The Origins of American Social Science.* Cambridge: Cambridge University Press, 1991.

Rossiter, Margaret W. *Women Scientists in America: Struggles and Strategies to 1940.* Baltimore: Johns Hopkins University Press, 1982.

Rouse, Joseph. "Policing Knowledge: Disembodied Policy for Embodied Knowledge." *Inquiry* 34 (1991): 353–64.

Rudwick, Martin. *The Great Devonian Controversy: The Shaping of Scientific Knowledge among Gentlemanly Specialists.* Chicago: University of Chicago Press, 1985.

Ryan, Judith. *The Vanishing Subject: Early Psychology and Literary Modernism.* Chicago: University of Chicago Press, 1991.

Santucci, Antonio. *Storia del pragmatismo.* Roma-Bari: Laterza, 1992.

Scarborough, Elizabeth, and Laurel Furumoto. *Untold Lives: The First Generation of American Women Psychologists.* New York: Columbia University Press, 1987.

Schaffer, Simon. "Astronomers Mark Time: Discipline and the Personal Equation." *Science in Context* 2, no. 1 (1988): 115–45.

Schirmer, Daniel B. *Republic or Empire: American Resistance to the Philippine War.* Cambridge, Mass.: Schenkman, 1972.

———. "William James and the New Age." *Science and Society* 33 (1969): 434–45.

Schmidgen, Henning. "Time and Noise: The Stable Surroundings of Reaction Experiments, 1860–1890." *Studies in the History and Philosophy of the Biological and Biomedical Sciences* 34 (2003): 237–75.

Schneewind, J. B. *Sidgwick's Ethics and Victorian Moral Philosophy.* Oxford: Clarendon Press, 1977.

Seigfried, Charlene Haddock. "James: The Point of View of the Other." In *Classical American Pragmatism: Its Contemporary Vitality,* edited by S. B. Rosenthal, C. R. Hausman, and D. R. Anderson, 85–98. Urbana: University of Illinois Press, 1999.

———. "On the Metaphysical Foundation of Scientific Psychology." In *The Philosophical Psychology of William James,* edited by M. H. De Armey and S. Skousgaard, 57–72. Washington, D.C.: University Press of America, 1986.

———. *Pragmatism and Feminism.* Chicago: University of Chicago Press, 1996.

———. *William James's Radical Reconstruction of Philosophy.* Albany: SUNY Press, 1990.

Serjeantson, R. W. "Testimony and Proof in Early-Modern England." *Studies in the History and Philosophy of Science* 30 (1999): 195–236.

Shapin, Steven. "The House of Experiment in Seventeenth-Century England." *Isis* 79 (1988): 373–404.

———. "The Philosopher and the Chicken: On the Dietetics of Disembodied Knowledge." In *Science Incarnate: Historical Embodiments of Natural Knowledge,* edited by Christopher Lawrence and Steven Shapin, 21–50. Chicago: University of Chicago Press, 1998.

———. *A Social History of Truth.* Chicago: University of Chicago Press, 1994.

Shapin, Steven, and Simon Schaffer. *Leviathan and the Air Pump: Hobbes, Boyle, and the Experimental Life.* Princeton: Princeton University Press, 1985.

Shook, John R., ed. *Pragmatism: An Annotated Bibliography, 1898–1940.* Amsterdam: Rodopi, 1998.

Schultz, Bart. *Henry Sidgwick Eye of the Universe.* Cambridge: Cambridge University Press, 2004.

Simon, Linda. *Genuine Reality: A Life of William James.* Chicago: University of Chicago Press, 2000.

———. Introduction to *The Correspondence of William James,* vol. 5.

Sklansky, Jeffrey. *The Soul's Economy: Market Society and Selfhood in American Thought, 1820–1920.* Chapel Hill: University of North Carolina Press, 2002.

Sklar, Martin. *The Corporate Reconstruction of American Capitalism, 1890–1916: The Market, the Law, and Politics.* New York: Cambridge University Press, 1988.

Skrupskelis, Ignas K. Introduction to *Essays, Comments, and Reviews.* The Works of William James.

———. Introduction to *Essays in Philosophy.* The Works of William James.

———. Introduction to *Manuscript Essays and Notes.* The Works of William James.

———. Introduction to *Manuscript Lectures*. The Works of William James.

———. "James's Conception of Psychology as a Natural Science." *History of the Human Sciences* 8, no. 1 (1995): 73–89.

Smith, Crosbie, and Jon Agar. *Making Space for Science: Territorial Themes in the Shaping of Knowledge*. New York: St. Martin's Press, 1998.

Smith, Crosbie, and M. Norton Wise. *Energy and Empire: A Biographical Study of Lord Kelvin*. Cambridge: Cambridge University Press, 1989.

Smith, M. Brewester. "William James and the Psychology of the Self." In Donnelly, *Reinterpreting the Legacy of William James*, 173–87.

Sokal, Michael M., ed. *An Education in Psychology: James McKeen Cattell's Journal and Letters from Germany and England, 1880–1888*. Cambridge, Mass.: MIT Press, 1981.

———. Introduction to *Psychology, Briefer Course*. The Works of William James.

———. "Origins and Early Years of the American Psychological Association, 1890–1906." *American Psychologist* 47, no. 2 (1992): 111–22.

Sprigge, Timothy L. S. *James and Bradley: American Truth and British Reality*. Chicago: Open Court, 1993.

Star, Susan Leigh, and James R. Griesemer. "Institutional Ecology, 'Translations,' and Boundary Objects: Amateurs and Professionals in Berkeley's Museum of Vertebrate Zoology, 1907–1939." *Social Studies of Science* 19 (1989): 387–420.

Starr, Paul. *The Social Transformation of American Medicine*. New York: Basic Books, 1982.

Stern, Sheldon M. "William James and the New Psychology." In Buck, *Social Sciences at Harvard*, 175–222.

Suckiel, Ellen Kappy. *The Pragmatic Philosophy of William James*. Notre Dame: University of Notre Dame Press, 1982.

Super, R. H. "The Humanist at Bay: The Arnold-Huxley Debate." In *Nature and the Victorian Imagination*, edited by U. C. Knoepflmacher and G. B. Tennyson, 231–45. Berkeley: University of California Press, 1977.

Swijtink, Z. G. "The Objectification of Observation: Measurement and Statistical Methods in the Nineteenth Century." In *The Probabilistic Revolution*, edited by L. Krüger., 260–85. Cambridge, Mass: MIT Press, 1987.

Taves, Ann. "The Fragmentation of Consciousness and *The Varieties of Religious Experience*." In *William James and a Science of Religions: Reexperiencing "The Varieties of Religious Experience,"* edited by Wayne Proudfoot, 48–72. New York: Columbia University Press, 2004.

Taylor, Eugene I. "New Light on the Origin of William James's Experimental Psychology." In Johnson and Henley, *Reflections on "The Principles of Psychology,"* 33–62.

———. "Radical Empiricism and the New Science of Consciousness." *History of the Human Sciences* 8, no. 1 (1995): 47–60.

———. *William James on Consciousness beyond the Margins*. Princeton: Princeton University Press, 1996.

———. *William James on Exceptional Mental States: The 1896 Lowell Lectures*. Amherst: University of Massachusetts Press, 1984.

Taylor, Eugene I., and Robert H. Wozniak, eds. *Pure Experience: The Response to William James*. Bristol, UK: Thoemmes Press, 1996.

Thayer, H. S. Introduction to James, *The Meaning of Truth*. The Works of William James.

———. "James and the Theory of Truth." *Transactions of the Peirce Society* (Winter 1980): 39–48.

———. *Meaning and Action: A Critical History of Pragmatism*. 1968. 2nd ed., Indianapolis: Hackett, 1981.

Thomas, John L. *Alternative America: Henry George, Edward Bellamy, Henry Demarest Lloyd, and the Adversary Tradition*. Cambridge, Mass.: Belknap Press, 1983.

Trachtenberg, Alan. *The Incorporation of America: Culture and Society in the Gilded Age*. New York: Hill and Wang, 1982.

Treitel, Corinna. *A Science for the Soul: Occultism and the Genesis of the German Modern*. Baltimore: Johns Hopkins University Press, 2004.

Turnbull, D. "Constructing Knowledge Spaces and Locating Sites of Resistance in the Early Modern Cartographic Transformation." In Paulston, *Social Cartography*, 53–79.

Turner, Frank M. *The Greek Heritage in Victorian Britain*. New Haven: Yale University Press, 1981.

———. "The Victorian Conflict between Science and Religion: A Professional Dimension." *Isis* 69 (1978): 356–76.

Turner, James. *Without God, Without Creed: The Origins of Unbelief in America*. Baltimore: Johns Hopkins University Press, 1985.

Tzuzuki, Chushichi. *Edward Carpenter, 1844–1929: Prophet of Human Fellowship*. Cambridge: Cambridge University Press, 1980.

Vailati, Giovanni. *Epistolario, 1891–1909*. Torino: Einaudi, 1971.

Veysey, Laurence. *The Emergence of the American University*. Chicago: University of Chicago Press, 1965.

Warwick, Andrew. *Masters of Theory: Cambridge and the Rise of Mathematical Physics*. Chicago: University of Chicago Press, 2003.

Weindling, Paul. "Dissecting German Social Darwinism: Historicizing the Biology of the Organic State." *Science in Context* 11 (1998): 619–37.

———. "Theories of the Cell State in Imperial Germany." In *Biology, Medicine, and Society, 1840–1940*, edited by C. Webster., 99–105. Cambridge: Cambridge University Press, 1981.

Westbrook, Robert B. *Democratic Hope: Pragmatism and the Politics of Truth*. Ithaca: Cornell University Press, 2005.

———. "Lewis Mumford, John Dewey, and the 'Pragmatic Acquiescence.'" In *Lewis Mumford: Public Intellectual*, edited by Thomas P. Hughes and Agatha C. Hughes, 301–22. New York: Oxford University Press, 1990.

White, Paul. *Thomas Huxley: Making the "Man of Science."* Cambridge: Cambridge University Press, 2003.

Wiebe, Robert H. *Self-Rule: A Cultural History of American Democracy*. Chicago: University of Chicago Press, 1995.

Wiener, Philip P. *Evolution and the Founders of Pragmatism*. Cambridge, Mass.: Harvard University Press, 1949. Reprint, Philadelphia: University of Pennsylvania Press, 1972.

Willey, Thomas E. *Back to Kant: The Revival of Kantianism in German Social and Historical Thought, 1860–1914*. Detroit: Wayne State University Press, 1978.

Wilson, Daniel J. "Professionalization and Organized Discussion in the American Philosophical Association, 1900–1922." *Journal of the History of Philosophy* 17 (1979): 53–69.

———. *Science, Community, and the Transformation of American Philosophy, 1860–1930*. Chicago: University of Chicago Press, 1990.

Wilson, R. Jackson. *In Quest of Community: Social Philosophy in the United States, 1860–1920*. Oxford: Oxford University Press, 1968.

Winsor, Mary P. *Reading the Shape of Nature: Comparative Zoology at the Agassiz Museum*. Chicago: University of Chicago Press, 1991.

Winter, Alison, "A Calculus of Suffering: Ada Lovelace and the Corporeal Constraints on Women's Knowledge in Early Victorian England." In *Science Incarnate: The Physical Presentation of Intellectual Selves*, edited by Christopher Lawrence and Steven Shapin, 202–39. Chicago: University of Chicago Press, 1997.

———. "The Construction of Orthodoxies and Heterodoxies in Early Victorian Life Sciences." In *Victorian Science in Context*, edited by Bernard V. Lightman, 24–50. Chicago: University of Chicago Press, 1997.

———. *Mesmerized: Powers of Mind in Victorian Britain*. Chicago: University of Chicago Press, 1998.

Wise, M. Norton. "Precision: Agent of Unity and Product of Agreement." In Wise, *Values of Precision*, 221–36.

———, ed. *The Values of Precision*. Princeton: Princeton University Press, 1995.

Wise, M. Norton, and Robert Brain. "Muscles and Engines: Indicator Diagrams in Helmholtz's Physiology." In *Universalgenie Helmholtz: Ruckblick nach 100 Jahren*, edited by Lorenz Krüger, 124–45. Berlin: Akademie Verlag, 1994. Reprinted in the *Science Studies Reader*, edited by Mario Biagioli, 51–66. New York: Routledge, 1999.

Withers, Charles W. J. "Encyclopaedism, Modernism and the Classification of Geographical Knowledge." *Transactions of the Institute of British Geographers*, n.s., 21 (1996): 275–98.

Wolffram, Heather Mary. "On the Borders of Science: Psychical Research and Parapsychology in Germany, c. 1870–1939." PhD diss., University of Queensland, 2005.

Woodward, William R. Introduction to *Essays in Psychology*. The Works of William James.

Woodward, William R., and Mitchell G. Ash, eds. *The Problematic Science: Psychology in Nineteenth-Century Thought*. New York: Praeger, 1982.

Wright, T. R. *The Religion of Humanity*. Cambridge: Cambridge University Press, 1986.

Yeo, Richard. *Defining Science: William Whewell, Natural Knowledge, and Public Debate in Early Victorian Britain*. Cambridge: Cambridge University Press, 1993.

———. *Encyclopaedic Visions: Scientific Dictionaries and Enlightenment Culture*. Cambridge: Cambridge University Press, 2001.

Young, Frederic Harold. *The Philosophy of Henry James Sr*. New York: Bookman, 1951.

Zammito, John H. *Kant, Herder, and the Birth of Anthropology*. Chicago: University of Chicago Press, 2002.

Zimmerman, Andrew. *Anthropology and Antihumanism in Imperial Germany*. Chicago: University of Chicago Press, 2001.

# INDEX

*Italic page numbers refer to illustrations*

Abbott, Andrew, 26, 278n, 285n
Aesthetics, 126, 150, 168, 185, 186, *169*;
  aesthetic preferences, 126, 186, *185*,
  250
Agar, Jon, 278n
Agassiz, Louis, 31
Agreement, 140–141, 143, 150
Altruism, 196, 208, 209
American Philosophical Association,
  1–4, 17, 22, 26, 48, 50, 51, 56,
  137–138, 144, 177–180, 247, 265, 270
American Psychological Association,
  22, 26, 55–56, 84, 169, 177, 178, 180,
  247–248, 270
American Society for Psychical
  Research, 100, 198
Anarchism, 193, 194, 249; anarchist
  communities, 215
Anatomy, 31, 32, 66, 67, 69, 158
Anderson, Douglas R., 265, 334n
Anesthesia, 197, 198
Anesthetics, 201
Angell, A. R., 56, 57, 168–169, 171, 178
Annan, Noel, 281n

Anthropology, 11, 62, 66, 75, 156, 157,
  182. *See also* science of human nature
Anthropologism, 157, 161
Anti-imperialism, 195
Anti-Imperialist League, 193
Aristotelian Society, 37, 41, 170
Arnold, Matthew, 64
Aristotelian Society, 37, 41, 170
Ash, Mitchell, 47, 55, 158, 161–162, 284n,
  285n, 287n, 310n, 311n, 312n, 327n
Associationism, 70–71, 80, 84–85, 149
Atomism, 243
Atoms, mental, 87
Attention, 110, 150, 203, 204
Automatic writing, 33, 104, 190, 198, 200
Automatism, 102, 104, 199
Avenarius, Richard, 47; James on, 47

Bain, Alexander, 36, 149, 150; James on,
  36
Baker, G. P., 311n
Bakewell, Charles Montague, 137
Barberis, Daniela, 42, 282n, 283n, 328n
Barua, Dilip Kumar, 323n

Barzun, Jacques, 6, 276n, 279n, 318n
Beard, George, 114–115; James on, 115; on the science of human testimony, 103–104
Becher, Tony, 276n
Beisner, Robert, 195, 319n, 320n
Belief, 116–117, 140, 150, 151, 165, 175, 186, 200
Bellamy, Edward, 189, 191, 211, 216
Ben-David, Joseph, 33, 280n
Bercovitch, Sacvan, 318n
Bergson, Henri, 43, 106, 214, 215; James on, 43
Binet, Alfred, 198
Biology, 31, 32, 169, 66, 169, 173, 176, 183–184, 234, 238; and philosophy, 44, 253
Bird, Graham, 5, 303n, 304n, 306n
Bjork, Daniel, 6, 276n, 277n, 279n, 300n, 333n
Bledstein, Burton J., 275n
Blindness, ancestral, 194, 215; of hysterics, 197; "selective" blindness, 198
Blondel, Christine, 298n
Blood, Benjamin Paul, 201
"Border ground," of philosophy and psychology, 59
Boundaries of disciplines and discourses, 4–6, 8, 11, 15, 17, 27, 28, 34, 40, 45–46, 60, 69– 70, 74, 76–77, 84, 87, 88, 92, 95, 120, 129, 130, 139, 153, 157, 160, 168, 163, 172, 176, 177, 179, 181, 183, 184, 252, 267, 269, 270; of the self (see Self)
"Boundary" (or "boundaries"), 16, 54, 69, 94, 129, 163, 167, 168
Boundary objects, 267
Boundary work, 7–10, 13, 27, 29, 34, 48, 58, 60, 69, 78, 81, 87, 95, 129, 136, 157
Bourdieu, Pierre, 7, 276n
Boutroux, Émile, 43, 143, 256; James on, 43
Bowditch, H., 74, 96
Bowen, Francis, 60–61, 67
Bradley, Francis, 180, 209, 213
Brain, 29, 69, 70, 80; brain localization theory, 71–72; brain physiology, 62, 66, 72, 74, 79; Ferrier's experiments, 72–74
Brain, Robert, 296n
Braun, M., 298n
Brooks, John I., 328n
Brown, Alan, 281n
Brown, Edward M., 298n
Browning, Don, 319n
Buchner, Edward F., 56

Büchner, Ludwig, 44, 63
Bucke, Richard Maurice, 211

Camfield, Thomas M., 287n
Caplan, Eric, 321n
Capshew, James H., 287n
Carl, Wolfgang, 158, 310n
Carpenter, Edward, 210–212, 266
Carpenter, William Benjamin, 37, 62, 65, 76– 77, 102–103, 114, 146; James on Carpenter, 76–77; theory of unconscious cerebration, 102
Cattell, James McKeen, 56, 121, 122, 124, 234, 246
Chambers, Ephraim, 232
Character, 65, 78, 101, 102, 203, 204
Charcot, Jean-Martin, 197
Charts of knowledge. See Taxonomies of knowledge
Clairvoyance, 200
Clarke, Edwin, 298n, 309n
Clendenning, John, 323n
Clifford, James, 26, 278n, 333n
Clifford, William Kingdon, 98, 116; James on, 116–117
Coady, C. A. J., 93, 296n
Cognition, 5, 98, 138, 144, 148–152, 165, 169– 170, 174–176
Cohen, Morris R., 53
Colapietro, Vincent, 323n
Collini, Stefan, 280n, 282n
Collins, Randall, 33, 280n
Communities, of inquirers and/or of knowledge, 7, 8, 10–12, 14–15, 18, 51, 52, 90, 92, 95, 126, 132–134, 181, 187, 221, 230, 258, 272; social communities, 194, 196, 207–209, 211, 213–215, 220, 243, 271
Compounding: of feelings and mental states, 85–87, 213; "self-compounding" of consciousnesses, 12, 90, 213–215; of selves, 12, 213
Comte, Auguste, 39, 40, 232, 234, 245
Conant, James, 184, 317n, 332n
Consciousness, James on, 88, 90, 97–98, 183, 198, 214; cosmic consciousness, 208, 214; pulses of consciousness, 214; synthetic unity of consciousness, 84–85; threshold of consciousness, 207
consistency, 142, 144, 145, 146, 149
conversion, 204, 266

Coon, Deborah J., 17, 93, 97, 118, 122, 130, 193–194, 215, 277n, 297n, 298n, 299n, 300n, 302n, 303n, 319n, 320n, 324n, 331n
Cooper, Wesley, 319n, 303n, 304n
Cotkin, George, 2, 4, 195, 205, 275, 299n, 303n, 318n, 319n, 320n, 322n
Crary, Jonathan, 204, 283n, 322n
Creighton, E. G., 137, 169, 176–177, 179; on philosophy, 50, 51, 53; on pragmatism, 138, 144, 261
Cresswell, Timothy, 276n, 325n
Croce, Paul J., 6, 276n, 279n, 286n, 288n, 289n
Crookes, William, 101–103
Cross-disciplinarity, 277n

Danziger, Kurt, 298n, 309n, 312n, 327n
Darnton, Robert, 327n
D'Arsonval, Jacques, 106
Darwin, Charles, 46, 67
Daston, Lorraine, 10, 92, 95, 276n, 285n, 296n, 302n, 335n
Davidson, Arnold I., 296n
Davidson, Thomas, 19, 41; James on, 263–265
De Certeau, Michel, 276n, 333n
Decleva, Fernanda Caizzi, 278n
de Rijcke, Sarah, 288n, 294n, 297n
Desmond, Adrian, 281n
Dewey, John, 53, 138, 149, 150, 168, 171, 175, 195, 226; architectonic conception of philosophy, 53
Diamond, Solomon, 285n
Discipline, mental, 65, 259. See also philosophy; science
Disciplines, 7–8, 26
Dissociation, 190, 198, 199
Dogmatism, 10, 117–119, 122–123, 135, Draaisma, Douwe, 288n, 294n, 297n
Dresser, Horatio W., 202
Dreyfus, affair, 118
Double, projection of, 33, 190, 200, 201
Du Bois-Reymond, Emil, 96
Durand de Gros, J. P., 232, 233

Ebbinghaus, Hermann, 47; James on, 45, 178; on philosophy and psychology, 45
Economy, moral and social economy of science. See Scientific ethos
Ecstasy, 191, 211, 212

Education, 242; debates on, 16, 43, 49, 60, 62–66, 69, 77–78
Egoism, 208
Eliot, Charles W., 30, 31, 66, 67, 68, 69, 74, 75, 247
Emerson Hall, 15, 18, 21, 22, 24–26, 29, 30, 35, 56–58, 246–250, 256, 248
Emotions, 104, 111, 116, 124, 125, 128, 129, 142, 144–146, 148, 152, 153, 169, 171, 175, 202, 285; aesthetic emotions, 145; intellectual emotions, 145; and knowledge, 98, 149, 152, 168, 175; and temperament, 185–186; and truth, 165
Encyclopedia, 223, 225, 231, 234, 240, 241; encyclopedic mind, 219, 232, 242, 254, 256; "Encyclopedic sage," 254
Endosmosis, 214–215
Energy, 1–5, 14, 146, 180, 202, 205, 265–268
Epistemology. See Theory of knowledge
Erkentnisstheorie. See Theory of knowledge
Ethos of science, 10, 16, 17, 50–52, 58, 92–93, 95, 96, 97, 99, 100, 106, 128, 179
Evans, Rand B., 79, 293n, 297n, 301n
Evidence, 17, 46, 93–94, 99–106, 110, 113–119, 121–123, 125–129, 132–135; absolute evidence, 117, 119; faggots of evidence, 105, 115, 121, 132, 133, 134; "objective evidence," 117

Fabiani, Jean-Jacques, 43, 283n, 289n, 328n
Fechner, G. T., 85, 207–208, 221, 254, 256; as "cross-roads" of truth, 219; James on, 219
Federation, 243
"Federal republic," 215
Feinstein, Howard, 6, 276n, 279n, 319n
"Fence": between lots in the field of knowledge, 69–70; separating individuality from cosmic consciousness, 208
Ferrier, David, 36, 72–74
Ferrier, James Frederick, 38
Fisch, M. H., 279n
Fletcher, Horace, 202, 203, 206
Flournoy, Theodore, 206
Fontinell, Eugene, 319n, 320n
Forgan, Sophie, 278n
Frank, Robert G., 296n
Franzese, Sergio, 11, 75, 181–182, 277n, 292n, 317n
Fujimura, Joan H., 276n
Fuller, Steve, 7, 276n

Fullerton, G. S., 84, 86
Furumoto, L., 301n
Fye, W. Bruce, 296n

Gale, Richard, 6, 12, 214, 276n, 277n, 319n, 320n, 322n, 324n, 333n
Galison, Peter, 95, 241, 296n, 302n, 328n, 335n
Galton, Francis, 27, 36
Gardiner, Harry Norman, 138
Gauld, Alan, 281n
Geiger, Roger, 285n
Gieryn, Thomas, 7–8, 28, 276n, 278n, 326n
Gilman, Daniel Coit, 69, 74
Ginzburg, Carlo, 296n
Giri, Ananta Kumar, 276n
Goldstein, Jan, 207, 282n, 288n, 289n, 321n, 322n, 328n
Good, James, 7, 276n, 277n, 287n
Gray, Asa, 31
Green, Thomas Hill, 40, 41; on philosophy, 38; on the science of man, 63
Gregory, Frederick, 283n, 288n
Griesemer, James R., 276n, 335n
Guarneri, Carl J. 318n, 324n
Gurney, Edmund, 105, 120–121, 130, 133, 198–199; James on, 37

Habit, 8, 14, 19, 64, 65, 92, 99, 146, 151, 203, 206, 259–261, 263–266, 272; and consistency, 146; habitual train of thought, 152; number habits, 124; of the philosopher, 64; scientific, 65, 102; of thinking, 102–103
"Habit-neuroses," 14
Hacker, P. M. S., 311n
Hadot, Pierre, 278n
Haeckel, Ernst, on philosophy and the sciences, 44
Hagner, Michael, 291n
Hale, Matthew, 327n, 329n
Hall, G. Stanley, 35, 49, 51, 54, 55, 56, 57, 62, 96; on Gurney, Myers, and Podmore, 130; on James's Of Psychology, 99, 261; on philosophy, 62; on the psychologist, 51; on psychology, 49; on the relationship between philosophy and psychology, 57
Hallucinations, veridical, 105, 120–121, 200
Hamilton, Sir William, 61
Hansen, F. C. C., 124
Harper, R. S., 277n
Harrington, Anne, 242, 291n, 329n

Harvard Psychological Club, 2
Harvard Natural History Society, 67, 96
Harvard Psychological Laboratory, 21–22, 23, 29, 33, 56
Haskell, Thomas, 134, 285n, 286n, 302n, 335n
Hébert, Marcel, 142
Helmholtz, Hermann von, 59, 62, 70, 75; James on, 70, 75
Herzog, Max, 284n, 315n
Heyck, T. W., 280n
Hibben, John G., 137, 177
Hilgard, Ernest R., 308n
Hodgson, Richard, 126–128, 132, 133
Hodgson, Shadworth, 33, 36, 37, 39–41, 43, 81, 176, 253; James on, 36, 41, 281n; on philosophy, 38; on the relationship between philosophy and science, 40; on pragmatism and humanism, 176
Höffding, Harald, 150, 253
Hollinger, David A., 6, 10, 17, 92, 93, 196, 275n, 276n, 286n, 295n, 297n, 299n, 302n, 303n, 316n, 320n, 333n, 335n
Hoopes, James, 320n
Hoorn, Willem van, 285n, 312n, 327n
Howe, Daniel Walker, 288n, 322n
Howells, W. D., 189–191, 193, 209, 216
Howison, George Holmes, 53–54, 163–164, 179, 230, 236, 243, 246, 250, 257; boundaries of philosophy, 54; on logic and psychology, 163; on philosophy and the sciences, 53–54
Hull, David, 276n
Humanism, 169, 170, 176, 177, 179
Hume, David, 101
Husserl, Edmund, 159–163, 165–166, 171–172
Huxley, Thomas H., 37, 39, 63–64, 65, 75, 98, 116, 151; James on, 116–117
Hygiene, mental, 18, 33, 202
Hypnosis, 96, 100, 103, 120, 106, 123, 147, 198, 200, 204, 266
Hypnotism. See Hypnosis
Hyslop, James, 100
Hysteria, 197–199
Hysteric subject, 267, 317n

Idealism, 54, 85, 164, 209, 212, 213, 230, 243
Ideo-motor action, 146–147
Imitation, 206
Imperialism, American, 118, 193, 195, 215

Individualism, 194, 196, 201, 209, 211, 242, 263, 264
"Individuality," 208, 211
Institut Général Psychologique, 106
Instruments, self-recording, 16, 95–97, 105–107, 109, 111, 113–115
International Congress of Arts and Sciences, St. Louis, 18, 54, 222–223, 226, 227–229, 230, 231, 236, 243–250
International Congress of Experimental Psychology, 120
Interpenetration, 14, 214–215

Jacyna, L. S., 298n, 309n
James, Henry, 200, 247
James, Henry Sr., 210–211
James, William: boundary work, 8, 10, 13, 34, 58, 60, 69, 78, 81, 87, 94–95, 129–131, 136, 157, 260, 269–270; Cognition, 5, 98, 138, 144, 148–152, 165, 169–170, 174–176; Communities (social), 13–15, 134, 181, 196, 208, 212, 214–21, 270–271; Communities of inquirers, 10–12, 14, 133–134, 181, 187, 221, 258, 272; on consciousness (see Consciousness, James on); courses taught at Harvard university, 31, 32, 33, 67, 280n; dreams, 201, 214; early academic career, 32–33; on evidence, 17, 93–94, 99–100, 114–117, 119, 121–122, 125–129, 132–135; ethos of science, 119, 128–130, 131, 135; on the knowing of things together, 87–88; insomnia, 31, 202–203; James's library, 256; on logic, 173–174; on the MD title, 31, 122–123; on metaphysics, 16, 80–81, 87, 88, 89, 254; on objectivity, 97–98, 150; on philosophy (see Philosophy, James on); pragmatism, 17, 98, 135, 137–141, 152, 157, 172, 173, 184–185, 251–252, 256–257; and psychical research, 6, 12, 17, 33, 37, 94, 100–104, 106, 115–116, 120–121, 123, 125–128, 132, 198, 200, 208; on the "psychologist's fallacy," 99; on psychology, 16, 58, 79, 80, 82; and the science of man, 11–12, 16, 62, 66, 68, 69, 74, 78, 82, 181, 183–184, 187, 192, 267; on the "scientist" or the scientific investigator, 16, 91, 95, 97, 100, 117, 118, 119, 130–131; on the self, 7, 12–15, 17–18, 191–192, 194, 196–197, 199, 200–205, 207, 211–213, 215, 217, 270–271; social and political vision, 192–196, 206, 215, 216, 221; and spaces of knowledge, 18, 19, 252, 256–257; on truth, 17, 135, 137–147, 157, 172–175, 184, 249; unification of knowledge, 9–11, 89–90, 221, 245–246, 254, 256–258; on the university, 249–250; writing style, 260–263 WORKS: "Aesthetics" seminar, 150; "Briefer Course, 82; Diary, 1, 32; "The Energies of Men," 1–5, 14, 19, 180, 260, 265–69; "Gospel of Relaxation," 206; Human Immortality, 207; Introduction to Little Book of Life after Death, 254, 277n; Johns Hopkins Lectures, 69–70, 74; "The Knowing of Things Together," 84, 88, 180; Lowell Lectures, "The Brain and the Mind," 69–70, 72, 74; The Meaning of Truth, 171; "Moral Equivalent of War," 216–217; "Philosophical Conceptions and Practical Results," 169; "A Plea for Psychology as a Natural Science," 82; Pragmatism, 138, 147, 171, 184–185, 251, 253, 256, 262; Principles of Psychology, 16, 32, 58, 60, 74, 79–85, 87, 96, 99, 117, 147, 150, 152, 174, 175, 185, 196, 203, 213, 260–261, 263; public lecture at Harvard's Sanders Theater, 73; reviews of Carpenter, 76–77; "Sentiment of Rationality," 152, 175; Some Problems of Philosophy, 253, 254, 255, 277n; "The Teaching of Philosophy at our Colleges" 77, 277n; "The Will to Believe," 96, 99–100, 116, 117, 119; The Will to Believe, 118–119, 262
Janet, Paul, 42, 240–241, 253
Janet, Pierre, 106, 197–199
Jasanoff, Sheila, 7
Jastrow, J., 103–104, 115
Jerusalem, Wilhelm, 159, 171, 172
Joseph, W. B., 166
Jurkowitz, Edward, 10, 276n, 292n, 295n, 329n

Kant, Immanuel, 138, 155–158, 160, 182, 250, 254
Kim, Mimi, 327n
Kittelstrom, Amy, 275n
Klein, Julie, 7, 276n
Kloppenberg, James T., 12, 193, 319n
Knowledge: as an encyclopedia, 240; spaces of knowledge, 8, 9, 27–28, 220–222, 232–234, 246, 252, 255–257; theory of knowledge (Erkenntnistheorie), 157, 159, 163–164, 171, 235; unity of knowledge, 9–11, 18–19, 46,

Knowledge (*cont.*)
    53, 226, 236–243 (*see also* James, William;
    Münsterberg, Hugo)
Kuklick, Bruce, 5, 49, 67, 139, 174, 175, 275n,
    276n, 278n, 285n, 290n, 302n, 303n, 305n,
    315n, 323n
Kuklick, Henrika, 285n
Kusch, Martin, 46, 47, 158, 161–162, 284n,
    285n, 286n, 310n, 311n, 312n, 314n, 327n

Laboratory, 39
Ladd, George Trumbull, 54, 56, 80–82, 84,
    149, 150; on philosophy and science, 52–53
Lamberth, David C., 5–6, 87–88, 276n, 285n,
    295n, 299n, 303n, 304n, 321n
Latour, Bruno, 272, 276n, 284n
Lears, T. J. Jackson, 190, 318n
Leary, David E., 5–6, 276n, 279n, 287n, 297n,
    301n, 318n, 319n, 320n, 321n, 322n, 324n,
    327n
Leehman, A. G. L., 124
Lenoir, Tim, 276n, 278n
Lentricchia, Frank, 319n
Lewes, George Henry, 36, 40; on philosophy
    and science, 39
Leys, Ruth, 301n
Lipps, Theodor, 158, 161, 171
Lipton, Peter, 93, 296n
Livingston, James, 13, 191, 194, 206, 277n,
    303n, 318n, 319n, 320n, 322n, 323n
Livingstone, David N., 276n, 325n
Lloyd, Brian, 319n
Lloyd, Henry Demarest, 208
Logic, 155–158, 161–162, 169, 174; and
    biology, 176; and psychology, 160,
    163–166, 168–173, 180
Lotze, Hermann, 75, 182
Lovejoy, Arthur O., 143
Lutoslawski, Vincenti, 268
Lutz, Tom, 319n
Luys, Jules, 72

Mach, Ernst, 46; James on, 46–47
Mantegazza, Paolo, 149–150
Marey, E.-J., 96, 106
Marshall, H. R., 149–150
Maudsley, Henry, 62, 71–73, 80
McClay, Wilfred M., 318n, 323n, 325n
McCosh, James, 75
McDermott, John, 204, 319n, 322n

McLachlan, James, 329n
Menand, Louis, 6, 31, 272, 276n, 279n, 286n,
    335n
Metaphysical Club, 32
Mill, John Stuart, 36, 61, 259
Miller, Joshua I., 319n, 320n, 325n
Mind: faculties and functions of the mind,
    138, 148, 151, 169; philosophical mind,
    46, 77; science of the mind (*see* mental sci-
    ence); training of the mind (*see* Discipline;
    Philosophy; Science)
*Mind*, 35, 42
Monism, 212, 215
Montague, W. P., 146–147, 176
Moore, Addison Webster, 175
Morawski, Jill, 93, 99, 276n, 287n, 296n, 297n
Müller, Detlef K., 289n
Müller, Johannes, 74
Münsterberg, Hugo, 15, 18–19, 21–22, 24–26,
    29–30, 33, 56, 58, 171, 222–226, 230–231,
    244–252, 261; on boundaries of science,
    129; and Emerson Hall, 15, 21–22, 24–26,
    246–250, 256; on experimental psychol-
    ogy, 22, 24; his house, 25; James on, 130,
    250; on the PhD title, 24; on philosophy
    (*see* philosophy, Hugo Münsterberg
    on); on pragmatism, 167; on psychical
    research, 103–104, 106, 129–130; on psy-
    chology's place in the map of the sciences,
    235; on relativism, 255; on Schiller, 244;
    on science, 129; social and political vision,
    244–245, 252; spaces of knowledge, 18,
    25, 245, 255; taxonomy of knowledge,
    18, 222–226, *224–225*, 235; on the unity
    of knowledge, 19, 24–25, 221–222, 226,
    230–231, 246; on the university, 24–25; on
    the use of questionnaires in psychology,
    134
Myers, Gerald E., 5, 12, 150, 276n, 303n, 313n,
    319n, 320n, 322n, 324n
Myers, F. W. H., 93, 120, 130, 131, 200–201;
    James on, 131
Mugwumps, 193
Munford, Lewis, 194
Mysticism (and mystical experiences), 12, 14,
    18, 90, 191–192, 194, 201, 210–211, 214,
    217

Neary, Francis, 280n
Neurology, 103–140, 169

Newcomb, Simon, 222, 230
New Thought, 202, 266
Nitrous oxide, 201, 212
Noakes, Richard, 298n

Oakes, Guy, 311n, 327n
Objectivity, 95–98, 125, 133, 135, 272–273
O'Donnell, John, 55, 287n
"Officialism," 249
Olesko, Kathryn M., 51, 81, 285n, 294n
One and the Many, 12, 14, 214, 243, 257
Oppenheim, Janet, 281n, 296n, 298n, 321n
Organisms, colonial, 199
Ostwald, Wilhelm, 234
Outram, Dorinda, 278n
Owen, Alex, 281n, 296n, 298n, 318n

Palladino, Eusapia, 106–107, 107, 108, 109–110, 111, 112, 128
Pandora, Katherine, 335n
Papini, Giovanni, 2, 186–187, 215, 256–257, 262, 272; on philosophy, 186
Passmore, John, 282n
Paulsen, Friedrich, 47, 241, 246, 253, 254, 256, 272; James on, 45, 262; on the philosopher as an encyclopedic mind, 46; on the philosophical unity of knowledge, 242; on philosophy's relationships to the sciences, 45–46; social vision, 242–243
Pearson, Karl, 223, 246
Peirce, Charles S., 32, 50, 52, 120, 134, 139–140, 169, 261, 262; on the community of inquirers, 52, 134; on philosophy, 50, 52; on veridical hallucinations, 120
Perry, Ralph Barton, 143
Persona, scientific, 92
Personality, 189–191, 198–199, 201, 203–204, 209, 212
PhD, 24, 249, 250, 269
Phantasms of the living. See Hallucinations
Phenomena, psychic, 6, 17, 90, 93–94, 100, 102–106, 120–121, 125–126, 132–133, 135, 200, 214; as the "Dreyfus case of science," 123
Philosopher, 33, 44, 46, 50, 162, 242. See also James, William; Münsterberg, Hugo
Philosophy: architectonic conception of, 9, 40, 52–53, 78; in Britain, 35–42; as the "circle" of knowledge, 53, 240; in France, 42–44; as the "germ and crown" of the sciences, 40;

in Germany, 44–48; as mental discipline, 9, 65, 77, 78, 264, 265; as mental disposition, 46; moral and mental philosophy, 16, 38, 48–49, 61–63; a new era of philosophy, 25, 76; perceived crisis of philosophy, 9, 26, 44, 236; place of philosophy, 50 (see James, William; Münsterberg, Hugo); positive philosophy, 39; professionalization of philosophy, 40–41, 45; and psychology, 17, 45–46, 54–58, 155–157, 159–168, 176–180, 235 (see also under James, William; Logic; Münsterberg, Hugo); as a science, 47, 50–51; and the sciences, 9, 15, 27, 34, 38–40, 42–43, 44–54, 62–66, 71, 156–157, 160, 236–238, 240–243, 245–246 (see also James, William; Münsterberg, Hugo); as scientia scientiarum, 9, 18, 40, 237; as "the sum-total of all scientific knowledge," 45; as a system of completely unified knowledge, 238 (see also under James, William); in the United States, 48–54, 62
philosophy, Hugo Münsterberg and: its function and position at the university, 24, 247, 249; on the philosopher, 18, 245; on philosophy, 24, 225; on philosophy's relationships with psychology, 22, 24, 56, 179; on philosophy's relationships with the special sciences, 18, 24–25, 247, 249; on the place of philosophy or of the philosopher, 18, 225–226, 255–256
philosophy, William James and: 32, 89, 186, 253, 268, 279n; architectonic conception of philosophy, 16, 78; "general philosophy," 10, 18, 89–90, 187, 221–222, 245–246, 255, 258, 270; inductive philosophy, 16, 76, 270; on "introspective" philosophy and physiology, 69, 70–71, 73–74; on the philosopher, 11, 19, 181, 187, 219, 221, 246, 254, 256, 258, 260, 264–265; philosophy as expressive of a person's "total reaction against the presence of the universe," 185; philosophy as a "forum" for a discussion of the sciences, 89, 270; philosophy as mental discipline, 76, 77, 78, 260, 264–265; on philosophy and psychology, 16, 58, 82–84, 87–88, 180, 253; on philosophy and science (or the sciences), 16, 64, 75–76, 77–78, 83, 89–90, 221, 253–255, 258, 269–270; philosophy as a system of completely unified knowledge, 10, 221, 246, 254–255;

philosophy, William Jame (*cont.*)
　place of the philosopher, 18–19, 219, 221,
　245, 246, 256, 258; as the "reflection of
　man on his relations with the universe,"
　182
Physiology, 69, 70, 71, 72, 73, 74, 81, 85, 96, 98,
　139, 146, 149, 169, 184, 234
Piper, L. E., 100, 126–127, 131, 133, 199
Pireddu, Nicoletta, 308n
Pittenger, Mark, 323n
Pleasure, 145–146, 149, 150; of consistency,
　149
Podmore, F., 120
Pols, Hans, 321n
Porter, Noah, 65, 75
Porter, Theodore, 295n
Pragmatism, 124, 136, 143–144, 148, 155,
　165–167, 169–171, 175–177, 179, 181, 261.
　*See also* James, William
Pratt, James Bissett, 143, 172–173
Presumption, 121–122
Psychologism, 17, 57, 153, 155, 157–166, 170,
　175, 180
Psychology: act psychology, 124; as "an-
　techamber of metaphysics," 74; behav-
　iorism, 124; as a child of philosophy, 56,
　56; empirical psychology, 156, 157; func-
　tional, 3, 124, 168–169; its position among
　the sciences, 235; and logic (*see* logic); as
　a natural science, 57, 74, 79–80, 82, 88,
　162, 178, 183; "philosophical" psychology,
　55, 164; and philosophy (*see* philosophy);
　and physiology (*see* James, William); pro-
　fessionalization of psychology, 54–55;
　"rational" psychology, 156, 164; "sci-
　entific" psychology, 54–55; structural
　(structuralism), 124, 168; transcendental
　psychology, 156
Putnam, Hilary, 139–140, 143, 184, 303n,
　304n, 317n
Putnam, J. J., 74, 96
Putnam, Ruth A., 184, 317n

Rabinbach, Anson, 275n
Radical empiricism, 33, 43, 58, 87–88, 119, 135,
　140, 252
Rationality, 152; sentiment of, 151
Reflex action, 38, 97, 102, 145, 150, 151, 175,
　183, 185, 190, 203; of the brain, 151; and
　temperament, 185

Relations, metaphysics of relations, 6, 140
Relationships, social and human, 13, 14, 18,
　191–192, 195, 206, 208, 214–215, 220, 221,
　271. *See also* "communities"
Relativism, 157, 160–161, 167, 171, 255
Renouvier, Charles, 42, 43, 253; James on, 42
Reuben, Julie, 61, 288n, 289n
Ribot, Théodule, 42, 199; James on, 42
Rice, Charles E., 287n, 289n
Richards, Graham, 293n, 298n
Richards, Robert J., 307n, 319n
Richter, Melvin, 281n, 282n
Rickert, Heinrich, 159–163, 166–167, 235;
　James on, 171
Ringer, Fritz K., 283n, 289n, 328n, 329n
Roberts, Richard H., 277n, 287n
Robertson, George Croom, 35, 36, 37
Rodgers, Daniel, 34, 280n
Roosevelt, Theodore, 2, 193, 230, 231
Rorty, Richard, 272, 278n, 297n, 303n, 304n,
　335n
Rosmini, Antonio, 241, 246
Ross, Dorothy, 285n, 287n, 296n, 302n
Rossiter, Margaret W., 301n
Rouse, Joseph, 276n
Routine. *See* Habit
Royce, Josiah, 91, 119,145, 180, 213, 257; on
　the relationships between philosophers
　and scientists, 53, 201; on the self, 201, 209
Russell, Bertrand, 166, 170
Russell, John Edward, 166, 169, 172
Ryan, Judith, 318n

Santayana, George, 67, 260
Satisfaction (and / or satisfactoriness), 139,
　141–146, 149–150, 175
Scarborough, Elizabeth, 301n
Schaffer, Simon, 93, 276n, 296n, 298n, 301n
Schiller, Ferdinand Canning Scott, 32, 67, 123,
　138, 175, 179–180, 212, 244, 259, 261, 265,
　272; on logic and psychology, 168–171
Schinz, Albert, 155, 177, 261
Schirmer, Daniel B., 319n
Schmidgen, Henning, 277n, 301n
Schneewind, J. B., 278n
Schurman, Jacob Gould, 50, 53; on
　philosophy, 50
Science: definitions of, 50; of human nature,
　11–12, 16, 17, 49, 54, 60–65, 68–69, 74–75,
　82, 156–158, 181–184, 187, 192, 267, 272;

as mental discipline, 65, 77, 92, 102; mental science, 61, 66, 67,75; philosophical science of human nature, 49, 60, 61, 64–66, 78, 183; relationships of science with philosophy (*see under* Philosophy)

Sciences, special, 24

Scientific temper, 131

Scientist (and "man of science"), conceptions of, 51, 77, 82, 91, 97, 99, 124, 125, 237. *See also under* James, William

"Scratch Eight," 37

Scripture, E. W., 81–82, 180

Seigfried, Charlene H., 5, 11, 12, 125, 126, 174, 181–182, 276n, 277n, 297n, 301n, 303n, 304n, 305n, 308n, 309n, 315n, 317n, 318n, 319n, 320n

Self, 17, 187, 188–192, 201, 208–212, 221 (*see also under* James, William); absolute self, 212, 213; boundaries (margins) of the self, 13–15, 191, 192, 200–201, 208, 209, 212, 214, 215, 217, 270; division of the self, 190, 192, 196–199, 201, 202, 204; social self, 191, 196, 208, 209, 217; subliminal self, 130, 199, 201; unity (and unification) of the self, 13, 197, 199, 203–206, 217

Selfishness, 202. *See also* Egoism

Sentiment: of rationality, 151, 175; of truth, 149

Serres, Michel, 234, 327n

Shapin, Steven, 10, 92–93, 276n, 278n, 284n, 295n, 296n, 298n

Sidgwick, Henry, 35, 37, 39–41, 61, 101, 104, 121, 123, 133, 253; on philosophy, 40, 61; on philosophy and science, 40

Sidgwick, Eleanor, 121, 123

Sigwart, Christopher, 158, 160, 161, 171, 172

Simon, Brian, 289n

Simon, Linda, 31, 279n, 280n, 287n, 289n, 290n, 296n, 299n, 302n, 319n, 321n, 322n, 331n

Sklansky, Jeffrey, 205–206, 318n, 319n, 322n

Sklar, Martin, 319n

Skrupskelis, I., 16, 75, 84, 87–88, 277n, 280n, 289n, 290n, 291n, 293n, 294n, 295n

Small, Albion, 49, 222

Smith, Crosbie, 278n

Smith, M. Brewester, 319n

Socialism, 194, 211, 216, 242

Society for Psychical Research, 37, 100,101, 106, 116, 121, 125, 126, 132, 200

Sociology, 49

Sokal, Michael, 55, 177, 287n, 294n, 316n, 333n

Specialist, 94, 102, 104

Spencer, Herbert, 9–10, 36, 40, 62–65, 67, 98, 149–150, 174, 221, 223, 237–238, 246, 254, 255, 258, 259

Spiritualism, 93, 102, 115; philosophical and psychological, 42, 70, 75, 80, 85

Sprigge, Timothy L., 5, 12, 209, 214, 276n, 277n, 303n, 319n, 323n, 324n

Stanley, Hiram S., 149–150

Star, Susan Leigh, 276n, 335n

Stephen, Leslie, 37, 117

Stern, Sheldon, 280n

Stewart, Balfour, 103

Stratton, George, 163–164, 177, 178

Strong, Charles Augustus, 44–45, 137, 148, 166

Stumpf, Carl, 33, 46–47, 158, 161, 162, 171; James on, 46; on philosophy and its relationships to psychology and other sciences, 46

Sully, James, 37, 79, 149–150

"Supernormal," 93

Taves, Ann, 321n

Taxonomies, of knowledge, 8, 18, 29, 220–222, *224–225*, *227–229*, 231–232, *233*, *238–240*, 234–241, 245, 250. *See also* James, William; Münsterberg, Hugo

Taylor, A. E., 56–58, 148, 164–166, 170, 176–178, 261

Taylor, Eugene, 5, 33, 96, 130, 276n, 280n, 292n, 293n, 295n, 296n, 302n, 320n, 321n

Telepathy, 5, 17, 93, 122, 123, 200

Temper of science, 92, 95, 131

Temperament, 184–186, 250, 252; and aesthetic preferences, 185; hysterical temperament, 204

Testimony, 17, 93–94, 100–105, 112, 120, 129, 133–134

Thayer, H. S., 303n, 309n, 313n

Thilly, Frank, 56

Thomas, John L., 318n, 322n

Titchener, E. B., 123–125, 130, 168

Trachtenberg, Alan, 318n

Trance, 17, 100, 102, 104, 126–127, 133, 190, 198–200

Treitel, Corinna, 296n

Trees of knowledge, 18, 221, 253–254, 258. *See also* taxonomies of knowledge
Trine, Ralph Waldo, 210, 212
Truth, 17, 135, 137–139, 140–147, 150, 157, 159–161, 164–167, 169, 171, 172–175, 184, 224
Turner, Frank M., 278n, 289n
Turner, James, 278n
Tyndall, John, 64
Tzuzuki, Chushichi, 323n

Unconscious cerebration, 102, 114
Unity: of knowledge (*see* knowledge); of the universe, 243, 257–258

Verhave, Thom, 285n, 312n, 327n
Veysey, Laurence, 285n
Vivekânanda Swami, 202, 204
Volition (and will), 73, 144, 147–151, 153, 176

Ward, James, 36, 148, 149, 150, 253; on philosophy and science, 39
Ward, Lester, 234
Warwick, Andrew, 280n
Webb, Beatrice, 211
Weindling, Paul, 329n
Wells, H. G., 216, 266
Westbrook, Robert B., 193, 175n, 319n, 320n
White, Paul, 39, 281n, 289n
Whitman, Walt, 211

Whittaker, Thomas, 238, 240, *240*
Wiebe, Robert H., 318n
Will. *See* "Volition"
Wiener, Philip P., 308n
Wilson, Daniel J., 5, 48, 51, 178, 276n, 285n, 286n, 316n
Wilson, R. Jackson, 323n
Windelband, Wilhelm, 159, 162, 163, 171, 235, 246
Winsor, Mary P., 279n
Winter, Alison, 27, 278n, 293n, 298n
Wise, M. Norton, 294n
Withers, Charles W. J., 327n
Wolffram, Heather Mary, 299n
Woodward, William, 287n
Woodbridge, Frederick J. E., 176
World-soul, 208
Wright, Chauncey, 32
Wright, T. R., 280n
Wundt, Wilhelm, 46–47, 52, 59, 74–76, 85, 158, 161–162; on philosophy and the sciences, 47, 52; chart of the sciences, 235; James on Wundt, 46–47, 75
Wyman, Jeffries, 31

Yeo, Richard, 231, 326n, 327n
Youmans, Edward, 63, 65

Zammito, John H., 156, 182, 310n
Zimmerman, Andrew, 276n, 295n